A Race on the Edge of Time

Previous books by David E. Fisher

Novels
Crisis
A Fearful Symmetry
The Last Flying Tiger
Variation on a Theme
Katie's Terror

Science
The Creation of the Universe (YA)
The Creation of Atoms and Stars (YA)
The Ideas of Einstein (juv.)
The Third Experiment (YA)
The Birth of the Earth

A Race on the Edge of Time

Radar—The Decisive Weapon of World War II

David E. Fisher

McGraw-Hill Book Company
New York St. Louis San Francisco Bogotá Hamburg
Madrid Mexico Milan Montreal Panama Paris
São Paulo Singapore Tokyo Toronto

Copyright © 1988 by David E. Fisher. All rights reserved. Printed in the United States of America. Except as permitted under the Copyright Act of 1976, no part of this publication may be reproduced or distributed in any form or by any means or stored in a data base or retrieval system, without the prior written permission of the publisher.

1 2 3 4 5 6 7 8 9 D O C D O C 8 7

ISBN 0-07-021088-8

LIBRARY OF CONGRESS CATALOGING-IN-PUBLICATION DATA

Fisher, David E., 1932–
   A race on the edge of time.
   1. World War, 1939–1945—Radar.   I. Title.
D810.R33F57   1987   940.54'86   87-3764
ISBN 0-07-021088-8

Designed by Kathryn Parise

For Leila L.,
who has somehow managed to extend the edge of
time into an infinite plane,

and, with gratitude,
To the memories of A. F. Wilkins and
Hugh C. T. Dowding

Acknowledgments

Many people shared their reminiscences with me, too many to list individually, but in particular I would like to thank Sean Swords of Trinity College (Dublin), for sending me his technical history of radar before publication, and E. G. Bowen, for stimulating letters and unpublished papers; Mrs. Nancy Wilkins, for her own thoughts and for permission to quote from her husband's papers; my British agent, Gerald Pollinger, for his RAF expertise and ability to find paths to solutions; the RAF Community Relations staff at Coltishall, Bentley Priory, and Bawdsey, and their counterparts at Raytheon and Boeing; the late Johnny Kent for sharing his experiences and thoughts on the Battle of Britain; Zola Mae Blakeslee of the Dade County Public Library; and Owen Dinsdale of the Duxford Aviation Society.

Contents

Introduction ix

Part One
The Death Ray
1

1. In God We Trust — 3
2. Origins: The Death Ray — 15
3. Origins: Radio — 32
4. Beginnings: Red Sky at Sunset — 50
5. The Opposition: A Lion in Winter — 64
6. The Battle of Barking Creek — 84

Part Two
Toil and Trouble
93

7. The Opposition: 'Round and 'Round the Mulberry Bush — 95
8. Technical Years: 1939–1940 — 110
9. The Spectre and the Terror — 127

Part Three
The Battle
153

10. In the Still of the Night — 155
11. The Human Element — 171

12.	Attack of the Eagles: *Adlerangriff!*	196
13.	London Calling . . .	218
14.	Pride and Petulance	233

Part Four
The Blitz and the *Boot*
239

15.	The Blitz	241
16.	Microwaves	266
17.	*Das Boot*	281

Part Five
The Band Plays On
313

18.	Armageddon	315
19.	The Death Ray: Finale	332
	Appendix	349
	Notes	352
	Bibliography	356
	Index	359

Introduction

The thrust of history that pushed us out of savagery, past barbarism, and into a tenuous sort of civilization is generally thought of as a varying mixture of religious, economic, political, and sometimes moral factors. But science and technology have also, if only occasionally, exerted overwhelming influences. The discovery of iron lifted us out of the stone age, that of bronze pushed us ahead another notch, and the technology of uranium fission may yet bring us back full circle.

But the most important scientific advance of all, in terms of its influence on our standard of living today—indeed on the very nature of today's civilization—is not generally recognized as such. The discovery of iron and bronze, of penicillin and the polio vaccine, the invention of the airplane and of television were all inevitable—and it wouldn't make much difference to most of us today if they had appeared a bit earlier or later than they did. But the invention of radar in the late nineteen-thirties in England was the one thing that allowed us to defeat Nazi Germany, and if its utilization had been delayed by only five or six months, the outcome of the Second World War would have been reversed and the story of our world today would be vastly different.

It seems to me that the most important thing about our collective memory of the Second World War is that we have forgotten how we so nearly lost it. Hitler seems to us so evil, the Nazi government so horrible, its excesses so brutal, that we have come to believe that God in his mercy would never allow such an abhorrence to be visited upon Americans. But God has no mercy, and this abhorrence and all its associated terrors were in fact visited upon millions of people. But for the grace of a half-dozen chance occurrences (in which you may, if you like, see the mercy of God), it could easily have happened to us.

The lessons of history are often lost in the telling of it. The organization of a historical story into a beginning, middle, and end—into a cause, a conflict, and a resolution—often lulls us into subconsciously accepting it as a well-made play with an ending predestined from the

Introduction

beginning to come about in accordance with the moral principles and dramatic sense inbred in the playwright. But history does not always have a happy ending or one in which justice, however fearful and strict, is meted out to those deserving of it. The Persian hordes were indeed turned back at Thermopylae, the barbaric armies of the Mongol Khan did indeed pause and lose interest before Vienna—but that was small solace to the thousands caught by the Persians before Thermopylae or the millions between Mongolia and eastern Europe who were overrun and slaughtered—or 'forced to submit to indignities worse than death'—by the soldiers of the Khan. And the Goths, remember, *did* sack Rome, and a twilit nightmare of chaos *did* descend over Europe for a thousand years.

In the run of history anything can happen, and frequently does. Though we look back through the mists of centuries to see a world seemingly orchestrated in an orderly fashion, this is only because of the skill and malfeasance of historians who insert the thread of order into the random and chance happenings of mankind in order to make sense of it all. In so doing we often lose the most basic sense of what has happened.

Nazism spilled out of Germany and enveloped continental Europe with not one major setback until it tried to cross the English Channel. It failed there, and it then turned its attentions to the east, where it came within the thinnest of margins of succeeding. Had it not been forced to maintain a defense against the English and American air forces operating out of England, and against the English and American armies attacking in Africa, it certainly would have won in Russia. Had Rommel and his Afrika Korps been available at Stalingrad, the knife edge of victory on the Russian front would have thrust the other way. It is impossible to predict what might have happened beyond that, but an America isolated by a Nazi Europe and Africa and by a Japanese Pacific and Asia would not be the America we know today. We might even speculate about the trans-Atlantic rockets that were on von Braun's drawing board for testing in 1945; armed with the nerve gases that Germany was inventing and perfecting in the 1940s, they would have been more destructive than our atomic bomb which we had no effective way to deliver across the Atlantic Ocean.

The atomic bomb is generally considered to be the most important single scientific/military invention of all time, but in terms both of our history and of our life today, it is not. If we had not put together the bomb in 1945 we would have lost more than a million lives in the invasion of Japan, but we still would have won the war, whereas if we had not made radar work in 1940, we would have lost it. The entire subsequent history of the world rested on the fulcrum of the English islands from the summer of 1940 until the invasion of Europe in 1944. Once that invasion was well established and our forces were moving inland, there was little doubt about the ultimate outcome, only about how long it would take and the price we would have to pay. But until that point, the winning of the war hung daily in the balance, and victory would not have been possible at all if England had succumbed to invasion in the summer of 1940, to the night bombings of 1940–1941, or to the U-boat blockade of 1940–1943. In all these crises there was one scientific weapon which meant the difference between victory and defeat.

Taken all in all, radar must be the most important scientific/political/military invention of them all, bar none. Today it is the basis of our defense system, and thus the bulwark of our system of government and way of life—for who can doubt that, whatever its potential for terror and catastrophe, it is the deterrent effect of mutually assured destruction that so far has prevented the outbreak of World War III? We think that our defense system is based upon nuclear weapons, but this is not so. In their absence nerve gases and biological cannisters would serve an equally destructive purpose, but the other component of the system—early warning—is and can be fulfilled by radar alone.

Our current notion of security is based upon the knowledge that if a preemptive strike should be launched by another nation we could retaliate in somewhat better than kind. Given the destructive energy coiled within the military structure of the superpowers, this retaliation is possible only if each nation is capable of receiving instant knowledge of rockets let loose without warning or of bomber fleets or cruise missiles suddenly taking to the air. Surrounded and protected by its radar screen, each nation can afford to relax sufficiently to carry on its own business, its daily life. But without this screen, the pressures and threats of international tensions would long since have pushed one or the other of us over the edge into the roiling maelstrom of nuclear and biological war.

Never before in our history has one invention borne such a distinct and significant role in the outcome of such a pivotal war, and in the stability of the world empires which followed in its wake. Radar was a device invented not just once, but again and again in many countries at many different times, but in only one of these countries was it implemented at the precise time that would make all the difference. It began with a death ray, a cup of coffee, and a milkman's jangling bottle.

Part One
The Death Ray

> When can their glory fade?
> O, the wild charge they made!
>   All the world wonder'd.
> Honor the charge they made!
> Honor the Light Brigade,
>   Noble six hundred!
> —*Alfred, Lord Tennyson*

> There were no instruction manuals, no maintenance staff, no test equipment, no nothing....
> —*E. G. "Taffy" Bowen,
> radar scientist*

Chapter One
In God We Trust

> —but the bomber will always get through.
> *Stanley Baldwin,*
> *Lord President of Great Britain,*
> *1932*

In the spring of 1939, when Hitler was playing his geopolitical games which would shortly lead to the Second World War, Germany had a working radar system named Freya, capable of detecting aircraft at a distance of fifty miles. The Germans were aware through their intelligence services that neither France, Poland, nor the Soviet Union had anything remotely comparable; but they were a bit worried about the British. During the previous year a series of 350-foot steel towers crowned with what were obviously radio antennae had been observed under construction along the eastern coast of England, although the dimensions of the antennae did not correspond to what the Germans had determined were the optimum parameters for a radar system. Still, whatever the towers were, the Luftwaffe director general of radio signals wanted to know.

General Wolfgang Martini argued with the *Oberkommando der Luftwaffe* that when war came it would be his job to counteract any Allied radio activity; at this stage of the political/military process it was his responsibility to determine what dangers existed on the other side of the Channel. For the past year the only surveillance had been by the crews of Lufthansa airliners, together with some abortive attempts by Admiral Wilhelm Canaris's secret intelligence service, the *Abwehr*. Binocular observations from the airliners had revealed nothing more than the size and obvious radio-related complexity of the masts; the reports of spies were more intriguing but hardly more illuminating.

The center of this radio construction, it had been determined, was

Bawdsey Manor, the ancestral home of Sir Cuthbert Quilter, which had been purchased for 79,000 pounds the previous year by the Air Ministry when its Telecommunications Research Establishment outgrew their home at Orfordness. Bawdsey obviously had been selected with one eye to secrecy; it was stationed at the mouth of the river Deben on the Suffolk coast just northeast of London, in rather lonely and desolate country on a promontory separated from the mainland by a wide though shallow water. Shallow, but deep enough; the site was available only by boat. A regular ferry service was run by Charley Brinkley, a one-armed man with a metal hook for fingers, who took no one across without a military pass.

The area was not one of Britain's premier garden spots, but neither was it entirely unknown as a vacation resort, and since the previous summer when the first signs of activity had become obvious the neighborhood had seen a succession of German tourists camping out in the surrounding woods. The radio apparatus they carried on their backs were of necessity weak and ineffectual, and revealed nothing beyond the fact of strong radio emissions from the towers and a host of radio activity within the compound. The stories they picked up from the locals were more interesting. Nearly everyone they met had a story to tell which consisted of only minor variations on one central theme. The person would have been driving home late at night and, perhaps influenced by the local ale, would have decided to drive up and take a look by moonlight at what was going on in the old Manor. As this person drove up the slight hill, at the top of which one would be able to look out over the Wash, the car's engine would stall. Nothing the driver could do would get it started again. Eventually, as the starter was wearing out, a soldier would appear out of the dark woods and ask what the trouble was. The person would tell him about the stalled engine. The soldier would nod, look at his watch, and advise a wait of ten minutes—or five or eight or fifteen—after which the engine, he said, would be all right. He would then stand there with the driver in silence, periodically glancing at his watch, and at the appointed moment he would motion to start the car. With no trouble at all the engine would turn over. The soldier would then watch as the car turned around and drove away.

Everyone had this story to tell; it had either happened to them or to someone they knew.

Such stories, dealing with mysterious rays that could cause automobile engines or watches to stop, or could even stun or kill, that could reach out and drop airplanes out of the sky or start fires in remote forests, were a recurring theme of the 1930s—much as reports of animals glowing in the dark would become a staple anecdote around atomic research laboratories after the war. The "death ray" had become such a standard feature of British scientific folklore that the government had established a permanent prize of 1000 pounds to anyone who could demonstrate a ray capable of killing a sheep at 100 yards. There were so many applicants that the office in charge wanted to set up a five-pound fee, but the native affection for eccentricity would not allow this. The German military scientists were well aware of this British trait, and accordingly they didn't take the stories seriously.

An electromagnetic ray capable of interfering with the ignition system of an automobile or airplane engine was theoretically possible, but totally out of line with power requirements; simply put, General Martini reported, it was impossible. The stories of stalled and unstartable cars were explainable by the circumstances—local inhabitants, none too bright to begin with, sneaking up the road after an evening of heavy drinking, daring themselves to see what the military camp was up to. Easy to see how they might stall the car, even easier to see how they would flood the engine in their frantic attempts to start it and get moving again before they were discovered. When the sentry found them their problem would be clear to him; he would have told them to wait a few minutes not because of any electromagnetic emissions from the research establishment but simply to give the carburetor time to drip clear. Then when he gave them the signal the car would start easily. It would have nothing to do with any mysterious rays from Bawdsey Manor.

But, Martini went on, those 350 foot towers did in fact exist, and it was his responsibility to find out why. Hikers could not carry the necessary heavy radio equipment, and airplanes like the Lufthansa airliners could not hover motionless in the vicinity long enough to take the necessary readings. Accordingly, Martini requested authorization and funds to equip a squadron of twelve Zeppelin-type airships which would sail up and down the English coast, staying within international waters, and would monitor the radio signals from the Bawdsey establishment. With sufficient data the German scientists could interpret the English activity, and suitable precautions could then be taken in "the unlikely event" of war between the two countries.

Goering was unimpressed. All available personnel, money, and equipment were urgently needed to increase the offensive capabilities of the Luftwaffe as quickly as possible. But after long argument, with Martini supported by Goering's rival, Secretary of State for Air Erhard Milch, the Luftwaffe chief agreed to let Martini have use of the only two dirigibles that were then in airworthy condition; if significant results were obtained from preliminary reconaissances by these two airships, the situation could be reviewed. Which was as good as Martini could have expected. He brought his technical staff to Frankfurt-am-Main, where the L.Z.127, the *Graf Zeppelin,* had been mothballed for a year, and got to work.

The Death Ray

The *Zeppelin* was the child of Count Ferdinand von Zeppelin, who had retired from the German Army in 1891 to give his full attention to an attempt to fly by designing and constructing rigid, hydrogen-filled airships. He succeeded so well that during the Great War of 1914–1918 a fleet of eighty-eight "Zeppelins", armed with machine guns and carrying hundreds of pounds of bombs, caused panic in the streets of England with repeated raids. Both before and after the war, these huge airships carried thousands of passengers hundreds of thousands of miles in comfort and luxury and absolute safety; not one life was lost in more than two dozen years.

In 1928 the L.Z. 127 *Graf Zeppelin* was launched. The next year the ship flew around the world in the remarkable time of twenty days and four hours, stopping only in Lakehurst, New Jersey; Los Angeles; and Tokyo. During the following nine years she made nearly 600 flights, covering more than a million miles and crossing the Atlantic Ocean 139 times without incident.

And then on May 6, 1937, the *Zeppelin*'s sister ship, the L.Z. 129 *Hindenburg* burst suddenly into flames while landing at Lakehurst; the awful flammability of hydrogen gas was all too vivid to see. In that one moment airships became obsolete. Since then the L.Z. 127 had been stored in its huge shed at Frankfurt-am-Main, in size and color as well as usefulness a true white elephant.

But now General Martini's radio technicians began to work on her, bringing her back to flight readiness, installing and testing their equipment—one of the technicians fell to his death while leaning out of the cupola to tighten an aerial in flight—and on the evening of May 7, 1939, the 3-million-cubic-foot monster was hauled out of her shed and, with General Martini in command, set sail for England.

She came across the North Sea from Frankfurt at eighty-eight miles an hour and headed southeast, directly for Bawdsey. Ten miles off England, the captain turned her abruptly in heavy clouds and began to cruise slowly northward along the coast. Inside the cupola attached to the underside of the huge blimp was a mass of electronic instruments, radio receivers of one type or another. They were listening to the emissions that would come from the transmitting towers on shore, and they would analyze and catalog them according to frequency or wavelength and according to intensity, directional characteristics, modulation, polarization—anything at all that they could diagnose.

When the Zeppelin had come just within a hundred miles of England it had been picked up on the radar screens—for indeed the tall masts were radar transmitters. At Bawdsey it was the biggest single blip they had ever seen. At first they thought it must be a large formation of aircraft, but as they followed it for several minutes they were able to estimate its speed; at eighty-eight miles per hour it was slower than any airplane that could have crossed the North Sea. By the time it had reached the coast, slowed even more, and turned on a dime to the north, they realized that it was a dirigible, but of course they had no idea what its purpose was. And so they followed it by radar on its course up the coast.

Inside the *Zeppelin* things were not progressing too smoothly. The radio equipment gave out nothing but static. Either there were no radio transmissions coming out of the Bawdsey towers, or the airship's radios were not working. Martini had no option but to continue cruising up the coast and see if he could pick up anything.

As they slowly sailed northward, they were plotted by each of the radar stations in succession. Over Yorkshire they radioed a report back to Germany, giving their position as a few miles off the coast where the Humber empties into the North Sea. In Bentley Priory, the suburban London headquarters of Fighter Command, where all the radar reports were filtered and plotted, a burst of laughter swept around the table when the report of the German transmission was received. The radar plot showed the *Zeppelin* to be well inland, over the city of Hull in fact, though the Germans couldn't know this as they were still in the middle of the thick clouds which blanketed the eastern half of the country. Flight Lieut. (later Air Marshal) Walter Pretty was on duty that day; "We were sorely tempted to radio a position correction to the airship, but this would have revealed we were seeing her on radar, so we kept silent."

In any event, their secret would have been revealed had the German radio equipment been working. But as the blimp continued to sail slowly northward, its radios emitted nothing but squeals and squawks, cacophonous static, and the myriad dials and registers jerked spasmodically and randomly, recording nothing of interest. Finally Martini gave up and they returned across the North Sea, emerging from the clouds for a few hours before lowering through the early morning mists onto their station at Frankfurt.

They tried once more. On the night of August 2, just one month before the blitzkrieg was to be unleashed against Poland and the long dark night would begin, the *Zeppelin* set sail again. This time General Martini, convinced beyond any doubt that his radio detection equipment was working perfectly, remained behind and left the airship under the command of *Oberstleutnant* Gosewisch. Because of a full moon and a clear sky he kept her fifteen miles offshore, again making landfall near Bawdsey and cruising northward along the east coast.

This time Martini was correct in that the German radio receivers worked perfectly, but he had not reckoned on another possibility: This night it was the turn of the British equipment to malfunction. The entire radar network was down, and not a peep came out of the antennae lining the coast. The British had no idea that the *Graf Zeppelin* was out there, and Gosewisch received no hint of any radio emissions from the

The Death Ray 8

mysterious towers. Frustrated but unwilling to give up, he kept the airship cruising just out of sight of land all through the dawn and the following daylight hours. Finally, in midafternoon, she came within sight of a Coast Guard lighthouse off Stonehaven. Two fighter planes of No. 612 Squadron were scrambled from nearby Dyce aerodrome to take a look. They found her nearly twenty miles off the Kincardineshire and Aberdeenshire coast, and as they circled around her, making positive identification, *Oberstleutnant* Gosewisch finally accepted defeat and turned for home, having learned nothing.

Two nights later the Royal Air Force's summer air exercises began. For the first time in the three years that the radar equipment had been integrated with the maneuvers, it worked perfectly.

Had the *Zeppelin's* radios been working on her first flight, or had the British radar been working on her second flight, or had there been a third flight, the Luftwaffe would have known that England's defense rested on those frail 350-foot towers lining her coasts.

And the outcome of the Second World War would have been quite different.

Today we spend a considerable percentage of our lives working to generate income which is taken from us in taxes for the purpose of designing and building bigger, more destructive atomic bombs. We live our lives under the constant threat of these bombs. In part we accept such a threatened existence as normal, and in part we think of ourselves as cursed and damned above all the generations of mankind; but in truth our situation is neither inevitable nor unique. It all began in 1932 when Stanley Baldwin, Lord President of England, using the then-new technique of direct communication with the people via radio, warned the British public in a BBC broadcast that no matter what anyone could do, the bomber would always get through.

Aerial bombardment of civilian populations began in the First World War when on Christmas Eve of 1914, barely eleven years after Orville Wright became the first man in history to fly under his own power, one tiny, lone German bomb fell from the skies into a suburban garden at Dover. The Kaiser, grandson of Queen Victoria, was horrified. He forbad any further adventures of this type; warfare was not honorably conducted against helpless women and children, nor could such cowardly actions have any conceivable effect on the battlefields, where victory or defeat would finally be decided.

Actually the Italians had begun to experiment in aerial warfare even earlier, when in 1911 they had used airplanes to bomb the Sanusi tribes in Africa; there was, of course, no way to distinguish between spear-carrying warriors and baby-carrying women, but this action was only barely noticed (with a combination of slight amusement and disdain) in Europe. Actions taken against African blacks had little to do with European civilization. The bombing of England was something else entirely, something quite beyond the pale of accepted behavior.

But once something is technologically possible, one might as well command the tides to roll back as command it not to happen. The German Imperial Naval Staff began to submit memoranda urging the limited bombing of coastal military targets, and in January 1915 the first airships cruised the English coasts at night looking for such targets. They couldn't find any, but reported that the docks of London were clearly visible in moonlight, and what could possibly be more suitable as a naval target? They did *not* report on their total lack of bomb-aiming equipment, nor did they lecture their superiors on the tactical necessity of remaining high above the range of any possible rifle or machine-gun fire and of the resulting scatter in the bombing pattern. Permission was granted to attack the docks, but the bombing raids that followed damaged more homes—even suburban homes—than they did military targets on the docks.

Escalation: One of the nastiest words in our language, it does not belong to our generation alone. The German Army staff could not let this new military tactic fall by default to the navy, and accordingly ordered its own airships to bomb London, seeking not only the docks—which the navy could argue belonged to *them*—but any likely "targets of opportunity." Again the Kaiser protested, but the arguments against him were too strong, the chances of success too high, the airships both too expensive to waste and too unsuitable for any other military operation. On May 30, 1915, he capitulated to the extent of authorizing aerial bombing of targets to the east of the Tower of London. The first bombs were dropped on May 31.

And still the process escalated. The German naval staff now proposed attacking military targets "of interest to naval strategy" within London, targets such as the Admiralty offices, the railways which brought goods to the docks, and the Bank of England which provided the financial support for all naval operations. Soon the German militarists were arguing for the destruction of *all* railways, *all* offices, *all* administrative and logistical support facilities. Of course, these were scattered throughout the city, interspersed with civilian offices and residences, and so one had to accept the necessity of civilian casualties. Once again—for the last time—the Kaiser balked. But English aircraft had already bombed Karlsruhe, and though that single raid had been more ineffective than the German attempts to hit the docks of London, the future was clear or so the German naval staff argued. The enemy would build bigger airplanes and more destructive bombs, and once he had them, what would hold him back from using them? The only defense was to rain destruction on England so dramatically that they would not dare to challenge warfare in this new realm. The Kaiser conceded. Years before unrestricted submarine warfare would rouse the indignation of the western world and bring the United States into the war, the unrestricted aerial bombing of London began.

And, of course, it escalated. Soon the airships were wandering all over the southeast of England, reaching as far as the Midlands, and by 1916 German airplanes as well as airships were dropping their bombs on homes and farms. In May of 1917 the giant Gotha bombers made their first appearance over London, dropping hundreds of pounds of bombs each, traveling at heights (12,000 feet) and speeds (nearly ninety miles per hour) beyond the reach of the Royal Flying Corps fighters or the newly designed antiaircraft cannon. In June of that year a force of twenty Gothas attacked London in daylight, dropping two tons of bombs directly onto the Liverpool Street railway station and one 110-pound bomb directly onto a nursery school, killing sixteen babies and severely injuring a score more. On February 16, 1918, Hauptmann Richard von Bentivegni, of *Reisenflugzeugabteilung* 501, a squadron formed specifically for the purpose of bombing England—the name means long-distance-aircraft-organization—dropped a specially designed one-ton bomb on Chelsea. Just weeks before, this same commander had attacked London with the heavy bombs standard at that time, 660 pounds in weight. Aiming for the Admiralty offices, he hit the Odham printing works—which had been designated an air-raid shelter because of its heavy construction, designed to hold the gigantic printing presses beneath which, on this evening of January 28, 1918, 500 Londoners huddled in the basement. The bomb shattered the foundations, weakening them so that moments later the whole building trembled as the printing presses collapsed and brought the upper floors crashing down into the basement. Thirty-eight people died and a hundred others were injured. Had Hauptmann von Bentivegni carried the one ton bomb that night, all of the building's occupants would have been killed.

The future looked promising.

In 1923 a committee set up in England to study defense requirements recommended, and the government accepted, that a minimum Royal Air Force strength of fifty-two squadrons was necessary to secure the land from attack. But a decade later there were barely forty squadrons, and instead of talking about building more, the emphasis now was on discarding airplanes altogether. In that same year of 1923 the League of Nations had recommended an international peace conference; a decade later, on February 5, 1932, the World Disarmament Conference finally opened in Geneva. The annual sessions of its preparatory commissions, meeting every year for that decade, had produced only a preliminary proposal that was fully acceptable to none of its members. The Germans, still enslaved by Versailles, wanted equality among the nations; the French would fight to the death to keep the Germans enslaved, but would accept equality among all other nations and in fact demanded compulsory arbitration, enforced by an international police force, of all international problems; the British were determined not to accept any further international commitments; the United States had refused to join the League and now reiterated its refusal to cooperate with the League against any aggressor; and the Japanese, lending a bit of humor to the proceedings, were still insisting on racial equality. In the background, the Russians plotted their world revolution.

For the first time in history, air power was the center of attention. On November 10, 1932, the past and future prime minister of England, Stanley Baldwin, told the House of Commons and the listening world that the bomber was a weapon that could not be resisted: "I think it is as well for the man in the street to realize that no power on earth can protect him from being bombed. Whatever people may tell him, *the bomber will always get through.*"

Germany by this time had walked out of the Geneva Conference and the talks were stalled; Baldwin's speech was intended to shock the participants back into action. He went further in private, insisting that all nations must abolish the military airplane or "the world would be destroyed by this type of warfare."

His warning was seized upon by the people of the world, as it seized their imagination. It was backed up with official statistics, and we all know how frightening they are in their authority: the Air Ministry revealed that their studies had shown that the French *Armée de l'Air* was capable of attacking London and producing nearly 7000 deaths and 12,000 casualties within the first week. (Yes, it was the French the English worried about, and of course the Russians; the Germans were still a broken people and no threat was foreseen from that direction.) In Parliament Mr. Chuter Ede pointed out in visual terms how the noble Thames was an arrow that shone by moonlight, leading potential French bombers straight from the coast into the heart of London. Mr. Nobel-Baker described the new incendiary bombs, and told how just twenty bombers armed with these could reduce London to a blazing inferno in just one night. Stories of the Great Fire of 1666 paled by comparison.

London was particularly defenseless because of her geographic position just fifteen miles from the coast. Paris, and every other major city of every other major nation, lay deep inland; if British bombers were to attack Paris, she could be defended by hordes of fighter aircraft which would have ample time to destroy or turn back the armada of bombers as they made their way over the French countryside. But if French bombers were to come against London, there could be no warning of their approach until they were within sight of the coast. And from there, flying at 200 miles an hour at 15,000 feet, they would reach London in just four and a half minutes. The British Bulldog fighter, which in 1932 was only beginning to reach the operational squadrons, would need fifteen minutes merely to reach that altitude. There was no possibility of even intercepting an enemy raid before it reached London, let alone destroying it or turning it back.

Small wonder that the British position at the disarmament conference was dedicated to the dissolution of the world's air forces. And yet there was a voice of nagging dissent at home. It came from the offices of the Colonial Secretary and the Chancellor of the Exchequer.

Before the 1914 war, the management of Britain's colonial empire required the expenditure of vast sums of money for the support of standing armies along the Northwest Frontier of India, in Khartoum and Sudan, in Johannesburg and Singapore. Wherever there were native populations to be kept under control, wherever the "white man's burden" included the keeping of the peace in alien lands, it was necessary to keep an army garrisoned, clothed, fed and housed. But by the end of that great war a few cheap squadrons of mobile bombers were replacing these large, expensive armies. If a tribe along the Northwest Frontier began attacking its neighbors or the local missionaries, a couple of bombers could fly up from Bombay within hours and destroy the offending village from the air, while all the natives could do was stand helplessly by and watch the bombs fall. If another rebellion were to take place the very next day hundreds of miles away, the same squadron could deal with it. A miniscule number of men and machines could control all of India, scouting the country in continuously scheduled reconnaisances and nipping insurrections in the bud before they ever rose beyond the nuisance stage. By the boatloads, Britain's merchant fleet brought the soldiers home and turned them loose on the streets of London to join the ranks of the unemployed, where they were no expense to the Crown. And throughout the world a few squadrons of bombers patrolled the skies and kept the King's peace.

So it wasn't quite cricket to suggest the League of Nations prohibit the use of military airplanes, in the manner that poison gas was prohibited. What was wanted was something a bit subtler—a resolution suggesting that the civilized nations of the world agree not to use air power against each other, but permitting each nation to utilize it in internal affairs as they saw fit. To put it bluntly, it was seen as morally permissible to bomb the natives in the hills of Africa, but reprehensible to use those same airplanes to bomb the citizens of Berlin or London.

Such an argument, swollen and cluttered with dangling participles and subordinate clauses, could not have the simple moral suasion of those that had been used successfully to argue for the prohibition of poison gas, particularly when the subtleties of international law were presented in opposition. Could His Majesty's Government accept with surety an international agreement prohibiting the use of aerial bombers on the grounds that the civilian population of countries at war should not be made to suffer, when they themselves depended on their Royal Navy to impose blockades as their first form of offensive action? Did not these blockades, whose object clearly was to starve the offending nation into submission, themselves constitute martial action against civilians?

Of course they did. Clearly all that a nation suffering such a blockade had to do was question in the court of international opinion, or in such courts of international law that the League might set up, the legality and morality of the British blockade, and either the blockade must be removed or they themselves would be assured of the right to counterattack with aerial bombardment against London. His Majesty's Government could not press for an international agreement against air forces unless it was prepared to give up its own navy, or at least its navy's most effective weapon. And this it was not prepared to do.

The arguments went even further, seeping into the chinks in the armor of each one of the powers. Fear and mistrust were, then as now, basic concomitants of international relations. One nation could trust another not to build military airplanes only insofar as it built no airplanes at all, for civil airliners could be modified overnight to carry bombs instead of passengers. (The German Focke-Wulf Condor airliner was in fact so modified at the beginning of the Second World War, and became Germany's only four-engine bomber.) It was suggested that an international agreement might prohibit government subsidies to airlines and airline manufacturers, so that they might not design and build airplanes suitably constructed for quick modification to bombers, but such an agreement would be impossible to enforce, and would not be particularly effective even if enforced. Airliners and bombers share much the same characteristics: they both strive for the best possible speed, range, and carrying capacities. And even if these characteristics should be limited by government agreement, with cataclysmic repercussions for the burgeoning civil aircraft industries of the European nations, there would still be no safeguard against aerial bombardment; there would only be a shift from gigantic lumbering bombers to great hordes of smaller, faster bombers. The Secretary of the Air Ministry, Sir Christopher Bullock, pointed out in a memo to the Chief of the Air Staff that even such a craft as the deHavilland Puss Moth—a small biplane intended for civil use, as meek and unwarlike as its name—could be modified overnight to carry 350 pounds of bombs over distances of hundreds of miles.

All these arguments were real and true, and they led inescapably to the conclusion that no nation could realistically rely on international agreements, public opinion, and the basic good will of humanity to protect it against destruction from the air. The only possible recourse was to pursue the opposite course: to build for oneself an air force capable of deterring attack by imposing the threat of instant massive retaliation.

In 1933 Hitler came to power and pulled Germany out of the Geneva Conference, which thereupon collapsed. The German diplomats went home to find Goering already starting to build a Luftwaffe; the Americans came home to read newspapers which reported that the Europeans were wrangling again and that this time we shouldn't send our own boys to save their decadent hides; the Japanese went home and continued the invasion of China they had begun two years before and which they had put on hold pending the outcome of the Conference: if there was to be no rule of international law and no declaration of universal racial equality, they would take what they could by force of arms; the British retired across the Channel, the French went back to Paris, and each stood looking with mistrust at the other, forgetting their recent alliance of 1914 and remembering instead Henry V and Joan of Arc; the Poles and the Czechoslovaks and the Bulgarians and the Yugoslavs continued sniping at each other in midnight cross-border raids. The possibility of a system of worldwide international law and cooperation, snarled and entangled in a mass of contesting and constricting self-interest and mutual distrust, died aborning. Each nation went home, and bigger bombs and better bombers began to stream out of the dark caverns of its factories, to sit gleaming and deadly in the sunlight, awaiting the order to take off and fly.

And so was born the world we live in today.

In 1939, with Germany poised to launch its *Blitzkrieg* against Poland and France, RAF Fighter Command still didn't have its fifty-two squadrons. It had just five squadrons of Spitfires and a dozen of Hurricanes, plus another couple of dozen either still equipped with obsolete biplanes or just beginning a transition to the modern machines. And it had radar—incomplete, difficult to operate, and prone to malfunction, yet the one factor that could nullify the disadvantages Britain was heir to, that could provide its one chance of survival.

Air Chief Marshal Sir Hugh Dowding, Commander-in-Chief, RAF Fighter Command was asked if their only hope was to pray to God and trust in radar. He replied, "At this stage I would rather pray for radar, and trust in God."

Chapter Two
Origins: The Death Ray

> Quite entertaining, but probably a hoax.
> —*H. E. Wimperis,*
> *Director of Scientific Research,*
> *Air Ministry*

On September 19, 1939, Hitler made a speech in Danzig claiming—according to the British Foreign Office—to have a secret weapon against which all defenses would be useless. This raised again worries which had been circulating privately through official channels for the past several months. All during the previous spring and summer, when it appeared that war might be likely, Joachim von Ribbentrop, the German ambassador in London, had been spreading rumors that the Nazis had a new explosive of unprecedented power. What he was referring to was actually no secret: in December of 1938 the tremendous release of atomic energy in the fission of uranium had been discovered—in Berlin. The chemists who had done the experiment did not realize its significance; the first person to understand what had actually happened in the Berlin experiments was the theoretical physicist in their group, Lise Meitner, who came to the conclusion while she was fleeing Germany to escape the Gestapo, which had ordered her arrest because she was Jewish. Thus the secret was lost by Germany before its military potential was understood. But since the experiments had been done in Berlin, no one in England was quite sure how far the German scientists might have progressed toward the possibility of a bomb.

Winston Churchill, at that time still out of the inner government circles, conducted his own investigation and reported to the Secretary of State for Air that "while at first sight this might seem to portend the appearance of new explosives of devastating power," his scientific friends were quite sure that nothing of any practical importance could be done with it for some time, at the very least. Which, looking back on it, turned out to be quite true: It would be another six years before the Bomb exploded. What von Ribbentrop had being doing in those last few months of peace was trying to spread whispers of Germany's military superiority, and so to influence the British not to take a chance on war but rather to continue to accept Hitler's territorial demands without physical opposition.

Other scientists supported Churchill's conclusion that summer that the threat of fission bombs was merely illusory, a bogey in the night. When Germany invaded Poland on September 3, Britain duly announced its ultimatum that Germany should immediately leave Polish territory and cease offensive operations. When Hitler ignored this demand, Britain resolutely declared war. But now, two weeks later, Hitler's public claim of a secret weapon brought back all the old fears. The Prime Minister, Neville Chamberlain, immediately asked the Secret Intelligence Service for information. Dr. R. V. Jones, a scientist and new SIS employee, was assigned the task of sifting through the prewar files to attempt to find some indication of what Hitler might be talking about. There were numerous suggestions from the military, with each service plumping for a choice that would provide new funds for counter-development of its own pet schemes: aside from the atomic bomb, other suggestions came from the Navy who wanted the secret weapon to be a magnetic mine and the Air Force who wanted it to be a rocket-propelled fighter. The SIS itself considered the death ray most likely.

The concept of a ray that could cause physical, mental, or mechanical incapacity, or even death, was a popular one in the 1930s; in America the cartoon-strip character Buck Rogers had not only a death ray but a disintegrator ray, a heat ray, and a stun ray, and his popularity was based in large part on the belief that there was more truth than fantasy in his gadgets. This feeling was reinforced a few days after Hitler's Danzig speech when the Foreign Office translators revealed that he had evidently let slip something of the nature of the weapon, referring to its effect as "blinding and deafening" its victims.

But none of this made scientific sense. As he continued to draw a blank from the SIS files, Dr. Jones began to wonder about the speech and the German language; he asked a friend, also working for SIS, to listen to the BBC tape of Hitler's speech. Frederick Norman, normally Professor of German at King's College, London, reported back that Hitler's use of the German language was barbaric, and that the phrase he had used was actually capable of two intepretations—one translation might be "a weapon against which all defenses were useless," while another could be "a weapon with which Germany could not be effectively attacked"; the latter phrase presumably could have referred to the Luftwaffe's overwhelming superiority over the RAF, and not to any secret weapon at all. As for the potential of rendering the victims blind and deaf, that was clearly a misinterpretation of a German idiom similar to the English term "thunderstruck." Jones's conclusion was that Hitler's ranting was certainly a threat of military violence, but not of any secret technology.

Yet it is easy to see how any threat of scientific gadgetry was to be taken seriously. In the early years of the twentieth century the concept of the death ray, in particular, was a curiously satisfying amalgamation of fantasy with the recent discoveries of science that seemed even more fantastic, and which promised a new world unconstrained by previous notions of reality.

The existence of mysterious rays can be traced back at least as far as the year 1602, when a Bolognese cobbler and part-time alchemist, Vincenzo Cascariolo, was climbing Mt. Pesara and noticed a heavy mineral reflecting the sun's rays. He brought it home and gave it to his wife, who placed it on the mantel in their bedroom. That night he was awakened by his wife's screams: she had got up to use the pot kept under the bed, and had opened her eyes to see the devil's mineral shining at her in the dark.

Nearly 300 years later this phenomenon of fluorescence, the ability of certain natural materials to store sunlight and emit it later, "to shine in the dark," was still not understood. On November 8, 1895, Konrad Wilhelm Roentgen, professor of physics at the university in Wurzburg, was in his laboratory studying the phenomenon by correlating it with the passage of electricity through air. The latter phenomenon was just as puzzling as the former, since it was well known that air does not conduct electricity. Yet in a "cathode tube" it sometimes did.

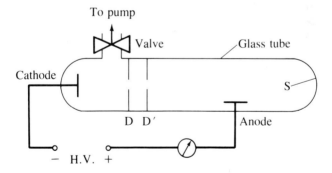

If the two electrodes, cathode and anode, were connected to an electrical source such as a battery or an induction coil, no electric current flowed under normal conditions; this was understandable since the two electrodes were connected on one side (the external wire) but not on the other (through the tube). But when the air in the tube was gradually pumped out, something curious happened; a current would begin to flow, even though there was still no connection through the tube. In fact, the current would grow in intensity as the pressure was decreased.

Though this curious phenomenon and that of fluorescence were different, they seemed to be related by an observation that had only recently been made: with sufficient current flowing, a spot on the glass tube directly opposite the cathode would begin to fluoresce. It was thought that the current through the tube was carried by mysterious "cathode rays" which could only penetrate from cathode to anode when the interfering air molecules were sufficiently pumped away. Where these rays missed the anode and hit the glass wall they induced the fluorescence. It was this possible connection that Roentgen was studying.

His method was simple; he systematically varied the applied conditions one at a time, noting the effect of each. It was the standard drudge work of science. Because the fluorescent effect was mild, dropping off as the current was reduced, his laboratory was darkened. Because of this dim lighting he noticed a faint glow coming from the other end of the room; investigating, he found a sheet of paper coated with glowing barium platinocyanide crystals. This is a compound that he had used in previous work, since it fluoresced when exposed to sunlight. The fluorescence was transitory, however, disappearing within minutes; and the crystals had been stored in this darkened room for days. Curious, Roentgen went back to the cathode tube and turned down the current. As he watched, the barium platinocyanide slowly began to dim; in ten minutes the room was totally dark. He turned the current up, and again the crystals began to glow. Obviously the cathode rays were escaping from the tube and exciting fluorescence in the crystals. It was an interesting observation; he jotted it down in his notebook and then, getting back to his original work, he moved the crystal-coated paper behind a screen to shield it from the cathode rays and cut off its glow so that it wouldn't interfere with his observations of the tube.

He worked for another couple of hours, varying the pressure of gas within the tube and the electric potential applied across the electrodes, and noting how the fluorescence varied with changes in these parameters. Needing a screwdriver, he left the apparatus, crossed the room, and as he returned he noticed the barium platinocyanide still brightly glowing behind the screen.

He paused, puzzled. He turned off the tube, and the glow died away; he turned it on again, and the glow returned. Clearly some rays from the tube were causing the fluorescence, but how could these rays pass right through the screen? He tacked several sheets of thick black cardboard to the screen, and still the crystals glowed. He reasoned that the rays must be coming out of the tube at all angles, bouncing off the walls, and coming at the crystals from behind. He wrapped the cathode tube in the black cardboard, so nothing could get out. And still the crystals fluoresced.

Roentgen had forgotten the original experiment now; this observation had all his attention. Certainly it was possible for one object to go through another—a bullet could easily pass through the black cardboard—but it would leave a visible hole behind. Light, on the other hand, could pass through glass without causing damage, and so he realized that what he had in this experiment was proof of another kind of electromagnetic ray, one with the ability to pass through opaque substances and carry sufficient energy to induce fluorescence.

He set up further experiments, one after the other. He put different types of material between the cathode tube and the paper covered with fluorescent crystals, and found that the mysterious rays passed easily through glass, wood, and paper; not so easily through light metals such as aluminum; and least of all through heavy metals such as iron or lead. He measured the precise thicknesses of each of the materials through which the rays could travel. Then one day the following week he reached out to take a sheet of wood from between the tube and the crystal-coated paper; as he did he noticed a shadow pass over the paper. Cautiously, he held his hand directly in between the two, and clearly on the fluorescent paper he could see the faint shadow made by his hand—not by his whole hand, but only by the bones inside it.

Skin and muscle are made of the light atoms carbon, hydrogen, and oxygen, the same atoms that comprise wood and paper, through which the rays passed easily. But bones are made of compounds of calcium, a much heavier element; and the heavy elements stopped the rays.

In the next few weeks Roentgen found that the mysterious x-rays—he emphasized their mystery by naming them after the familiar algebraic notation for the unknown quantity—had the property to expose photographic film as well as to excite fluorescence, and he used that property both to make more detailed studies and to keep a permanent record of his observations. He took pictures of the bones in his wife's hand and of metal objects inside closed wooden boxes. On New Year's day 1896, not two months after he'd first been aware of the x-ray, he mailed copies of his pictures and an accompanying paper to physicists all over Europe. It caused an immediate sensation. Newspapers in Vienna reported that a German physicist had invented a camera that could take pictures of the bones inside living people—getting the story right, but misspelling the name of the physicist as "*Routgen.*" An American newspaper got the name right, but the story just a bit wrong: they reported that Roentgen had invented a camera that could see through things, and speculated that if people such as those living in Paris got hold of it they could point it at fully clothed people and take pictures of them naked. More importantly, the discovery stimulated research all over the world. The medical profession was taking x-rays of broken bones within weeks of Roentgen's announcement, while in England

J. J. Thomson identified the cathode rays as streams of electrons and in Paris Becquerel began his investigations which were to lead to the discovery of radioactivity. Within ten years Marie Curie had discovered radium, Rutherford had identified alpha, beta, and gamma rays, and the world was suddenly filled with all sorts of mysterious emanations. On lecture tours Pierre Curie carried a small vial of the glowing yellowish crystals of radium salt; he was fond of setting it down on a block of ice so that everyone could watch as it melted its way through. He showed people the red welt on his skin under his vest pocket where he carried the vial; it was all very amusing.

Madame Curie claimed that the radium rays could cure cancer; others saw them as more deadly. H. G. Wells wrote of atomic bombs which would spew out rays that would kill the populations of entire cities. The military of all nations asked its scientists if they could produce enough radium to use in a death-ray gas. A St. Louis newspaper predicted the end of the world. In the midst of the first Great War of the new century, the belief in the existence of still more deadly but as yet undiscovered rays was sufficiently advanced that people tried to swindle the government with false claims. Field Marshal Lord Haig, Commander-in-Chief of the British Forces, witnessed a demonstration in 1916: two rabbits in a cage were irradiated with a newly-invented deadly ray, within minutes they were twitching on the floor; in another few minutes they were dead. The ray looked promising, as much as or more so than the Boer War's machine guns or the monstrous tanks that would soon see action on the Somme. But further investigation of the demonstration showed that for the "death ray" to be effective the rabbits had first to be secretly poisoned.

In 1924 H. E. Wimperis, newly appointed Director of the Air Ministry Laboratory, visited the home laboratory of a Mr. Grindell Matthews, who purported to demonstrate a working death ray. "Quite entertaining," Wimperis reported later, "but probably a hoax." Probably? The Air Ministry was sufficiently impressed to offer a reward of 1000 pounds for any such device that could pass a strict scientific test.

Matthews was never able to rev up his equipment to pass such a test, the basic requirement of which was to kill a sheep at a distance of a hundred yards, but others took his place. A corporal in the Royal Artillery invented a ray that would kill mice, which he proved with a series of convincing photographs—convincing enough so that the army promoted him to lieutenant and built a laboratory for his experiments. When a year went by with nothing but more photographs, they finally called in a scientist, who quickly found evidence of doctored pictures and total fraud.

There is a mathematical theorem which states that no non-zero integers exist which satisfy the relation $x^n + y^n = z^n$, for $n > 2$. The theorem is thought to be true, but the proof has escaped the mathematicians.

In 1665 the great mathematician Pierre de Fermat died, and when his disciples went through his possessions they found scribbled in the margin of a book the statement, "I have discovered a truly remarkable proof (of this theorem) which this margin is too small to contain." But despite more than three hundred years of searching for it the proof was never found, and from that day to this, mathematicians are still trying to prove Fermat's Last Theorem. Similarly, after Marconi died in 1937 the rumor spread that he had discovered the death ray, though no written description was ever found. But rumors and fantasies feed on their own flesh, and the various government ministries in England were kept busy throughout the thirties investigating claimants to the 1000 pound reward, together with other interesting inventions—such as the scheme for solidifying clouds and using them as sites for antiaircraft weapons, or for freezing the English Channel with liquid nitrogen so that troops could cross. It was difficult to reject such claims out of hand; the story still circulated throughout Britain's ministries of the drawings for a proposed tank that before the Great War of 1914 had been relegated to the War Office's dusty archives with the scribbled comment, "The man's mad."

Professor da Costa Andrade of the University of London was a frequently used consultant. One morning a general broke in upon him with the exciting news that someone had invented a ray which could detect hidden stores of dynamite. The general explained that he had carefully tested the man, who by passing his ray over a map could determine where any explosives were hidden. The general had ordered a sergeant to bury a box of explosives in a field and had given the man a map of the area. The fellow had been able to locate approximately where the box was buried. Professor Andrade replied that he could do the same, and immediately did. He simply looked at the map and picked a spot in the center of the field: "Naturally that is the place a sergeant told to bury an explosive somewhere in a field would select."

Andrade's death ray investigations were similar. He was called one day to witness what seemed to be a successful device. Going to an office in London, he found the equipment set up and working; since there were no sheep in London the inventor demonstrated the effect by putting out an electric light bulb with the ray. Clearly it worked, but "I was convinced," Andrade said, "that it was being worked quite simply by a flex that went invisibly under the carpet and could be manipulated as required by the inventor." And so while the death ray was turned on, he diverted the inventor's attention for a moment and quickly stepped directly into the path of the death ray. When the inventor turned back he saw him standing there, grinning happily, though the electric light was still 'mysteriously' doused.

* * *

The Death Ray

Governmental interest in the death ray peaked in 1935, sparked by a chain of inquiries that began with Hitler's accession to power in Germany in 1932 and the RAF summer air exercises of 1934. The object of the air exercises was to test the defenses of London and the surrounding provinces which housed the aircraft industry; the results showed that, without exception, there were in effect no defenses at all. The exercises consisted of night attacks, and even though on some nights bad weather made navigation difficult and kept the bomber speeds down to sixty miles an hour, more than half the bombers reached their targets without being intercepted, just as if they were airliners sweeping through peaceful skies. But to most people the dreadful implications of the exercises, combined with Hitler's announced intention of ignoring Versailles and building a Luftwaffe 4000 bombers strong, were lost in competing angers and anxieties about the cost of the exercises to the taxpayers and about the disturbing noise. The next morning in Parliament a Member asked the Air Minister, "Is the right honourable gentleman aware that many inhabitants in Regent's Park and elsewhere were awakened last night, including the hyenas at the Zoo, which in turn awakened several others?"

It was time when preparations for war were unpopular in democracies and were largely ignored until conflict actually broke out. Today our military budget is 300 *billion* dollars which, even allowing for inflation, is orders of magnitude greater than defense expenditures in the first half of this century. The massive pendulum swings from one extreme to another, and it has always been very difficult for men to argue it into submission using nothing but logic against the contrary claims of fear (today) or complacency (yesterday).

> *Yes, makin' mock o' uniforms that guard you while you sleep*
> *Is cheaper than them uniforms, an' they're starvation cheap;*
> *An' hustlin' drunken soldiers when they're goin' large a bit*
> *Is five times better business than paradin' in full kit.*
>
> *Then it's Tommy this, an' Tommy that, an' "Tommy, 'ow's yer soul?"*
> *But it's "Thin red line of 'eroes" when the drums begin to roll....*
> *The drums begin to roll, my boys, the drums begin to roll,*
> *O it's "Thin red line of 'eroes" when the drums begin to roll....*

It was the same in California and Hawaii and the Philippines in 1941, when the war came rolling in over the Pacific swells; it was the same in Korea and Vietnam, the same as it had been a hundred and fifty years earlier when General Wolfe wrote from Quebec: "I think our stock of provisions for the siege full little, and none of the medicines for the hospital are arrived. No horse or oxen for the artillery, etc. Our clothes, our arms, our accoutrements, nay, even our shoes and stockings, are all improper for this country."

And so it was that while the Luftwaffe took shape in Germany, the call in Parliament was to allow the citizens of London, and the hyenas in the zoo, to sleep more soundly. But a few thousands of pounds had recently been expended in creating a new position in the Air Ministry, a Director of Scientific Research. The first Director, appointed after much civil service haggling ten years before, was forty-seven year old Dr. H. E. Wimperis, who had joined the Naval Air Service in 1915 and was at the time running the small Air Ministry Laboratory in Imperial College. He had one assistant, a young man named A. P. Rowe, who had begun his scientific career as a meteorologist. The new Directorate was much less than it appeared to be from its title. It was not to interfere in the development of dirigibles, radio, or even armaments—flight alone, not bombs or guns, was to be its concern. "Above all," Rowe later wrote, "there was the curse of secrecy. Although secrecy in defence science is often essential, it must be admitted that the more secret a project the less efficient it is likely to be.... The few workers in this field rarely if ever had to encounter the criticism of greater or even other minds."

In science, it is often said, the most important thing is to ask the right questions. In 1934 the directorate asked the right question. Shortly after the summer air exercises, Rowe journeyed to the Biggin Hill aerodrome, half an hour south of London, to view a demonstration set up by the army's Air Defence Experimental Establishment. The problem that group had been working on was how to provide advance warning of an aerial attack on London—"the greatest target in the world, a kind of tremendous fat cow, a valuable fat cow tied up to attract the beasts of prey," as Churchill had described it. The method they had devised was sound location, in which sensitive acoustic systems were set up to gather the sound of any aproaching airplanes long before they could be heard by the human ear. The center of this system was a giant, acoustically molded wall, 200 feet long and 25 feet high, imbedded along its length and breadth with the most sensitive microphones. (It stands to this day in splendid abandoned isolation in Romney Marsh). The wall was immovable, of course, and designed to detect aircraft noises along a line from London to Paris—it had been designed and constructed a few years previously, when Germany was still unarmed and likely to remain so and when France was still England's traditional enemy. Now that the focus of enemy attack had switched to the northeast, it was impossible to move the wall; it stood then, as it stands today as a mute testimonial to archaic ideas. A part of the basic idea, but intended to be more mobile than the wall, was a system of directional microphones mounted on movable bases. It was intended that a fleet of these be disposed around the coastline, to pick up the sound of airplane engines and follow them in their flight, passing accurate information by telephone to the antiaircraft guns with which the army was concerned, and perhaps also letting the RAF know what was going on. The army was particularly pleased that they were so far ahead of the RAF in aerial defense measures, and was happy to give this demonstration to the Air Ministry's Directorate of Scientific Research, in the person of the director's assistant, A. P. Rowe, and to the RAF's Air Member for Research and Development, Air Vice-Marshal Hugh C. T. Dowding. The test would consist of a bomber flying a predetermined route toward Biggin Hill, whereupon its noise would be picked up long before the plane itself came into view.

As the equipment was being set up, "there appeared in the distance an ordinary milkman, driving his horse and cart, with his milk-churns jangling away," Rowe recalled. "The road along which he was coming should have been closed, but for some reason this had been overlooked. The milkman was stopped, reprimanded, and told that under no circumstances would he be allowed to pass along the road that afternoon."

The reason, it was immediately obvious to both Rowe and Dowding, was that the sound locators would be confused by any extraneous noises—even those as innocuous and unavoidable as those of a milkman making his rounds. God forbid the Germans should attack when anyone in England was up and moving, or if an English bird was chirping!

Just as serious were the innate limitations placed by the laws of physics on such devices. Sound travels at about 700 miles an hour, which is not so tremendously greater than the speeds bombers were expected to attain in a few years. If a sound echo were received from forty or fifty miles away it would take several minutes to reach the microphones, and in that space of time the bombers would have flown so far that they would no longer be where the sound detectors indicated.

Rowe began to wonder if there was anything that anyone could do to avert the coming catastrophe. Official estimates of the damage that a modern bomber fleet could inflict on London were horrendous: thousands upon thousands killed and dismembered and burned to death, hundreds of thousands made homeless, all within the first days of war. With the failure of the Geneva disarmament conference there seemed no way to prevent the approach of Armageddon; people hunched their heads between their shoulders and were afraid to glance up into the baleful skies.

* * *

When the last great ice sheet melted 10,000 years ago, it dumped enough water into the ocean to raise sea level and flood the low-lying valleys on the edges of the continents. In this way the English Channel was created, and since then England has been, for all practical considerations, an island separated from the political vicissitudes of the continent by that formidably wet military barrier to the passage of arms and armies. Secure behind her wall of water she could send forces to France or not, as she chose; the European armies that decimated the peoples and fields of France, Germany, Holland, and Italy never touched her shores. But with the coming of the bomber, all was changed. Instead of the clash of armies on foreign soil, the sounds of war would now be the scurrying feet of women and children, the sobbing anxious breathing of old men and women hurrying to their underground shelters as their homes exploded behind them. "War, which used to be cruel and magnificent," Winston Churchill complained, "has now become cruel and squalid. In fact it has been completely spoilt.... I wish flying had never been invented. The world has shrunk since the Wrights got into the air; it was an evil hour for poor England."

He was among the first to realize the consequences of this new form of warfare, and he spent the 1930s railing at the attitude of His Majesty's Government, from which he was largely excluded, roaring at them to increase their defenses. But when they sporadically did, they only further exposed the weaknesses of the RAF whose leaders squabbled among themselves with each new pound to be spent. Should they buy more bombers to avoid war by the threat of destroying any enemy homeland, or should they buy more fighters to try to knock down the enemy bombers? Little by little the bomber enthusiasts won; led by Lord Trenchard, they formulated the concept of peace through the threat of mutually assured destruction of civilian populations.

At the Air Ministry, after the demonstration of the nonefficacy of the army's sound-location detectors, A. P. Rowe took it upon himself to study all scientific proposals that had been made for improvements in air defense since the Great War. He was at the time the only staff member of the Directorate of Scientific Research of the Air Ministry to be employed on armament problems, and even this was probably illegal since the Directorate itself had been told that armaments were none of its affair. Nevertheless, he now searched through the files and found only fifty-three, among the many thousands of Air Ministry files, concerned with air defense. Fifty-three only, and not one of them was of the slightest use; not one of them could even remotely shake one's faith in Baldwin's pronouncement that "the bomber will always get through." There seemed to be nothing to do but accept the Trenchard strategy and build more bombers, but clearly that way led to destruction. With all of her population crammed into a small island having no avenue of escape; with London, her greatest city, tethered fast at the very southeastern corner of the island, too close to the coast for any effective defense to be mounted; with her only real strength floating on her warships, which were increasingly being shown as vulnerable to aerial attack, England was bound to lose in the coming games of war. "You came into big things as an accident of naval power when you were an island," Churchill told England. "Through an accident of airpower you will probably cease to exist."

Rowe took the line of any conscientious civil servant, performing the sort of action that is so often ridiculed by men of action: he wrote a memo. And his memo started the chain of events that was to save England and the rest of the western world. He wrote to his superior, H. E. Wimperis, Director of Scientific Research, claiming that everything else the directorate was engaged in amounted to nothing compared to this impending doom by bomber forces, and urging him to tell his superior, the Secretary of State for Air, of the dangers ahead—that "unless science can find some way to come to the rescue, any war within the next ten years is bound to be lost."

Rowe received no reply that summer of 1934, and took his annual vacation in a dispirited sense of mind: "My holiday companion had to listen to gloomy predictions regarding the coming war. That a few others were equally worried I have no doubt; but whatever their worries or mine, there was no sign in that summer of a drive to solve the problems of air defense." Wimperis, however, did in fact take Rowe's warning seriously. He was in a unique position, having been head of the Air Ministry's sole scientific research office since its establishment nearly ten years before, and yet never having had a free hand. He was at the service of the Air Ministry, which was in turn fighting for its existence against the other services which held it in no great regard. As Air Marshal Sir John Slessor was later to write:

> In 1907 the Admiralty were offered Wright's patents; Lord Tweedsmuir replied on behalf of their Lordships, "I have consulted my expert advisers with regard to your suggestion as to the employment of aeroplanes. I regret to have to tell you, after the careful consideration of my Board, that the Admiralty, while thanking you for so kindly bringing the proposals to their notice, are of opinion that they would not be of any practical value to the Naval Service."

The army was no more impressed. A member of their Council, on being urged to consider airplanes as suitable reconaissance machines, replied that since aircraft could not fly at less than forty miles per hour, it would be impossible for an observer in the plane "to see anything of value." Airplanes were rather reluctantly introduced into military maneuvers in the years just before the First World War, but were not a conspicuous success; more than one autobiography of the period reports superior officers ridiculing the idea that they could ever be of any use. The British chief of the Imperial General Staff (CIGS) regarded aviation as "a useless and expensive fad, advocated by a few individuals whose ideas are unworthy of attention." Even after the First World War, the other services totally denigrated the RAF. As Slessor wrote, "Every single advance in the use of airpower had to be fought through tooth and nail against usually stubborn and often intemperate opposition by officers of the older Services."

With the Air Ministry under continuous external attack from the army and navy, and with the scientists within the RAF under continuous internal attack because of the competition for the few monies available, it is little wonder that Wimperis had to keep a low profile. But faced with Rowe's memorandum, he realized that this was the time to forge full speed ahead. He took the action of inviting an old friend to lunch at the Atheneum, his London club. The old friend was Professor A. V. Hill of University College. Though a distinguished scientist, in fact a Nobel Laureate, Hill was a physiologist; he was not at all sure what advice he could give on air defense, although during World War I he had been director of the Anti-Aircraft Experimental Section, dealing in questions of the effect of blast on the human frame. Wimperis, in fact, wanted to talk to him about the death ray; telling him about Rowe's memo, he confessed that he could think of nothing else that science could do about the problem, and he wanted to know if, in Hill's opinion, such a weapon might be possible. Hill thought not, but wondered if more minds might not find more solutions.

Accordingly, Wimperis took the next step, one about which he must have felt much trepidation: he wrote to the Secretary of State for Air, Lord Londonderry, and to the Air Member (of the RAF Air Staff) for Research and Development, Air Marshal Sir Hugh Dowding, proposing the formation of a committee to "consider how far recent advances in scientific and technical knowledge can be used to strengthen the present methods of defence against hostile aircraft." This in itself may not have been terribly courageous, but Wimperis took the extra step of suggesting that the committee be composed not of Air Ministry personnel but of outsiders. He was suggesting, in effect, that the atmosphere inside the Air Ministry was not such as to be efficacious for the flowering of men of independent minds, and that a truly effective examination of potent possibilities could not come from such men as existed within the ministry.

Such a suggestion does not strike one with the force of its courage in the same manner as Henry V's exhortation to his men, but it does exhibit courage of a different kind. At the time at which it was made, it surely meant that Wimperis was laying his job on the line; it would have been so much easier for him merely to suggest that the ministry itself look into the matter.

And it would have been easy for his superiors to ignore his note, particularly when one considers their own positions at the time. Two men less likely to be effectual are hard to imagine.

Londonderry was a most pleasant and personable man, and was wildly enthusiastic about the air—he was one of the few Air Ministers who ignored Sir William Gilbert's famous advice for getting on in governmental office ("Stick close to your desk/And never go to sea/And you all may be rulers/Of the Queen's Navy."); he actually learned to fly. But Gilbert was right after all: knowledge of and enthusiasm for the service were insufficient to carry the day in the corridors of power. A Conservative appointed in 1931 by MacDonald against Baldwin's advice, Londonderry proved to be probably the most ineffectual minister in the government. His pleasant manner was regarded as weak, his enthusiasm as childish; he carried no weight in the political infighting that was part and parcel of the job, nor in the more formal give-and-take in Parliament and the Cabinet. Baldwin at this time held the position of Lord President; deeply concerned with air defense, he tended to bypass Londonderry and make decisions himself.

Air Marshal Dowding was another kettle of fish entirely; known as "Stuffy" to his friends (and God knows what to his enemies, of whom there were many, including at the moment Lord Trenchard, head of the RAF, and later Winston Churchill), Dowding was a severe, iconoclastic, free-thinking individual in a service which included among its traditions neither severity nor iconoclasm, and certainly not free thinking. "Since I was a child I have never accepted ideas purely because they were orthodox," he once wrote, "and consequently I have frequently found myself in opposition to generally accepted views. Perhaps, in retrospect, that has not been altogether a bad thing." Perhaps not, but it was for this reason more than any other that he had been passed over for Chief of the Air Staff, and was put off to one side—in the position in which, as it turned out, he was to influence decisively the future of the coming war.

Londonderry's and Dowding's enforced isolation from the mainstream of governmental thinking worked both ways: the government was in turn isolated from their thoughts and paid little attention to what they did. And so there was in fact no opposition when they acquiesced to Wimperis's suggestion, forming the committee he had suggested under the title of the Committee for the Scientific Survey of Air Defence; they also accepted Wimperis's suggestion for chairman, Dr. H. T. Tizard, a pilot with the Royal Flying Corps in the first war who had gone on to become a leading scientist and chairman of the Aeronautical Research Committee; filling out the committee were Hill, Professor P. M. S. Blackett of Manchester, like Hill a Nobel Laureate, and Wimperis, with Rowe as secretary. (Soon Winston Churchill was to find out about the committee, and then would follow rumblings and reverberations that would nearly tear it apart and doom the concept of aerial defense before it matured. But for the moment it lived on in a mantle of obscurity and indifference.)

The committee met for the first time on January 28, 1935. The aging scientists who sat down around the table were about to alter the course of European history. Although scientists had been called on for military assistance since the days of Archimedes' catapults in Sicily and the even more ancient days when Chinese soothsayers predicted fortune or disaster on the basis of comets and falling stars, scientific help and advice was not consistently sought nor, when offered, always appreciated. When the Great War broke out in 1914 Napier Shaw, head of the British Meteorological Office and that country's most distinguished weather expert, had gone unsummoned to the War Office to offer his services in whatever capacity he might best serve, only to be rebuffed with the question, "Sir, do you suppose the British Army goes into battle carrying umbrellas?" But now, on January 28, 1935, for the first time in history the considered opinions of scientists were to become a firm platform on which future political options would be decided.

Before they began to consider what new under the sun might deflect the bomber forces, the committee members decided to reconsider all the old ideas, if for no other purpose than to finally dispose of them and get on with the new. So, then; was there any possibility of concentrating energy in a beam of electromagnetic energy sufficient to incapacitate a flying man or his machine? Wanting to get the committee off to a good start, Wimperis (who was also on the Radio Research Board of the Department of Scientific and Industrial Research) had already asked Dr. Robert Watson Watt, head of the Radio Research Laboratory at Slough, to meet with him at the Air Ministry. That meeting had taken place a week before, on January 18, and Wimperis had posed the death-ray question. Watson Watt thought it impossible, but declined to give a definite answer until he had carried out some calculations. By the time the committee met, Wimperis had received Watson Watt's answer, in which he reported that a death ray was definitely out of the question, but that he thought there might be "some help" he could give them. The result was "nothing less than a revolution in the science of air defence."

* * *

The Death Ray

Watson Watt is generally regarded as "the father of radar," but the paternity is open to question.

Winston Churchill presents the accepted version in *The Gathering Storm:* "In February, 1935, a Government research scientist, Professor Watson-Watt, had first explained to the Technical Sub-Committee that the detection of aircraft by radio echoes might be feasible and had proposed that it should be tested. The Committee was impressed."

This version was earlier published in A. P. Rowe's autobiography, *One Story of Radar,* when he wrote: "Watson-Watt's answer to Wimperis's death-ray question was a simple one. He said that, although there was no possibility of directing enough energy on to an aircraft to produce a lethal effect at useful distances, it should be possible to locate the plan position of an aircraft by measuring its distances from two points on the ground [by radio waves]."

Watson Watt—as he then was, only adding the hyphen after the war when he was knighted—accepted and promulgated this version in his own autobiography and in numerous talks around the world:

> The director of scientific research, Air Ministry (Wimperis)... asked me to come see him...to ask what I thought about the prospects for some form of....Death Ray. Back in my office I prepared a formalized problem which I handed for numerical solution to a junior scientific officer in whose knowledge, judgment, and discretion I, in turn, had the highest confidence. This skeleton problem did not explicitly disclose the objective, but as I had expected, A. F. Wilkins correctly inferred what it was all about. *We talked it over, and I wrote a memorandum....* (my italics).

In that memorandum Watson Watt suggested that though radio beams could not be used to destroy enemy bombers, they might quite reasonably be able to detect them. This memo is rightly considered the birth certificate of radar. But who was the author of the idea? It is certainly implied in the passage quoted above that it was Watson Watt himself, or at least he says nothing to disturb the accepted inference that this is so.

Not until recently did Arnold "Skip" Wilkins, Watson Watt's assistant, contest this story. In 1935 he was employed as Scientific Officer at the Radio Research Laboratory of which Watson Watt was Superintendent. "The station was quite small....The annual expenditure was correspondingly limited and the need to economise was constantly impressed upon us by the Superintendent, Robert Alexander Watson Watt."

It was, however, Watt's "economical Scot" upbringing rather than the exigencies of government-imposed economies, Wilkins suggests, "which caused him to use the leaves of a daily desk calendar to send requests and memoranda to members of the staff." One day shortly after Watson Watt's visit with Wimperis in London, Wilkins returned to his office from the hut in which he was carrying out some unrelated high-frequency radio experiments, and found there one of those calendar leaves on his desk. It was signed "S" (for Superintendent), and asked him to "calculate the amount of radio-frequency power which should be radiated to raise the temperature of eight pints of water from 98°F to 105°F at a distance of 5 km and at a height of 1 km."

In some of the early death ray discussions, emphasis had shifted from the ability to cause death by mysterious unspecified emissions to the more precise possibility of radiating enough energy into a bomber pilot's body to "boil his blood." Watson Watt was clever enough to realize that this was a classic case of overkill: boil a pilot's blood and he'd be able to fly without his bomber. It would be sufficient to raise his body temperature to the point where fever would knock him out or drive him into delirium, roughly 105°F.

And Wilkins was clever enough to realize that the human body contains about eight pints of blood—which, chemically speaking, is nearly pure water—and that raising it to 105° by radio-frequency radiation would be, in effect, to create a death ray:

> It seemed clear to me that it (the note) concerned the production of fever heat in an airman's blood by a death-ray and I supposed that Watson Watt's opinion had been sought about the possibility of producing such a ray.
>
> My calculation showed, as expected, that a huge power would have to be generated at any radio frequency to produce a fever in the pilot of an aircraft even in the unlikely event of his body not being screened by the metal casing of the fuselage....[As nothing remotely like the power required could be produced,] it was clear that no radio death ray was possible.
>
> I said all this to Watson Watt when handing him my calculation and he replied, "Well, then, if the death ray is not possible how can we help them?" I replied to the effect that Post Office engineers had noticed disturbances to VHF reception when aircraft flew in the vicinity of their receivers and that this phenomenon might be useful for detecting enemy aircraft.

It has been suggested that genius is the ability to see relationships between apparently disparate processes, and to connect them in new ways. If ever in the history of the world the birth of a great sequence of events can be traced to just such an inspiration, this was indeed the case on that January day in 1935.

Chapter Three
Origins: Radio

> In deciding whether a new theory might be true, the question is not whether it is crazy—but whether it is crazy enough!
> —*Niels Bohr,*
> *Danish physicist,*
> *Nobel prize winner*

The Post Office experiments to which Wilkins referred were not the first to notice interference of radio waves by aircraft. Curiously, the phenomenon had been observed, in one form or another, nearly as far back as the discovery of radio itself, but although every major world power was conducting research in the 1930s, it was only in Britain that the effect became a generally accepted part of the defense system prior to the initiation of hostilities. In the American military, for example, more than two years after the war began in Europe, radar was so little integrated within the system that when it gave warning of the Japanese attack on Pearl Harbor it was ignored, much as people in a large building treat any fire alarm as either a test or a malfunction. And in Germany Field Marshal Hermann Goering could never quite bring himself to believe that radar was an important part of the RAF defenses, despite repeated reports from returned pilots that no matter where or when they attacked they were always met by a bunch of Spitfires in the right place at the right time.

The concept of radio waves goes back to James Clerk Maxwell, but for an understanding of the underlying phenomena we have to think in terms of mysteries literally thousands of years old. In this all too real universe we are and have always been surrounded by invisible and mysterious radiations and emanations. Among these are the manifestations of radioactivity, gravity, solar neutrinos, and cosmic rays, but the first to be discovered were the Siamese twins of electricity and magnetism. Horace, in the first century B.C., mentioned the attractive properties of rubbed amber and the attraction of the lodestone for iron, but there was no attempt to explain these unseen forces until Roger Bacon in 1267 published his first theories. Following upon this, rumors began to spread concerning a certain "sympathetic needle" with which it was possible to converse over long distances. But the rumors remained without substance for several centuries. The first description of a "sympathetic telegraph" came nearly 300 years later: it consisted of two needle-shaped pieces of iron, each mounted at the center of a dial with letters equally spaced around the periphery. If both needles were magnetized by rubbing with the same lodestone, the two instruments could then be separated by any distance imaginable, it was said, and if the needle of one instrument were moved so as to point it to a particular letter, it would then cause a similar movement of the needle on the other instrument. Communication, it was thought, could thus be set up between two distant points. The idea was brilliant, but it never worked: The unknown force synchronizing the two needles died off quickly with distance, and communication by this device was limited to a matter of inches rather than miles. Fascinating, but hardly practical.

By the beginning of the seventeenth century the English scientist Gilbert listed a number of materials which he called "electrics"; when rubbed, they attracted light bodies but repelled similar materials similarly rubbed. In the middle of the next century the Leyden jar was invented by Musschenbroeck, who found that a charged body (a glass jar charged by rubbing) could store its "electricity" and transfer it over a distance by means of a metal wire conductor. A year later Watson in England demonstrated that an electric current could be transmitted through two miles of wire, and seven years later the first practical suggestion for an electric telegraph was made in Scotland by a writer to the *Scots Magazine* who signed himself C. M. and who suggested using an insulated wire for each letter of the alphabet. At the receiving end of each wire a light ball was to be suspended above a piece of paper marked with a letter. As a charge was sent along a given wire, the ball would attract the paper beneath it, and by observing the movements of the papers, words could be spelt out. "C. M." even suggested that bells might be substituted for the papers, the bells being struck in turn by the ball as a charge was sent along the desired wire.

This idea was in fact carried out in Geneva by a gentleman named Le Sage in 1774, and it worked, but its usefulness was limited by the infinitesimal quantity of electricity that could be created by rubbing and stored in the Leyden jar. It took another hundred years before Samuel Morse had the bright idea of merging these notions with Faraday's invention of induced electric currents to produce an instrument of revolutionary usefulness.

Michael Faraday in 1831 had discovered how to use the properties of magnetism to create a sustained electric current, culminating the scientific odyssey begun a dozen years before by Hans Christian Oersted who found that a magnetic needle orients itself at right angles to a wire carrying an electric current. This was the first demonstration that the invisible forces of electricity and magnetism were related, and was soon followed by Ampère's discovery that a magnetic field can act on and influence a current. The question then became whether a magnetic field could in some way *create* an electric current.

Faraday was born in 1791 in a small village near London. He was the son of the village blacksmith, but did not have the physique to follow in his father's occupation. He was bright, but an eighteenth century smithies' son did not go to school; instead he was apprenticed to a bookmaker. He learned to bind books, and to read them. He read everything, particularly books on science, and in his spare time began to carry out whatever experiments "could be defrayed in their expense by a few pence per week." This reading and these experiments were his only and entire scientific education, until at the age of twenty-one he managed to get into London to attend four chemistry lectures given by Sir Humphry Davy. Afterwards he submitted the careful notes he had taken "as proof of his earnestness" and applied to Sir Humphry for a position; to his great surprise he was hired as a demonstration assistant, and immediately proved himself the best Sir Humphry had ever had. Six months later he went with the Davys on a grand tour of the Continent, assisting with the lectures and meeting many of the great scientists of the day. He was as pleasant and modest as an assistant should be, but in addition he was uncommonly intelligent. "We admired Davy, we *loved* Faraday," one of the scientists reported.

So did Sir Humphry. Back in England he recognized more than an assistant in the young man and encouraged him to try his own experiments, using the apparatus available there at the Royal Institution (of which Sir Humphry was director). In the next three years Faraday published thirty-seven original papers, and six years later he succeeded Davy as director.

During this time Oersted had enlarged on his earlier discovery to show that not only could an electric current orient a magnetized iron needle, but it could actually induce magnetism in a virgin piece of iron. Faraday felt, anticipating Newton in a philosophic manner, that "every action must have its reaction"; he was sure that somehow magnetism in its turn could induce electricity. It took years of experiment and study, during which many false leads were followed, but finally, one day in 1831, he inserted "a cylindrical bar magnet into the end of a [wire wound in the form of a] helix, then quickly thrust it in the whole length." The ends of the wire formed a closed loop by connecting to a galvanometer, an instrument that reads the flow of electric current. When the magnet was "quickly thrust in the whole length," Faraday saw the galvanometer show "a mere momentary push"—a small flash of electric current—and then nothing. When the magnet was pulled out, "again the needle [of the galvanometer] moved, but in the opposite direction."

This was not what he had expected. If the magnet induced the current, why was there no current when the magnet was sitting motionless in the wire helix? And the effect was so small and so momentary that it could easily have been ignored. It is, after all, easy to ignore any small, momentary, unexpected effect; we brush them out of our minds every day as accidental, meaningless, coincidental. But Faraday repeated the pushing and pulling, and found that "the effect was repeated every time the magnet was put in or out, and therefore a wave of electricity was so produced from the *motion* of the magnet and not from its mere position within the wire." (We explain this today by noting that a metal wire is formed of atoms with electrons loosely bound to the nuclei, so that when a magnet moves into the vicinity it repels the electrons in the wire; they break loose from their nuclei and flow "downwind" along the wire, thus creating an electric current.)

Within two weeks Faraday had set up a copper disk on a crankshaft; by cranking the conducting copper between the poles of a magnet, he reversed the previous experiment and showed that it was only the relative motion of conductor and magnet (rather than the absolute motion of the magnet) that was needed to produce a current of electricity. Moreover, with this mechanism he was able to set up and maintain a continuous flow of electric current. He had constructed the first dynamo. (When he demonstrated his apparatus to His Majesty's Government, the prime minister, Sir Robert Peel, asked, "Of what use is it?" To which Faraday replied: "I don't know, but I'll wager that some day you'll tax it.")

In 1832 Morse was returning to America from Europe on the packet *Sully*; Faraday's recent work was being discussed, and a Dr. Watson of Boston remarked that one could thus create and pass a charge of electricity along a wire. Morse then thought that if one could also *detect* such an electric charge, one could send signals by opening the circuit—stopping the current—and then closing it, allowing the current to flow again. By the time the ship arrived in New York, Morse had the details worked out and had invented the dot-dash system for converting the resulting signals into letters. Within ten years an experimental line was operating between Washington and Baltimore, a distance of forty miles, but when Morse offered his system to the government for $100,000, it was turned down on the advice of the postmaster general, who was unsure about the demand for telegraphic services and thus "uncertain that the revenues could be made equal to its expenditures." Morse then raised private capital; by 1851 fifty companies were operating his system throughout the United States, and in another ten years throughout Europe as well.

At just about that same time, James Clerk Maxwell was graduating from the Edinburgh Academy with firsts in English and mathematics ("and nearly a first in Latin"). The son of a rich and distinguished family, he naturally went next to Cambridge and by the time Morse's first telegraphs were ticking away, he had graduated with high honors, particularly in physics. After a professorship at King's College, London, he retired at the age of thirty-four, then later came out of retirement to found the famous Cavendish Laboratory at Cambridge. It was while he was at King's, however, that he did his greatest work: In *A Dynamical Theory of the Electromagnetic Field* he formulated the general equations which led to the electromagnetic theory of light. That work, together with Newton's *Principia,* Einstein's *Theory of Relativity,* and the *quantum mechanics* of many authors, constitutes the theoretical basis of our present-day physical universe.

Maxwell showed that light consists of wavelike disturbances in a (hypothetical) universal medium called the aether, and is propagated through space with a velocity equal to the ratio of the electromagnetic to the electrostatic unit of charge. This was verified by experiment, and led to the most striking consequence of the electromagnetic theory: that other, as yet undiscovered, waves also propagate through space with the same velocity. These invisible rays—invisible because they do not stimulate the optic nerve as do the rays of light—must be produced in the same manner, by oscillatory electric currents, and differ from visible light only in the length of their waves. In fact, an entire spectrum of electromagnetic radiation became theoretically possible, from wavelengths infinitesimally small to infinitely large. Our eyes have evolved a sensitivity to radiations of only a limited range of wavelengths, on the order of hundredths of millionths of an inch (presumably because that is the magnitude of most of the radiation produced by our sun), but all other wavelengths are equally possible.

The search for these other radiations was immediately begun. It was later shown that Roentgen's x-rays are examples of very short wavelength radiation, and in 1887, ten years after Maxwell died, Heinrich Hertz in Germany discovered the longer wavelength radiation, the radio waves. By the end of the century their properties were understood well enough for Marconi to suggest the possibility of replacing Morse's conducting wires with these invisible rays of electromagnetism, thus sending telegraphic electrical messages via a "wireless."

Guglielmo Marconi was born in Bologna in 1874. He was one of those boys who are passionately fond of science, and in particular he was fascinated by the new science of electricity. At the age of twenty-one he had his vision of transmitting telegraphic signals through the very air. Imagine the breadth of that vision! It stunned him; the implications were so vast as to amount to nothing less than the transformation of the entire world. Instant communication with every place on earth, nowhere too remote to be lost from civilization, a means by which one people could speak to another, to understand each other—an end forever to enmity and misunderstanding between the peoples of the world, an end to war.

Now Marconi, you have to understand, was crazy. But not yet, not in this idea; here he was merely a man with a dream. The craziness came later. This first idea of his bore immediate fruit; he began to construct apparatus for sending and receiving the electromagnetic radiation waves—radio waves, for short—and within a year he was communicating over distances in excess of a mile.

He came to England in 1896 because Victorian England was the center of the expanding industrial and scientific universe. As painters in those years went to Paris, inventors went to England. He obtained a patent there for his concept of wireless telegraphy, continued to improve his equipment through experimentation, and demonstrated his results to Post Office officials. England was at the height of her powers as an empire; the virtually instantaneous transmission of information to colonies around the world without the expense and upkeep of laying suboceanic and intercontinental cables seemed to His Majesty's officials to be one more indication of a just God who had the interests of the Empire at heart. The first demonstration carried over a distance of only a couple of miles, but when Marconi sent a message across Salisbury Plain and then across the Bristol Channel, who could doubt that penetration of the English Channel was next, and then perhaps even the oceans? Visions of the money to be saved by canceling the cable to Australia then being planned danced in the officials' heads.

While they were mulling things over, the Italian government became interested in the enthusiastic newspaper reports. The next year Marconi returned to Spezia, where he set up a transmitting station and installed receivers in ships of the Italian Navy, which then communicated at distances of up to twelve miles. He successfully demonstrated this remarkable ability to talk to ships at sea in front of King Humbert and the Italian Chamber of Deputies. At about the same time he established in London the Wireless Telegraph Company, Ltd., setting up stations along the coast for shore-to-ship transmissions. The very next year the East Goodwin lightship was run down by a steamer; the captain radioed for help to the South Forland lighthouse and lifeboats were dispatched in time to save the men, who otherwise would have drowned.

In the next years the new science/industry of radio became the craze of its time. In 1898 stations began to communicate across the English Channel, naval warships stayed in touch at distances of nearly a hundred miles, and military communications were revolutionized in the South African war. (The first major use of the invention that was to eliminate war was in fact in the improved operations of war; and who today could conduct war without it?) In 1912 the *Titanic* struck an iceberg and went down. The world followed the disaster in newspapers that received the very latest information by radio; readers everywhere visualized the wireless radio operator in that doomed ship staying at his post, tapping out his message calling for help, giving exact position and details, bringing rescue vessels rushing to the spot; they visualized his finger tapping away in perfect rhythm as the ship listed, as the water reached his feet, as the radio cabin finally submerged and the ship sank under the waves, its last contact with the world being that final dot and dash from the heroic wireless operator's cabin.

A generation of schoolboys fell to sleep that night dreaming of becoming radio operators in ships at sea, but by that time Marconi's own dream—that of transmitting across entire oceans, of weaving a web of communication around the world and drawing it tight—had progressed from an idea to an obsession to a ridiculous craziness and finally to reality. In 1899 he had already begun to raise money to construct a transatlantic transmission station. This was the idea that was crazy, as was pointed out to potential investors by eminent scientists like Poincaré and Lord Rayleigh, because of a simple, incontrovertible, and overwhelming circumstance: The earth is round. Its surface is curved, while electromagnetic rays travel in straight lines; therefore Marconi's beams, as ingenious as he might be, could not follow the curved surface of the earth across the ocean to Europe, but must fly off on a straight-line tangent into space, irretrievably lost.

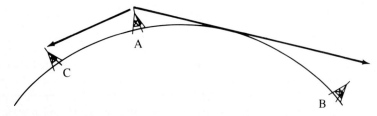

The transmitting station at A cannot possibly reach the receiving station at B, although it can easily talk to C. The distance radio waves can reach is fundamentally limited by the curvature of the earth. One could in fact calculate the useful range by applying the formula for the circumference of a circle to the earth, giving a result of about 200 miles. This was an absolute limit, and there was nothing to be done about it.

But Marconi was crazy. Having no answer to this objection, he simply ignored it. Most people with money have more of that stuff than brains, he found, and so he raised enough to set up stations on the coast of Cornwall in England and at St. John's in Newfoundland. On December 12, 1901, he successfully transmitted a radio message from Poldhu, Cornwall, across 3000 miles of the earth's curved surface to Newfoundland.

This raised a tremendous ruckus among scientists all over the world, and led to a series of experiments which tried to find out how the straight-line radio beams could have curved enough to cross the ocean. The problem was solved in the following year independently by an English physicist who had retired because of deafness thirty years before, and by a scientist at Harvard. The Englishman, Oliver Heaviside, had continued throughout his "retirement" to conduct theoretical investigations which no one believed because his methods were so unorthodox; little by little, after much arguing, he would get them published and accepted. The Harvard physicist, A. E. Kennelly, was an active worker using more usual methods; separately they came to the conclusion that there must exist high in the sky a layer of material that would reflect the radio beams back to earth. The transmitted beams fly away from the transmitter in all directions, each of them traveling in straight lines and therefore leaving the curved surface of the earth behind, but upon hitting the reflecting layer they would bounce back at an angle and thus appear to curve around the earth:

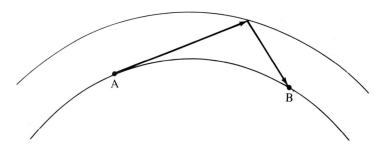

This layer, later studies of which would lead directly to the basic concepts of radar, and which was shown to consist of atoms ionized by solar ultraviolet radiation and cosmic rays, proved to be stable enough to permit consistent radio communication around the globe. The first indication to the general public of how the world was rapidly shrinking was the use in 1910 of a transatlantic wireless message to help capture the most famous English murderer since Jack the Ripper.

Dr. Hawley Harvey Crippen was neither a doctor nor English; an American graduate of the University of Michigan, he emigrated to London where he worked for a patent-medicine company and, part-time, as a dentist. His wife was also an immigrant (her family came from Poland); her maiden name was Mackamotzki, her stage name was Belle Elmore. She was a music-hall artiste; she was also domineering, loud-mouthed, extravagant, and bullying. Crippen was retiring, modest, self-effacing; the balding and bespectacled patent-medicine dentist was no match for his wife.

So he killed her. After a party on January 31, 1910, he fed her a hydrobromide poison and watched her die. That much is clear, as are his subsequent actions; what is not clear is his state of mind. He was either a "cold-blooded, premeditated murderer" or a "panic-stricken and wildly foolish" victim of a moment's craziness. Either way, "the wretched little man" carved the dead body up into pieces, filleted it, burned the flesh, and buried the bones in the cellar. He then told people that his wife had left him and gone to America. In time he brought home to console him a young secretary, Ethel Le Neve.

When friends remarked that it was strange that Mrs. Crippen never wrote to any of them, he became nervous and fashioned the tale that she had died in America of a fever. Well, of course people began to talk about the wife's disappearance and about the "secretary's" appearance, and before the year was out Inspector Dew of Scotland Yard entered the scene. He asked questions, returned to ask more, and returned once more to find the house deserted. Obtaining a warrant, he searched the house; finding signs of recent digging in the cellar floor, he dug up enough of Belle's corporeal remains to obtain an arrest warrant for Dr. Crippen, citing 'willful murder' of the missing wife (now only partially missing, to be sure).

But where were Crippen and Le Neve? They could have been anywhere; London itself was by then a multimillion-person warren into which fugitives could disappear as completely as if they had gone to Mars, and if that weren't enough, there were numerous other crowded cities on the island. Travel to the continent was easy enough, with no passport controls or other paperwork to provide barriers. It didn't look as if Inspector Dew had much chance of finding them.

But then Captain Kendall, sailing the Canadian Pacific liner *Montrose* from Antwerp to Quebec, was reading through his copy of the *Continental Daily Mail* that he had picked up before embarking. It carried photographs of Crippen and Le Neve. He had on board a Mr. Robinson and his son, who when they thought they were unnoticed were given to holding hands in an "unnatural way." Captain Kendall began to take an interest in Mr. Robinson and while engaging him in close conversation, noticed a light patch on his face that indicated the recent shaving of a mustache and marks on his nose that indicated customary use of spectacles which were never worn on the ship; the man seemed to be trying to change his appearance. Kendall invited the Robinsons to dine at his table and then, while they were waiting for him there, he searched their room—and found a woman's bodice. A thousand miles at sea, cut off from England, he was not as helpless as he would have been a year or two previously. He sent a wireless message to London:

> Have strong suspicions that Crippen London cellar murderer and accomplice are among saloon passengers. Accomplice dressed as boy; voice manner and build undoubtedly a girl.

Inspector Dew of the Yard responded by booking passage for Quebec on the much faster White Star liner *Laurentic*. As he raced across the Atlantic he coordinated his plans with Captain Kendall by wireless, while on board the *Montrose,* as Captain Kendall later recalled on a BBC radio program, "Mr. Robinson sat in a deck-chair looking at the wireless aerials and listening to the crackling of our crude spark-transmitter, and remarking to me what a wonderful invention it was." And all the while the London papers, kept informed by radio, reported on the chase.

When the *Montrose* pulled into the St. Lawrence River, a local pilot boarded her. As Captain Kendall recalled,

> The night before was dreary and anxious, the sound of our foghorn every few minutes adding to the monotony. The hours dragged on as I paced the bridge; now and then I could see Mr. Robinson strolling about the deck. When the "pilot" came aboard he came straight to my cabin. I sent for Mr. Robinson. When he entered I stood with the detective facing the door, holding my revolver inside my coat pocket. As he came in I said, "Let me introduce you."
>
> Mr. Robinson put out his hand, the detective grabbed it, at the same time removing his pilot's cap, and said, "Good morning, Dr. Crippen. Do you remember me? I'm Inspector Dew of Scotland Yard."
>
> Crippen quivered. Surprise struck him dumb. Then he said, "Thank God it's over. The suspense has been too great. I couldn't stand it any longer."

The suspense was over. The result of the trial back in England was a foregone conclusion. Crippen was hanged on November 23. The jury decided that Ethel Le Neve had genuinely believed in his innocence, and although she had knowingly fled with him, she was acquitted of complicity in his crime, either before or after the fact.

The new invention had captured its first arch-criminal. The age of radio was upon us.

*   *   *

When Heinrich Hertz in 1886 carried out the first experiments on the generation of the radio waves predicted by Maxwell's equations, he had noticed interference by surrounding objects. By the turn of the century such interference was well enough known for people to begin thinking about using it: In 1900 a frustrated and misunderstood genius named Nikola Tesla suggested his own ideas for a world system of wireless transmission. As part of this system he intended to use reflected radio waves to locate objects and even measure their distance. He explained the concept of radar even then:

> When we raise the voice and hear an echo in reply, we know that the sound of the voice must have reached a distant wall or boundary, and must have been reflected from the same. Exactly as the sound, so an electrical wave is reflected, and the same evidence [can be used to] determine the relative position or course of a moving object such as a vessel at sea.

But Tesla was a "visionary," a term at the time nearly synonymous with "anarchist" or "suffragette"; he never obtained financial backing and nothing came of his ideas. He was forgotten, and the concept was rediscovered and lost again several times before it reached fruition.

In 1904 a German inventor, Christian Hulsmeyer, the son of a farmer, was granted a British patent for a *telemobiloscop*, which was a "hertzian-wave projecting and receiving apparatus...to give warning of the presence of a metallic body such as a ship or a train." The idea here was to avoid collisions by warning ships sailing in fog of other ships nearby.

Hulsmeyer was one of the radio "freaks" who sprang out of the ground at the start of the century, building his own crystal sets on the farm in his teenage years and leaving home as soon as he was old enough to teach in Bremen, where he repeated Hertz's experiments and found for himself the phenomenon of interference. During this time a friend of the family was killed when two ships collided at sea, and the mother's grief somehow connected in his mind with his radio experiments to form a new idea: He could prevent such accidents. He met by chance a merchant from Cologne who agreed to back him with up to 5000 Deutsche marks, and with this he founded the *Telemobiloscop-Gesellschaft, Hulsmeyer und Mannheim*. Two years later, on May 18, 1904, he gave a successful public demonstration on the Hohenzollern Bridge in Cologne: His telemobiloscop was pointed out along the river, and as a ship approached, a bell on his apparatus began to ring. When the ship sailed away, out of the beam, the ringing stopped. All day he stood there as ships sailed by, and the system worked perfectly each time. The press was ecstatic, and he went home awaiting fame and fortune.

When he died in 1957, at the age of seventy-five, he was still waiting. The German Navy was not interested. A Dutch shipping company did invite him to give a demonstration, but though it went perfectly, they decided that they were not interested either. His English patent aroused no interest from the Admiralty; not even the Great War that broke out ten years later had any effect on naval curiosity in detecting ships through fog and smoke. (It is interesting to think what might have happened at Jutland had the German Kriegsmarine been equipped with the apparatus—even more interesting to think what Hulsmeyer must have thought as he read the newspaper accounts of that defeat.) Embittered, he gave up further experiments; during the terrible maelstrom of the postwar years in Germany, most of his equipment was lost, and although he made a living the rest of his life as an engineer, he did nothing else with his first invention. It was totally forgotten, although by the end of his life the stories of his early work were resurrected and the remains of his first telemobiloscop are now on display at the Deutsches Museum in Munich.

It seems that somebody really wanted Germany to win that first worldwide war, because two other Germans took up the crusade when Hulsmeyer retired from the field. In 1916 Hans Dominik and Richard Scherl, unaware of Hulsmeyer's work, unveiled their *Strahlenzieler* (ray-aiming) device, which directed electromagnetic waves onto unseen objects and measured their reflected beams. When, in May of 1916, they demonstrated their experimental device to the German Navy, the authorities were interested. "How soon can you have a working production machine?" they were asked. "Six months," they replied. "Too late," they were told, "the war will be over by then." And that was the end of that.

After the war—which Germany almost won with her U-boats before the United States got into things—there was for a while a great deal of interest in antisubmarine warfare, just in the unlikely event that there should ever be another war. Nikola Tesla popped briefly back into view with an "invisible electric ray" which, "when reflected by an unseen submarine hull, causes a phosphorescent screen to glow, giving warning that the U-boat is near." He never worked this idea out in detail, however, and the intrinsic problems of scattering, refraction, and attenuation of the rays by moving masses of water make this problem of submarine detection one of the most important military problems still facing us today.

In 1925 King George V visited an Admiralty research facility where the Royal Navy's Asdic apparatus for submarine detection was demonstrated to him. This is a system based on sound waves. A "ping" is generated and directed downward into the depths; when and if it hits a

The Death Ray

submarine it bounces off and an echo returns to the ships waiting on the surface, betraying the submarine's presence and giving a rough indication of direction and depth. The King asked if electromagnetic waves could not be substituted for sound waves; would they not travel faster and be less subject to interferences? The director of scientific research smiled and shook his head without bothering to use it. His Majesty did not understand these scientific problems, he murmured; it was quite impossible. King George retreated in royally muted embarrassment from the precipice of original thought; neither he nor the director ever imagined using the idea to detect airplanes.

By the early 1930s the idea of using interference with radio waves as a technique to locate metal objects had been forgotten; instead, the emphasis had switched to the elimination of these interferences and to the problems of providing longer range, more dependable, and clearer radio transmissions. In 1931 the British General Post Office (GPO) was conducting radio experiments with an eye to setting up communications between the Scottish islands and the mainland. Their engineers at Colney Heath, northeast of London, had carefully documented the various natural occurrences that affected the received signal strength: thunderstorms, winds, rain, and the occasional passage of an airplane from nearby Hatfield aerodrome. They documented these effects in GPO Report No. 233, *The Further Development of Transmitting and Receiving Apparatus for Use at Very High Radio Frequencies,* 3 June, 1932. Part V of that report is titled "Interference by Aeroplanes." At that time the effect was just that, an interference, a nuisance, and there is nothing in the report to suggest its use at any future time. But when Watson Watt asked Wilkins, in the winter of 1934, whether "there is anything we can do to help" defend Britain against air attacks, Wilkins suddenly remembered that GPO report. "I replied to the effect that Post Office engineers had noticed disturbances to VHF reception when aircraft flew in the vicinity of their receivers and that this phenomenon might be useful for detecting enemy aircraft."

The official history reads somewhat differently. The "official" father of radar is Watson Watt, not his assistant Wilkins. The historian John Terraine, for example, in his massively documented and excellent history *The Right of the Line: The Royal Air Force in the European War 1939–1945* writes simply that

> Watson Watt, with the aid of one of his staff, A. F. Wilkins, was able to dismiss the "death ray" once and for all; but he did not

leave it at that. He pointed out that even if a "death ray" could be devised it would be useless unless its target could be accurately located. And in the matter of locating targets in the air, he thought he could be of some help. He took his clue from a Post Office Report of June, 1932, in which it was mentioned that aircraft interfered with radio signals and re-radiated them; he considered the possibility of transmitting a radio pulse which would be reflected back by aircraft as a signal to the ground. Further calculations [by Wilkins] enabled Watson Watt to submit to the Air Ministry on February 12, 1935, an historic document.

...the document that was to be nothing less than a revolution in air defense.

Terraine, and most other historians, took their clue from the only authoritative account written about this early work. In the beginning, of course, nothing was written—or even talked about. Radar was then called RDF, "radio direction-finding," and in the years before the war when it was being developed, and on through the first years of the fighting, it was top-secret stuff. When the British night fighters, for example, began to knock down the German bombers in the darkened skies, the story was given out to the papers that the British pilots dined on raw carrots to improve their night vision. As a child I remember being told by my mother to eat my carrots so that I could grow up to be a night fighter; in vain I explained that I wanted to fly Spitfires—day fighters—instead. John Cunningham, the leading night fighter ace, was dubbed "Cat's Eye" Cunningham by the press, to his everlasting disgust. But there was nothing he could do about it; better to let the Germans think the RAF was breeding cat-eyed pilots on rabbit food than let them get hold of the idea of radar.

(In fact, they already had the idea. A few years before the war began, when England and Germany were still trying to maintain some sort of friendly relations, General Erhard Milch, Germany's Secretary of State for Air, visited the Royal Air Force. At a luncheon given at the headquarters of No. 11 Group (which was to bear the brunt of the Battle of Britain in 1940), the wine flowed so freely that Milch suddenly addressed the assembly in a loud voice: "Now gentlemen, let us all be frank! How are you getting on with your experiments in the radio detection of aircraft approaching your shores?" The resulting silence reverberated like thunder, punctuated only by Milch's drunken giggle....)

Still there was no point in telling them more than they already knew, and radar was kept under strict censorship for many years. The most up-to-date details still are classified, as even today radar forms the most vital cogpin of our own defenses.

After the war Watson Watt wrote his autobiography, and it was there that the official version of radar's origins began. This circumstance, together with the different temperaments of the two men involved and the difference in their relative positions, explains much. Watson Watt was, according to the distinguished scientific biographer Ronald Clark, "blunt, outspoken, quite confident and with a keen eye focused on the future; the man who discovered the secret and realized that he had done so by a mixture of intuition and hard work. Exuberant, ambitious, devoutly patriotic, Watson Watt was during the next few years to put the nation in his debt and to create a multitude of enemies by so doing."

At the time he was Superintendent of the Radio Research Laboratory. A. F. Wilkins was simply his assistant. A much more mild-mannered man, it was not until the 1980s, after Watson Watt's death, that Wilkins began to be sought out for interviews; somehow he had become the forgotten man. It is easy to see why:

> On one of his visits to Orford [where Wilkins was carrying out the research work on radar in 1935] Watson Watt told Bowen and me that the Air Ministry had advised him to take out a secret patent for RDF. This he thought was rather a nuisance but he would have to comply with the wishes of the Air Ministry. He went on to ask us whether we wanted to be named as co-inventors of the patent and there was something in the way he asked the question that gave me a strong feeling that he wanted to keep us out of it. If we had been considering a scientific publication rather than a patent, I would have felt insulted to have had no acknowledgement of joint authorship and would have pressed to be included. As we were merely considering a patent I told Watson Watt I would not wish to be included in authorship. Bowen, rather reluctantly, decided not to press his claim in the matter.
>
> At a later date Watson Watt also patented IFF [Identification Friend or Foe, a device for recognizing friendly aircraft on radar plots] before any trials had been made. His patent included the passive-keyed dipole, a method which I claim to have suggested to him although he denied it....

Later still, "when the existence of RDF was disclosed to the public in 1942 and Watson Watt was named as the inventor," Wilkins has said, "he told me he had thought of using radio waves for aircraft location *before* 1935...and had put his ideas to Dr. W. S. Tucker [Superintendent of the Air Defence Experimental Establishment at Biggin Hill]." But when Wilkins sought confirmation, "Tucker had no recollection whatsoever of it. [Other senior members of the ADEE research staff] also had no recollection of Watson Watt's suggestion." It is also interesting to note that during their first conversation, when Watson Watt asked about the death ray and Wilkins replied with his suggestion of using the GPO's 1931 radio effect for location of aircraft, Watson Watt "made no mention of ever having considered the matter previously."

The Post Office report of 1932 had not been a best-seller, even among technicians involved in the work. Like most such reports it had been filed away and forgotten. Wilkins was familiar with the work through a completely accidental set of circumstances. When he began work at the Radio Research Station in 1931, he was assigned to look into propagation of radio waves through ionospheric interactions. He first had to decide on a suitable receiver, so he collected a few different types and took them to Colney Heath to test them against the GPO receiver there. While this was being done, he learned, in casual conversation over a cup of coffee, about the troubles they were having with aircraft flying about from deHavilland's Hatfield aerodrome nearby, which were interfering with their beams. The airplanes were discussed as a nuisance, like mosquitos in summer; no one gave a thought then to making any possible use out of that damned interference. But when, a few years later, Watson Watt raised the question "What can we do to help?", the memory of that casual conversation over that now-historic cup of coffee came rushing back.

The GPO experiments were aimed at using the Heaviside ionospheric layer for radio communications. In order to maximize transmissions, they were trying to learn something about the layer and its interactions with radio waves, but how do you study something that is sixty-five miles above the surface of the earth? No airplane or balloon could travel that high, and there were no rockets in those days. The problem can be visualized by imagining an invisible wall postulated to lie beyond our reach. If it can't be seen or reached, how can we learn anything about it—such as its exact location, shape, structure, height, or even whether it actually exists?

The answer is to throw something, perhaps a tennis ball, at where we think it is. If the ball bounces back, we've learned that the wall does exist and where it is. From that beginning we can learn more by studying the characteristics of the bouncing ball: if it always comes back at a 180-degree angle, the wall has to be smooth and facing us; if it always comes back at an oblique angle, the wall is slanting; if it bounces at irregular angles, the wall must be rough. If the ball loses energy and just drops in a soft bounce, the wall is doughy or mushy. If the ball sometimes doesn't bounce back at all but keeps on going, the wall must have holes in it at least as big as the ball.

The principle is actually the same as when we see something. We walk into a dark room and turn on a flashlight. The beam of light shoots out in front, and when it hits something and bounces back we "see" the something. It was this principle that two naturalized American scientists used to prove that Heaviside's postulated layer actually exists. In 1925 Gregory Breit and Michael Tuve had shone a beam of radio-frequency electromagnetic pulses straight up into the empty sky, and found that it wasn't empty after all—the beam was reflected back at them. That beam was their electronic tennis ball, and as it bounced back it told them a lot about the Heaviside layer; for example, it gave the first measure of how high above the earth the layer was. From Maxwell's work and from many experiments it was known that light, and therefore all electromagnetic radiation, travels at a constant speed of 186,000 miles per second in vacuum. Its speed in air is not significantly different. They found that when they sent a pulse of radiation straight up, it bounced back to them in 0.0007 seconds. The distance it traveled was given by the time multiplied by the velocity, $0.0007 \times 186,000$, or 130 miles; since the beam had to go out and back again, the reflecting layer lay sixty-five miles above the earth. The layer would reflect only waves of certain frequencies, which is analagous to having tennis balls bounce back from a wall that lets ping-pong balls pass through: It told them something about the size of the ionized molecules that made up the "wall."

In the years that followed, this became the standard method of conducting such investigations in nearly every country in the world. In England the GPO investigators were irritated because they had spent a lot of time following false leads: their experiments showed that waves of certain frequencies that ordinarily passed through the layer without reflection occasionally gave freak returns. The time it took for the 'echoes' to come back was not 0.0007 seconds, and therefore the ghost layer that was reflecting them must be at a different altitude. They thought of it as a ghost layer because the time of reflection (and therefore the altitude) varied from time to time, and because the reflections appeared only occasionally. It was an exciting time in their lives, they thought they were on the threshold of an exciting new discovery, and then they began to realize that whenever the ghost echoes were observed there happened to be an aircraft flying by in the distance. It is not recorded which of them first made that observation, but airplanes in the early 1930s were not all that usual, and finally someone observed as they were studying the echo that there was an airplane out there and that the last time they saw the echo there had also been an airplane around....

Nuts. (Or more probably *Bugger it all.*) The ghost layer didn't exist after all: they were merely seeing reflections from those damned airplanes. The different "altitudes" they had been measuring for their ghost layer were simply due to the varying distances of the airplanes; the basic discovery they had thought they were making was just an experimental artifact, of no use at all. Well, it happens to all of us at one time or another. They grumbled about it, and had to be careful that when they were doing their experiments there were no airplanes flying about, and they forgot about it—except when making coffee conversation with an occasional visitor like Skip Wilkins.

Chapter Four

Beginnings: Red Sky at Sunset

> Ich weiss nicht was soll es bedeuten....
> —*H. Heine,*
> *Buch der Lieder*

Wilkins was undoubtedly the "father of radar"; he provided the germinal sperm, the initial idea. But Watson Watt was the mother, giving birth to the baby, nurturing it, and raising it to maturity. In fact, Wilkins's idea of radar was not the first proposed, even if we neglect those early visionary ideas of Hulsmeyer, Dominik, and Marconi. The idea had also been born in America—more than a dozen years previously, in fact, at the Aircraft Radio Laboratory of the Anacostia Naval Air Station, which was housed in a wooden shack on the east side of the Anacostia River in Washington, D.C.

Two radio researchers, A. H. Taylor and Leo C. Young, were working there on high-frequency communications, testing radio transmissions from a truck parked on the other side of the river. As they tuned in one day they heard the steady hum in their earpieces that they wanted, and then suddenly they heard it swell to a much greater loudness, and then very slowly begin to die away. They looked outside the shack and saw a steamer passing down the river between them and the truck. They realized they were getting interference and would have to wait for the ship to pass before they could resume their experiment. As the ship sailed between their radio and the transmitter the noisy hum continued, slowly dying down and then again swelling to its former intensity; after that it faded away, this time to its normal tone.

It was much the same situation as with the British Post Office engineers and their aircraft interference, but with a difference: Taylor and Young were navy men, not postmen; they had sailed on ships and were familiar with naval problems, one of the chief of which was how to see through fog and darkness. As they waited, they realized that this accidental interference could be useful: they thought in terms of guarding a harbor so that enemy ships couldn't sneak in by night. All that one had to do was set up a radio installation like the one they had here, with a transmitter on one side of the harbor and a receiver on the other, tuned to give a low, steady hum; if any unauthorized ship tried to sneak past, the radio set would immediately scream out its presence.

They got excited, and as they talked more ideas came to them. The sets could also be mounted on ships in convoy to defeat the submarine menace. The Germans had almost won the Great War of 1914 with their U-boats, which generally came to the surface to attack convoys in darkness since their underwater speed was too low to allow for the necessary maneuverability. Torpedoing one ship and slipping away unseen in the confusion, they would slide up through the rows of tankers in the convoy to torpedo another and another until the group was decimated, while the escorting destroyers would dash around, lost in the darkness. The large tankers and merchantmen sat high in the water and were visible against the stars and the night sky, while the low-lying subs were mere flitting shadows on the surface of the roiling, dark waters. But now if each merchantman had a radio set, any sub passing between them would reveal its presence with a howl of static.

As the river steamer passed on its way into oblivion, Taylor and Young sat in their shack and began to write a proposal. It was submitted to the commanding officer, NAS Anacostia, on September 27, 1922.

In countries all over the globe similar discoveries were being made. In Germany it happened in 1933, when Rudolf Kuhnhold, chief of the Kriegsmarine's Signals Research Division, who had been working on an echo sounder for detecting submarines and surface ships, decided to switch from sound waves to electromagnetic waves. He might have been amused to learn he was using King George's idea. (Or maybe not. He was German.) He quickly ran into technical problems and early the next year approached Dr. Runge of Telefunken, asking that that company take over the navy's research. But Dr. Runge's boss, Dr. Bohm, was Jewish, and in 1934 it was hard to find any German Jews interested in increasing their country's military capabilities. Telefunken officially refused to become involved.

Rebuffed, Dr. Kuhnhold set up his own company (*Gesellschaft für Elektro-akutische und Mechanische Apparate,* GEMA for short), and continued to work on his own. In October of 1934 GEMA demonstrated its equipment at Pelzerhacken (near Lübeck), locating the research ship *Welle* at a distance of seven miles. When a Junkers W34 airplane accidentally flew overhead, it produced its own blip, and the delighted observers followed it for a distance of 700 meters before it faded.

GEMA's work went on shrouded in military secrecy. The following summer, Dr. Runge, who by his own account had forgotten Kuhnhold's proposals, was working on a navigational system using radio-wave interference from coastlines. Independently he devised a scheme for locating aircraft instead of coastlines, set up the experiment (similar to the British GPO experiment of which he had never heard), and it worked. He reported the results to Dr. Bohm, who disapproved and refused financial support. Nevertheless, Dr. Runge continued to work on his own, and in 1938, with only "Aryans" in charge, Telefunken began its own radar work in earnest, concentrating on using the system to direct antiaircraft fire. The following year Runge was sued by Kuhnhold for infringement of patent.

In Russia there was great interest in aircraft detection in the 1930s. Efforts were first concentrated on sound location, but even though they went to such extremes as training operators who were blind, on the assumption that blind people have heightened powers of hearing, the system was not effective. In 1933 the *Protivovozdushnoi Voiska Oborony* (PVO), the Air Defense Forces, began a radar research program under the direction of Pavel Oshchepkov, an electrical engineer who had begun by working on antiaircraft sights. Supported by the army under the direction of Mikhail Tukhachevsky, the PVO made reasonable progress until Stalin's great purges of 1937–1938 decimated the military establishment. Marshal Tukhachevsky, by then vice commissar of defense as well as commander in chief of the Red Army, together with most of his top generals, was charged and convicted of spying for a foreign state. Oshchepkov was arrested and sent to prison for ten years. A politically acceptable group took over radar research, and although by the beginning of the war they had a rudimentary device in operation, they were following leads that could take them nowhere. All of Russia's radar technology of today has followed directly from the equipment brought through submarine-infested waters by Allied convoys from Britain, at the cost of many British and American lives.

In Japan, professors Okabe and Yagi of Osaka University demonstrated a workable radar device in 1936, developed totally without knowledge of the work anywhere else. In Italy Marconi demonstrated his predicted interference effect to the authorities in 1933, at which time a military radar program was inaugurated. In Holland, in 1934, *Jonkheer Ingenieur* J. L. W. C. Von Weiler noticed interferences with his short-wave radio experiments when a flock of birds flew between his transmitter and receiver. He was employed at the time by the Dutch armed forces, and immediately began to work on adapting the system for aircraft detection and antiaircraft gunfire. (Because the system gave an acoustic signal rather than a blip on a cathode-ray tube it was called the *Elektrische Luistertoestel,* or electric listening device.)

In 1928 French work was initiated by Pierre David, who proposed that radio waves might reflect from aircraft just as light waves do. His early experiments failed, but when he tried again in 1932 he succeeded, detecting an aircraft flying at 15,000 feet over Le Bourget. By 1936 *La Défence Aerienne du Territoire* (DAT) had set up a *ligne de guet électromagnetique* near Lorient; this "electromagnetic lookout line" successfully detected aircraft involved in the air maneuvers that summer.

But on September 4, 1939, when the first British bombers swept out of the clouds over Bremerhaven on the second day of that second great war, they saw German sailors hanging their wash out on the battleship *Admiral Scheer* and the battlecruiser *Emden,* looking up at them in absolute surprise. And when the Luftwaffe came roaring over the borders of France on May 10, 1940, to destroy the *Armée de l'Aire,* they caught virtually the entire French air force on the ground. No blips appeared on Dutch or Russian screens as the Junkers and Heinkels swarmed, and when the Japanese torpedo bombers and Zeros came soaring over Pearl Harbor on December 7, 1941, they saw the Army Air Force P-40s that could have shot them down lined up in neat rows on the tarmac at Hickam, while the Navy Wildcats were sitting serenely on their carrier decks miles away at sea.

Although radar eventually became a large part of the war-making apparatus of both Germany and Japan, as well as the United States, it was only in England that it became the weapon that changed the course of history. Its development in America and Germany, although important, was never decisive in the events that followed, but the work in England meant the survival of the western democracies and the defeat of Nazi Germany. Even there it was developed only in the very nickest of times. Had Skip Wilkins not gone to Colney Heath in 1931 to test his radio receivers, he would never have sat down for that cup of coffee nor heard of the Post Office experiments, and had he not recalled them when Watson Watt asked, "What can we do to help?" radar would surely still have been developed—but not in time. (When the Second World War broke out, it still wasn't working properly, as we shall see in the Battle of Barking Creek; it was barely ready for the decisive Battle of Britain the following summer.) Of all those experiments in all those countries, it was the British work alone that rechanneled the course of the war to come; and so this is the story that is of overwhelming historical importance to us today.

Wilkins and Watson Watt were the winners, but they weren't the first, not even in England. A few years previously, even before the GPO accidental discovery of the effect, two members of the Signals Experimental Establishment (SEE), Woolwich, put forward a method of detecting "ships from the coast or other ships, under any conditions of visibility or weather...by...an apparatus (which) depends on the reflection of Ultra Short Radio waves by conducting objects, e.g., a ship." W. A. S. Butement was a regular member of SEE, while P. E. Pollard was on loan from the Air Defence Experimental Establishment at Biggin Hill when they put their suggestion in writing both to the Royal Engineer and Signals Board and to the War Office. The official reply was that they were graciously to be "allowed to carry out experiments on a small scale in [their] own time," with the conditions that no chits for overtime pay nor for any other expense were to be submitted to the establishment and that their "real work" was not to suffer. Under these conditions they managed to progress to the point where they were able to achieve "a good signal from a mast about 100 yards away," but they soon gave up their work; neither Wilkins nor Watson Watt ever heard of it.

The ideas and concepts of radar were in the air, blowing in the wind. If it hadn't been worked out by Wilkins and Watson Watt, it would have been by someone else. It wasn't like the general theory of relativity which, without Einstein, might never have been formulated. At least that was Einstein's judgment. He felt that special relativity was due to be discovered by someone soon, and there were several physicists who might have done it, but general relativity was a more personal description of nature which without him might never have been formulated, just as the plays of Shakespeare would never have been written in their precise form by anyone else. Radar was more like Special Relativity or black players in major league baseball: If Jackie Robinson hadn't been the first, it would have been Larry Doby or Satchel Paige or Roy Campanella; the time was ripe and it had to happen soon. The difference between special relativity and black ballplayers on the one hand and radar on the other is that in the former the timing was not terribly important. If either had been introduced a few years earlier or later it would not have mattered overmuch to the world, but if radar had been worked out a few years earlier it would have been integrated into the German military mind by the summer of 1940 and Goering would have had no hesitation in attacking the high, vulnerable, flimsy transmitting and receiving antennae that lined the coast of England; while if the British hadn't got it working till just a few months later than they did, the Battle of Britain would have been over while the technicians were still fiddling with their dials, and Hitler would have been sitting in the catbird seat in Buckingham Palace.

When Wilkins suggested using the radio interference phenomenon to detect approaching enemy aircraft, Watson Watt immediately put him to work calculating the quantitative aspects on which practical success would depend. This meant calculating the strength of a return radio signal, given a broadcast signal of possible strength; and this in turn depended on choosing a suitable wavelength for broadcast.

The problem of wavelength choice is akin to the questions of why the sky is blue and why it turns red at sunset, why grass is green but dandelions are yellow, why light will go through a thick pane of glass but not a thin sheet of paper and why x-rays can pass through skin while visible light cannot. Its answer depends on the relationship between the size of objects suspended in the medium through which electromagnetic rays travel and the wavelength of the radiation itself.

"Everybody" knows that light consists of electromagnetic waves, and "everybody" is in a sense wrong. Newton thought that light was a stream of particles, and for a long time the wave-versus-particle controversy raged, settled at last, it seemed, by Maxwell's equations in which a wave description of light described perfectly nearly all known optical phenomena. But in the early years of the twentieth century, Albert Einstein gave scientific respectability to Max Planck's crazy quantum theory by showing, in his description of the photoelectric effect (in which a beam of light knocks electrons out of a metal) that light can indeed behave as if it were a stream of particles—the wave description of light simply does not allow enough energy to be transferred quickly enough from the light beam to the metal to knock out the electrons with the observed distribution of energy—and a few years later Prince Louis de Broglie of France won a Nobel Prize for his Ph.D. dissertation showing that particles such as electrons must sometimes be described as waves.

The resulting "wave-particle duality" of quantum mechanics states that some aspects of reality—such as light and electrons—are *both* waves and particles, which is nonsense because the two concepts are actually philosophic *opposites:* a particle is something which exists within definite boundaries, while a wave does not. A baseball exists within its surface and does not exist anywhere else; its position can be precisely defined (please, we won't go into the Heisenberg uncertainty principle); we can say with surety that the baseball is *here* and not *there*. But toss a pebble into the ocean and then tell me where the resulting wave ends; you might as well try to go and catch a falling star or get with child a mandrake root. The amplitude of the wave (its height) gets smaller and smaller as it travels away from the point of impact, but there is no one place in all that ocean that you can say, "*Here* the height goes to zero, or *here* the wave exists and *there* it does not." And so the wave-particle dual description of nature, while useful to theoreticians, to the rest of us only emphasizes that nobody knows what is really meant by the wave nature of any physical reality such as light.

Nevertheless, it is true that Maxwell's equations, which are of the same form as those used to describe well-known wave motions such as the vibration of a violin string or the waves generated by that pebble dropped into the ocean, describe extremely accurately most optical phenomena. We can then go on to *pretend* that light actually looks like a wave of water. (In a radio interview towards the end of his life, Einstein said that the whole theory of relativity was born out of his attempt to discover what a light wave would look like if we could move along at the same speed and observe it. The impossibility of ever doing this is what led him to his theory.) And if we pretend, we can think of the wavelength of light as the distance between corresponding points of any two successive electromagnetic waves, such as their crests, just as if they were waves in a pond caused by a thrown pebble.

This is the most important concept in understanding both visible light and the other forms of electromagnetic radiation: all such radiations are the same except for individual differences in wavelength. We see objects by the light they reflect to us from some external source (or from the light they themselves generate, in the case of stars or light bulbs). Differences in the *color* of objects are only differences in the wavelengths of the light they reflect. (Our eyes react chemically in slightly different ways to light of slightly different wavelengths, and this difference in reaction is interpreted by our brains as different *colors,* a concept intrinsic to our sensory description of nature rather than an extrinsic phenomenon characteristic of the object being seen. There are no colors where there is no light to be reflected or no eyes to see. A rose in complete darkness is a rose is a rose, but not quite.)

Sunlight, a mixture of light of all wavelengths (including those invisible to us, such as the ultraviolet light which gives us sunburn), falls upon a field of grass. Roughly speaking, the molecules of all substances are about 0.00000001 centimeters in diameter, roughly the same dimension as the wavelengths of visible light. But the molecules of the leaves of grass are of a precise size such that they absorb light of all wavelengths except the particular wavelength which corresponds to the color green, which they then must reflect; and it is by that reflected light that we see the grass, which therefore appears to us as green. The face of a dandelion has slightly different molecules of slightly different dimensions, so that it absorbs green along with other colors and radiates the wavelength corresponding to yellow.

In gases the constituent molecules are so far apart that light can easily travel through them without interruption, and so they are transparent, and we can see clearly through the air as if the molecules of oxygen and nitrogen were not present. (Visible smog is formed when small solid or liquid particles are floating in the air; it is these particles we see, and breathing them into our lungs is what makes smog so unhealthy. We see steam only when water vapor condenses; in its true state, steam consists of water molecules in the gaseous state, and they are invisible. In the same manner clouds are formed of condensing water vapor which scatters all wavelengths of visible light uniformly, and that is why they appear to us as fluffy white objects.) The "sky" consists of gaseous air and the empty space above it, and so it should not reflect light and should have no color: it should look black to us, as it does when seen from satellites or spaceships cruising above the atmosphere. The reason it has color when viewed from earth is that there are suspended in the air invisible particles of dust, with sizes again on the order of 10^{-8} centimeters. Although there is a distribution in their sizes, the average is closer to the wavelength of blue light than red, and so these particles scatter blue light preferentially to the other colors of the spectrum. If no light were scattered out of the sun's beam and one looked up at the sky, the sun would look white and the rest of the sky black (lightless and therefore colorless). But since blue light is scattered out of the direct beam, the color of the sun itself is shifted slightly toward red, thus appearing yellowish, and we see the rest of the sky by the scattered blue light. At sunset the sun appears much redder because it is sitting just above the horizon and its light must pass through a thicker layer of air (and dust). This geometry enhances the scattering of blue light out of the direct beam; since blue and red are the end members of the visible spectrum, and since we have lost the blue portion, the sun appears red.

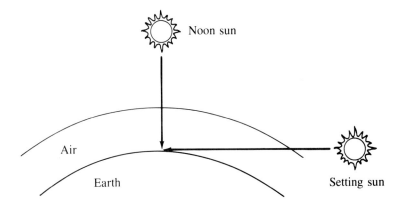

When Watson Watt and Wilkins sat down to make the calculations which would tell them if the radio interference effect could be of practical use in detecting aircraft, this same principle was their starting point: an object scatters and reflects electromagnetic waves most effectively if it is the same size as their wavelength. They had to start with the most effective reflective properties because they knew from the start they would always be fighting the problem of sensitivity, compounded because their electronic fingers would have to reach out for distance. For the device to be useful it would have to have a realistic range of at least several miles, and in this respect they would be fighting the undefeatable, irrefutable law of inverse squares, which states that as light, or any sort of electromagnetic emission, goes from point A to point B, it loses brightness according to the inverse square of the distance. So if a radio system is able to locate a bomber one mile away, in order to find one ten miles away it must have much more power: the radio beam reaching the bomber ten miles away will have only one-hundredth the power at that distance, and since the reflected beam must also travel that same distance it will also lose the same proportion of its strength. The total result is that the sensitivity of the system at ten miles will be only one *ten-thousandth* what it is at one mile (1/100 × 1/100). Since it would obviously be important to locate the bombers as far away as possible, it was essential to pick the most efficient frequencies to start with.

The form of the calculations, as first posed by Watson Watt to Wilkins, was something like this: "Consider that an aircraft is flying at a distance of ten kilometers from a radio transmitter T of one kilowatt radiated power and at a height of five kilometers. Calculate the field strength produced at the position T by energy reradiated (reflected) by the aircraft as a result of its illumination by waves radiated by T."

Bomber aircraft of the day had wingspans of about forty-five to seventy-five feet; allowing for moderate growth in a few years, Wilkins picked a wavelength of twenty-five meters. He next had to estimate the amount of radio broadcast power that might reasonably be expected to be attainable at that wavelength within the next few years; after throwing in possible efficiencies of reflection by the aircraft and detection by the system, he quickly came up with the following results:

> Let it be assumed that the typical bomber is a metal-winged craft, well bonded throughout, with a span of the order of 25 metres. The wing structure is, to a first approximation, a linear oscillator with a fundamental resonant wave-length of 50 metres and a low ohmic resistance. Suppose a ground emitting station be set up with a simple horizontal half-wave linear oscillator perpendicular to the line of approach of the craft and 18 metres above ground. Then a craft flying at a height of 6 km and at 6 km horizontal distance would be acted on by a resultant field of about 14 millivolts per metre, which would produce in the wing an oscillatory current of about 1 and 1/2 milliamperes per ampere in sending aerial. The re-radiated or 're-flected' field would be about 20 microvolts per metre per ampere in sending aerial.

The above description is taken from the memorandum Watson Watt immediately wrote to Wimperis, the point being that "it may be concluded that reflected fields on the order of a millivolt per metre are readily available..." and that such a reflected beam is "about ten thousand times [greater than that] required for radio communication''; in other words, the idea could work.

When Wilkins finished the calculations he couldn't believe that the desired effect was so obtainable; he checked his arithmetic again and again but could find no mistake, and so he gave it to Watson Watt. He, in turn, also found the results incredible and again checked the math. Finding no mistakes, at the end of January, 1935, just a couple of weeks after their first discussion, Watson Watt sent the above memo to Wimperis, together with a separate memo that showed that the radio intensities available were grossly insufficient for a death ray. The conclusion that the death ray was unworkable didn't surprise Wimperis, but this new idea did. He, in his turn, couldn't believe that God was so good; he checked the math himself and, still not quite believing, sent the calculations to *his* assistant, Albert Percival Rowe, who also checked them and found them good.

And it was evening, and it was morning, the first day.

The Committee for the Scientific Survey of Air Defence which had been set up on Wimperis's recommendation immediately became known as the Tizard Committee, after its chairman. It met for the first time on January 28, 1935, ten days after Wimperis had asked Watson Watt for his death ray recommendation. At that first meeting, Wimperis presented the committee with Watson Watt's recommendation that "there was some hope of [aircraft] detection by [radio] means," though any form of "death ray" was useless. The committee instructed Wimperis to ask Watson Watt for a detailed description of what might be done, and within two weeks they received from him a classic paper, "Detection and Location of Aircraft by Radio Methods." This document—the piece of paper "which made England an island once again" and which has been described as being as important to history as the Magna Carta—described with uncanny prescience the future development of the technique. Once the basic idea had been formulated, there was no one more capable than Watson Watt of carrying it to fruition; he himself more than a dozen years previously had designed a cathode-ray direction finder by which thunderstorms could be located using what was essentially this same technique. He now not only described the basic principles which would allow it to detect aircraft, he gave specific numbers as to range and height limits, and even discussed how a communication system should be devised to transmit the information directly to fighter pilots by radio. The committee, which had come to its first meeting prepared to spend its time discussing the death rays of cockeyed crackpots, suddenly had something solid, technical, *measureable,* to sink its teeth into.

Two days later Watson Watt came to London. He lunched with Tizard and Wimperis at the Atheneum, and the following day Wimperis went to the Air Ministry offices of the Air Member for Research and Development, Air Marshal Sir Hugh Dowding, and on the basis of Watson Watt's memo asked for 10,000 pounds to begin work.

Dowding said no.

"Stuffy" Dowding had been criticized, damned, vilified, and excoriated throughout his career by overlords, underlings, and onlookers, from the first "Boss" of the RAF, "Boom" Trenchard, to Winston Churchill, from fighter pilots and mechanics to members of Parliament and air ministers. He was passed over for Chief of Air Staff when he was the most senior member of the RAF and instead given the post of Head of Fighter Defence of Great Britain, at a time when no one but he thought that such a defense was possible. He became the "man who won the Battle of Britain" at a time when he was demonstrably and certifiably insane, and after he won it he was summarily fired and thrown out of the RAF. But all this was in the future; on February 15, 1935, he was the man in charge of allocating RAF monies for research, and 10,000 pounds was too much to give away without something more concrete than several pages of calculations. He wanted a demonstration of the expected effect.

"Dowding was no friend of the scientists," it has since been said, but in demanding experimental verification, he was being more scientific than the scientists who criticize him. He was no stranger to the military possibilities of new scientific inventions, in particular of radio itself. He had been the first ranking officer in the RAF to pounce on it when just a few years previously he had been in charge of fighter defenses in a summer air exercise. He had secretly sent radio vans out on the roads over which the bombers would have to pass; when observers in the vans saw the bombers, they radioed back to Dowding the news, telling him exactly when the bombers were on their way, and so he always had his fighters in position waiting for them—just as he would do with radar and the Luftwaffe bombers in the summer of 1940. He was so successful in intercepting these RAF bombers that the umpires changed the rules on him. He didn't complain; instead he came up with another idea, fitting one of his fighters with wireless—radios were not normally installed in airplanes in those days—and had it follow the bombers back to their base. When the bombers had landed and were being refueled, the tailing fighter radioed back the news, and Dowding's fighters attacked en masse, strafing the airfield and "destroying" the entire bomber force on the ground. The umpires were so disgusted with this unorthodoxy that they canceled the exercises and never published a summary report.

So Dowding was no military dinosaur who had no interest in new technical inventions. On the contrary, he knew better than anyone the need for something drastically new if Britain was to be defended against aerial attack; but he also knew better than anyone the strict limits to his research budget and the impossibility of getting any more from a government determined above all else not to raise taxes. He had to see something concrete before he could approve the disbursement of any funds.

Wimperis agreed; but the problem, he countered, was rather a chicken-and-egg situation: with no money, how could they build the apparatus to demonstrate the effect? Stuffy nodded; he saw his point. Nevertheless, 10,000 pounds was a lot of money (corresponding roughly to 50,000 dollars, which can be amusingly contrasted to the present American administration's defense appropriations of about one and a half *trillion* dollars), and he knew the tendency of scientists to damn the torpedoes and charge full speed ahead on a project that was scientifically exciting but might be useless on practical terms. As a military man he knew the foolishness of that, and he wanted to see results before he committed his small stock of money. Reminding Wimperis of the necessity for haste, he suggested a demonstration within a week or two.

Fair enough. Wimperis replied that he would have Mr. Watt modify an available radio transmitter and conduct the test within ten days; he passed this decision on to Watson Watt, without the faintest idea of how it could be done. Watson Watt accepted it equably; he also had no idea of how to do it, but he too had someone to whom he could pass the problem: Skip Wilkins, who was where the buck stopped, and who replied that "to modify a transmitter to operate suitably short pulses of high enough power to enable an aircraft echo to be displayed on a receiver was quite impossible in the time suggested." But just as he had done when he rejected the death-ray hypothesis, Wilkins came up with an alternative suggestion again: in investigating atmospheric effects on radio transmissions during previous years, he had "noticed...that the Daventry short-wave broadcasting station of the BBC...had a strength and quality [that] was quite high...and, in particular, that there was a station GSA on the forty-nine metre band working on an array of horizontal dipoles and directed southwards." Forty-nine meters was close enough to the twenty-five meters they had thought optimum that Wilkins thought this already existing transmission facility might be suitable for a test. Watson Watt agreed and left it to Wilkins to set up all the details.

It was arranged that an RAF Handley-Page Heyford biplane bomber would fly to Daventry on the morning of February 26, there to fly up and down a prearranged path while its radio interference was noted on the equipment to be set up by Wilkins. On the morning of the twenty-fifth, Wilkins loaded his equipment into a van and, together with a driver, set out to find a suitable location near the axis of the Daventry beam. They found a cleared field with an unobstructed view, set up the aerials just before an approaching thunderstorm hit, and left to find a local hotel and dinner. After eating, they returned to the field to test the equipment, using the transmissions from the radio station which was scheduled to close down at midnight. But they had arrived and set up the antennae in the afternoon and had not counted on having to work in the dark; there was no provision for lighting in the now dark—and muddy—field. They found an electric lamp in the van, but it wasn't working, and so they made all the million and one connections by the light of numerous matches held till they reached the tips of the van driver's frozen fingers. Wilkins managed to get the apparatus working at five minutes till midnight, and completed the tuning just as the station signed off.

When they tried to leave the field they found that the mud had frozen around the van's tires, locking it in, but they found a spade in the locker and managed to dig themselves out.

Early the next morning they returned to find the aerials still in correct position and the set working perfectly. They were soon joined by Watson Watt and his young nephew, brought along for the ride in the country, and A. P. Rowe, who would be the Air Ministry's official observer. For purposes of secrecy the van driver and the nephew were sent away to a far corner of the field, while the others settled down to wait for the bomber.

Squadron Leader (later Air Vice Marshal) R. S. Blucke was then Flight Commander of the Wireless and Electric Flight. He had been told only that he was to fly a bomber to Daventry, reaching there at a specified time at 1000 feet. He was to fire a Very flare, then turn for Wellingborough while maintaining a steady climb to 6000 feet. He was to circle there and fly back to Daventry on a reciprocal course, fire another Very flare, and then go home.

Shortly before 9 a.m. he took off, with W. T. Davies as his observer. "The morning was clear and crisp with a strongish southerly wind blowing," he said. "We reached Daventry (which in 1935 was for us quite a navigational feat), fired our Very light, and climbed up."

It was then that Blucke realized that the course led over the edge of his map sheet, and shortly after that he realized that the wind was stronger at height than it had been on the ground. They were lost. "We both had our doubts about being over the town we ought to have been over, and had some discussion about its being Wellingborough or Kettering... but there we were, over somewhere, so we flew back to Daventry, fired our Very light, and flew home."

In the field below, Wilkins, Watson Watt, and Rowe watched the bomber approach and then concentrated all their attention on the small cathode-ray screen. Precisely as calculated, a "small rhythmic beating of [a] small re-radiated signal was detected." Nearly out of their minds with joy, but exhibiting nothing much more than quietly satisfied smiles, they watched the signal slowly fade as the Heyford headed home.

And that was that. Watson Watt and Rowe jumped back into their car and headed for London, while Wilkins and the van driver loaded their equipment and drove off for Orfordness. Halfway to London Watson Watt noticed that the car was unusually roomy; looking around, he saw that his nephew wasn't there. In their excitement they had hopped into their car and driven off without him. They turned around and hurried back to the empty field where they found him worriedly walking around in the mud.

When Watson Watt went over the results that afternoon he noticed that they didn't quite conform to the expected position of the airplane. This was worrying. At a loss to explain it, he drove to the Farnborough aerodrome.

R. S. Blucke described the resulting scene: "That evening I was sent for by the C.O., and as I went to his office I wondered whether I had messed up the whole show [by getting lost]. I therefore owned up to the fact that I wasn't sure whether I had flown over the right town or not, and was very surprised when the civilian seemed absolutely relieved."

That same evening Rowe wrote to Wimperis, and the next day Wimperis went to see Dowding. The 10,000 pounds was authorized on the spot.

But at the same time as these first ideas and experiments were forming the basis of what was to become radar, as the first funds were being spent and the first aerials constructed, a formidable opposition was arising... in the form of a most unexpected person.

Chapter Five

The Opposition: A Lion in Winter

> We shall defend our island whatever the cost may be. We shall fight on the beaches, we shall fight on the landing grounds, we shall fight in the fields and in the streets, we shall fight in the hills; we shall *never* surrender!
>
> —*Winston Churchill,*
> *Prime Minister of Great Britain*

Winston Churchill was unquestionably the greatest leader of the Second World War and just as certainly one of the greatest of all time, but the man was a holy terror. He was the son of a second son of the great line of the Dukes of Marlborough, distinguished from the time the first Duke in 1704 brought his army through the marshy country bounding the rivers Nebel and Danube to win a crushing victory over the French and Bavarians at the conjugation of those rivers among the towns of Hochstadt, Lutzingen, Oberglau, Unterglau, and Blenheim. Parliament responded by appropriating funds to build one of the greatest palaces in England for him, which he named after the one town on his battlefield which sounded faintly English.

Churchill's lineage was thus distinguished, but his father being only a second son was merely Lord Randolph, and *his* son was plain (though the Honorable) Mr. Churchill. He was born at Blenheim palace in 1874, the son of the syphilitic lord and an American beauty whose promiscuity was kept under sufficient control to keep her free from such inconveniences as killed the lord; after his death she would marry again—and again—finally when she was sixty-three marrying a man of forty and living happily ever after (or at least for another fourteen years, until she died of a slight misstep, which led to a broken ankle and thus to gangrene, amputation, and finally an overwhelming hemorrhage). Winston took the traditional path of sons of second sons: education at Harrow and (failing to get into Oxford) Sandhurst and thence into the army. He wrote an account of his first action while it was going on, the last great cavalry charge in British history, and published it the following year under the title, *The Story of the Malakand Field Force;* it was received in 1898 as "brilliant." He served with the Tirah expedition and wrote *The River War* in 1899, another critical and commercial success. The income from the books enabled him to lead the life of an army officer without embarrassment, setting the pattern for the rest of his life. (He evidently saw his literary career as more important than the military, for in his second book he noted that during the action, "With the design of thereafter writing this account, I moved to a point on the ridge which afforded a view of both armies," a line which William Manchester points out affords a fascinating glimpse of both his view of himself as soldier/writer and of his aristocratic independence: the "lowly subaltern" felt himself perfectly free to move around as he pleased for the purpose of the book he intended to write.)

When the South African war started, he went as both army officer and correspondent for the *Morning Post;* he was captured by the Boers, escaped from prison and thence through the enemy lines, and wrote a glorious account of his adventures, published as *London to Ladysmith via Pretoria.* In the early years of the new century he was the highest paid reporter in the world.

But he was a gentleman in days when "one might with decency invite a doctor or even a solicitor to garden parties, but of course never to tea"—and "one" would not invite a reporter to anything at all. So he stood for Parliament in 1900 and was elected as a Conservative. His father was felled at an early age from the syphilis contracted in his youth, some said as a result of a prank when he was drugged by friends and dumped into a whore's bed, while most felt that since he frequented so many whores' beds of his own volition it was impossible to say where he had picked it up. He had been brilliant in politics early in his career and seemed slated to become the head of government. It was a tragedy when his speech slurred, his memory gave out, and his temper flailed; it was a relief when he died. His fate was tragic but not shameful, as it would have been in Victoria's reign, and his son was received by his father's colleagues with open arms and with high expectations, which were somewhat dampened as the first years rolled by. In reviewing his first speech, the Liberal *Daily Chronicle* saw him as an "undistinguished young man with an unfortunate lisp in his voice"; he obviously had no career ahead of him as an orator. The Conservative papers were more approving, which is surprising as his speech was definitely liberal in tone, presaging the dichotomy which would mark his life: he was liberal among Conservatives, and conservative among Liberals.

He had "heavy but not very mobile guns," Lord Balfour remarked, referring to his fine speeches but lack of repartee when challenged on the floor. It was a lack he felt heavily and trained himself to overcome, with devastating effect later in his career

He was not overly influenced by labels, which brought him the unwelcome reputation of political instability, but in judging his career one is reminded of Lloyd George, "whose practical instabilities rode lightly over a deep philosophical rigidity." When Churchill first left the Conservative Party in 1906 and won reelection as a Liberal, it was not for mere political expediency but because he felt strongly enough about free trade to risk his burgeoning career. It is easy to understand how he was quickly regarded as an opportunistic traitor by men such as Arthur Balfour, the Conservative prime minister for whom the leadership of England by the Conservative Party was more important by far than minor questions of economics. It is feared that Balfour's opinion was not revised upwards when both Churchill and the Liberals won the resulting election.

Winston was rewarded by the new prime minister, Campbell-Bannerman, with an appointment as Under Secretary for the Colonies. His subsequent treatment of the colonies reflected his basic philosophy, which was already strongly formed and had been clearly enunciated in the Boer War: A strong and firm Government, dissent from which was not to be tolerated, must react to any provocation strongly and swiftly but with humanity and justice, and after destroying dissent by force of arms must be magnanimous and forgiving, ready to right by its own policies whatever wrongs may have led the natives to protest. Churchill was to fight for this philosophy in the disputes concerning Ireland, India, South Africa, and Germany; in clinging to it he had often to change political parties or friends whose philosophies were conditional upon circumstances and who consequently denounced him as shallow and opportunistic. He was anything but; in his later days of glory it was his overwhelming honesty and belief in an ultimate right that convinced the people and dragged them along to "their finest hour." At the time of the Boer War, this Kiplingesque philosophy was dangerously liberal; at the time of the Indian fight for freedom it was dangerously reactionary; during the long Irish struggle it was often both at the same time, depending upon your point of view; at all times it was basic Churchillian.

As spokesman in the Commons for the colonies he did well, and entered the Cabinet a few years later as President of the Board of Trade, in which capacity he pushed through one of the first minimum wage acts. He was always, since his earliest days, much concerned with the poor and the working class, although his own life was so much of the upper class that it was nearly a prison. He never in his entire life, for example, rode on a bus; and on the one occasion when he took a subway train he is reported to have got lost. (William Manchester tells us that he was so totally lost that an expedition had to be sent after him; it found him wandering helpless and brought him home. I don't believe that for a moment. The man was patrician but not imbecilic. On the other hand, I find most believable another anecdote quoted by Manchester, according to which Churchill, upon arriving at Maxine Elliott's Riviera villa, told her: "My dear Maxine, you have no idea how easy it is to travel without a servant. I came here all the way from London alone and it was quite simple." She murmured, "Winston, how brave of you.")

The dark-red spectre of a bloody European-based world war first arose softly enough from the desert sands of Morocco in 1904, masquerading as just another of the familiar old colonial squabbles in far-off lands that meant nothing to the people at home beyond a joyful parade to send off another detachment of Tommies. In Morocco, Tangier had an international administration, and France was recognized as having a "friendly interest" there. When the Mohammedans were inflamed into rebellion and war, France moved in to suppress the dissidents militarily, with England's reluctant support. Russia, however, weakened by the Japanese conflict, was unable to help; and thus Germany made so bold as to oppose France. The Kaiser visited the Sultan at Tangier, greeting him as an independent monarch, and signed a secret treaty with Czar Nicholas for a war against England. But the bubbling pot simmered down when President Theodore Roosevelt, bringing the United States upon the European stage for the first time, privately induced the Kaiser to accept the verdict of an international conference, which was accordingly held the next year at Algeciras. At the same time, the Russian ministers learned about the Czar's secret treaty and abruptly backed out of it, and the Triple Entente of England, France, and Russia was born to keep the Kaiser under control.

Wilhelm II "half-despised and half-feared" his uncle, King Edward VII of England, and the people of England felt the same way about the Kaiser, without the 'half'. Every day there were new rumors of a war both feared and longed for. At Marienbad, where he went to take the cure, the King was photographed deep in conversation with his prime minister, Sir Henry Campbell-Bannerman. The caption in the newspapers read, "Is It Peace or War?" Campbell-Bannerman said afterwards, "Would you like to know what the King was saying to me? He wanted to have my opinion whether halibut is better baked or boiled."

In Morocco the natives continued to rebel sporadically, not in any great uprising but in recurring acts of terrorism which were not enough to uproot the French but were sufficient to give them an excuse to take further control as a means of establishing law and order (and where have we heard *that* before and since?). In 1911 the locals at Fez took arms against the puppet Sultan, and the French Foreign Legion moved in, causing public outcries in Germany (because of the Domino Theory, of course: let one sultanate fall to France and where will it all stop?). Announcing that the French move was a break with the Algeciras agreement, Germany declared herself free to move. She dispatched the gunboat *Panther* to Agadir, in the southern district of Morocco. France grew hysterical; Lloyd George announced that England would honor her obligations; the Great Beast began to stir her slow loins. But once again the diplomatic tables barred the armies from each other. At a conference in Berlin, Germany recognized France's right to move in and establish a protectorate over Morocco (protecting her from what? Herself?) in return for a portion of what had been the French Congo. The Congolese, of course, had no say in this, any more than the Czechs would in 1938 have any voice in the dismemberment of their country (done, of course, in the name of the same European peace).

War was averted, but at the cost of anger in France for giving away what was "rightfully" theirs and in Germany for having failed to use her growing military power to obtain a "place in the world in proportion to her importance." The Kaiser, in particular, was not amused.

It was during these events that Dr. Crippen was caught, tried, and executed. Churchill, in his new position as Home Secretary, signed the execution order, afterwards rinsing the "foul taste" of the murder from his mouth with champagne. Justice, as always in his philosophy, was to be swift and severe—the forgiveness in this case must come from Crippen's God, not man. There were other things to worry about; as the Edwardian Era faded into the Era of War there were signs, presages of things to come, the awful "baneful seething" of the huddled masses that was soon to erupt in Russia as revolution and in England and America as labor violence. In 1911 the new Home Secretary, dressed in a silk top hat and a long fur coat, led a small army into London's slums, armed with mobile cannon and a machine gun, to besiege a house in which were surrounded a group of armed aliens who had killed three policemen. The house was burnt to the ground, and the public was angrily jubilant or despairingly hateful, depending on which step of the social ladder they stood upon. In the air was distrust both of foreigners and of the different classes of Englishmen.

Competing with this overpowering miasma was the slight incense of a tiny, smoking sense of hope, fanned enthusiastically by a few men with visions of a more civilized future—men such as William Jennings Bryan who said, at a ceremony marking the hundredth anniversary of the Treaty of Ghent, "We know of no cause today that cannot better be settled by reason than by war.... I hope we have seen the last great war." His golden dream was not to be. The fairy-tale years were ending.

In 1904 Sir John Fisher had been appointed First Sea Lord. Under his command, the first of the new line of battleships had been launched, the *Dreadnought*. At 18,000 tons she was the largest warship ever floated; powered by the new steam-turbine engines, she was the fastest, and with ten 12-inch guns she was the most heavily armed; all other navies—including especially Germany's navy—were suddenly inferior. But there were continuing problems with Fisher. He once suggested to King Edward that the Royal Navy should use its new strength to attack Germany by surprise and wipe out her navy in one move. "My God, Fisher, you must be mad!" the King gasped. There followed a growing series of disagreements between the Navy and the War Office concerning the strategies to be followed in case of war, and as the disagreements and arguments with Germany increased, a concomitant sense of urgency arose. By 1911 there was a feeling that war—over one cause or another—was imminent.

Fisher by this time had retired, but the navy as it existed was his creation. Churchill was now transferred to the Admiralty, with secret instructions to "put the fleet into a state of instant and constant readiness for war" in case of attack by Germany, which meant replacing Fisher's men, who were opposed to cooperation with the War Office. He did it quickly and relentlessly, "knocking over the admirals like ninepins" and reshaping the navy into his own image. (When the admirals complained that he was ignoring their advice and undermining their authority, thereby destroying the basic traditions of the Royal Navy, he picked up his silk hat, started for the door, then paused and said: "And what are those traditions? I shall tell you in three words. Rum, sodomy, and the lash. Good morning, gentlemen.")

In his restructuring of the navy, Churchill managed to retain good ties with Fisher himself, punctiliously seeking his advice. If he didn't always follow it, he was still sure to explain his reasons beforehand and to obtain, if not the old man's blessings, at least his acquiescence. He did not, however, recall Fisher to his old post at the Admiralty; not yet.

Churchill did his work well, and at the outbreak of war the navy was ready. To his great disappointment, however, no noble actions seemed to be afoot; his navy transported the British Expeditionary Force to France without loss, cleared the oceans of sea raiders, and contained the German battle fleet in their harbors, and then simply waited. There was as yet no submarine menace, the German battleships stayed securely at home; there was nothing to do. It was a situation impossible for Churchill to accept, and his mind began to whirl and spin. It soon settled on the Dardanelles.

His original plan was to send the battleships of the Royal Navy through the Dardanelles into the Black Sea; on their passage through the narrow isthmus they would destroy the Turkish forts, and once they burst through into the Black Sea, they would overwhelm the Turkish naval forces stationed there. A follow-up force of ground troops would land on the mainland and occupy the battered forts, Turkey would fall into chaos and capitulate, and thus this great ally of the Kaiser would be driven out of the war.

The stategic plan was actually sound, even brilliant; unfortunately, its execution was not as finely honed as its conception. At the first attempt to force the straits, the Royal Navy was beaten back; but the losses the forts suffered were even more severe. Their ammunition stockpiles were sorely depleted and they were on the verge of capitulation when the British naval commander lost his nerve and telegraphed that he thought he ought not to continue the action. He suggested that instead a strong invasion force be prepared to attack the forts by land simultaneously with his renewed naval bombardment.

An exchange of telegrams began and continued through the following days. Fisher, who at the outset of war had been brought back to command the Admiralty as First Sea Lord, was in constant consultation with Churchill, offering his advice, sympathy, and encouragement. It was, however, impossible to overrule the commanding officer on the scene, and so, though Churchill strongly suggested an attack with all naval forces before the Turkish ammunition stores could be augmented, he did not insist, and the plan simply fizzled out. Instead, a large army force was prepared and sent off. Though that force was large in manpower, it was short in preparation: it went off ill-trained and ill-supplied, and it failed disastrously.

This became Churchill's first fall from the political ladder he had been climbing so successfully and steadily up till that moment. He was, most unfairly, blamed for the terrible failure that became known as the Gallipoli expedition, though the part that failed and turned into a massacre—the army's invasion against a line of entrenched fortifications—had never been a part of his vision and would never have been necessary had his naval commander carried out the original plan. But the admirals he had alienated turned now against him, led by the man who had become his friend, whom he had brought back to the Admiralty, Sir John "Jacky" Fisher.

Fisher resigned in protest against what he now called a "foolish and calamitous" plan which "had never held the confidence" of the navy—though he himself had been informed and consulted and had given his blessing throughout the planning and execution. The Liberal Government fell, and Churchill was replaced. Accused in Parliament of "unjust usurpation of power" in the Admiralty, he defended himself vehemently against Fisher's charges and then left politics and England, sailing for France and active service as commandant of the 6th Royal Scots Fusiliers.

In 1917 he was recalled by Lloyd George to serve as minister of munitions; quickly and emphatically he insinuated himself once again into the forefront of the conduct of the war. His love of the reckless throw of the dice upon which all might depend became part of the Allied war strategy, for when calamity fell and others instinctively drew back in indecision he just as instinctively stepped forward with one decisive plan or another. In early 1918 the Germans broke the English line and threatened to roll through to Paris and win the war; it was Churchill who stiffened Lloyd George's faltering resolve with the injunction: "Now is the time to risk everything; and to run risks for their very sake!" (Oh yes, he gloried in war; he reveled in the drama even as he wept at the tragedy.) And he was without mercy in the heat of battle: When the Red Cross tried to intervene to stop the use of poison gas by international agreement, he was furious—because the prevailing winds favored the Allies in the long run:

> I am much concerned at the attitude taken by the French representatives... on the Red Cross suggestion of the willingness of the German Government to abandon the use of poison gas.... I do not believe this is to our advantage. I hope that next year we shall have a substantial advantage over them in this field. Anyhow I would not trust the German word....
>
> I am on the contrary in favour of the greatest possible development of gas-warfare, and of the fullest utilisation of the winds, which favour us so much more than the enemy....

By the end of the war he was Secretary for War and Air, climbing the political ladder once again straight toward the summit. He efficiently demobilized the troops and sent them home, while at the same time beginning his first squabble with Lloyd George: he insisted that the communist government in Russia was England's ultimate enemy, and he wanted to destroy it before it got its claws into the Russian people. For two years England (and America) sent troops and supplies to aid the White Russians in their civil war against the Reds, but their aid was too desultory and the Whites were too incompetent, and by 1920 it was all over. As the only member of the Liberal Government that had supported all-out aid to the Whites, Churchill was in disfavor; when Bonar Law resigned the Chancellorship of the Exchequer it went to Sir Robert Horne instead of to Winston. He was disappointed, but turned his energies to the Irish question, in which he was in total agreement with Lloyd George that Home Rule was the only acceptable answer. In 1922, however, L.G.'s coalition government ran into trouble over a totally different question, the Greco-Turkish war.

Lloyd George and Churchill supported King Constantine of Greece, even though he had been pro-German in the last war; their attitude was determined by their opposition to any resurgence of the Turkish Empire, which might cut across their lines of communication and commerce to the "jewel in the English crown," India. But as the possibility of open war between Turkey, England, and France approached, Poincaré ("that horrid little man," as he was described by the foreign secretary) pulled out the French troops. Churchill and Lloyd George remained adamant and warlike, and it was only at the last moment (with French political help) that war was averted and a settlement reached.

The Conservatives were frightened at how close war had come. The Great War had been "the war to end war," and yet England had been fighting continuously ever since it was over, first in Russia and now nearly again in Turkey. They had had enough, they withdrew their support from the Government, and the coalition fell. This was to be the last time a British government stood up against aggression directed toward their allies until September 3, 1939.

In the ensuing election of 1922 Churchill was stricken with appendicitis; when he awoke from the anesthetic he found that both he and the Lloyd George government had been defeated. Estranged from both parties, he was, as he put it, "without an office, without a seat, without a party, and without an appendix."

Bonar Law took over direction of the Conservative Party and the government, but resigned the next year to die "of his fell affliction," as Churchill said, introducing upon the world stage Stanley Baldwin. Mr. Baldwin's government lasted only another year, to be followed by the nation's first Labour Government, which itself fell in less than a year. During that year Churchill stood as a Liberal in a Westminster by-election. The seat was contested by all three official parties (Labour as well as Conservative and Liberal), and Churchill lost to the Conservative candidate by forty-three votes out of 20,000. In the process, however, he had won a lot of Conservative support, and when the next election came in another two years, he won with their backing, though he ran officially as a "Constitutionalist." Once back in Parliament, once again a Conservative (though on the fringe of that party), he was surprised—

and, he reported, "the Conservative Party was dumbfounded"—when Baldwin chose to appoint him Chancellor of the Exchequer. One year later he formally announced his rededication to the Conservative Party which he had left twenty years before. Once again he was a powerful force in British politics.

Baldwin's government lasted until 1929, when it was brought down by economic issues and Labour took charge. It was clear that no party, no person, and no economic philosophy could cure the woes of a sick nation in such a sick world, and it was to be expected that the Conservatives would soon be back; but when they came, Winston was no longer one of their leaders.

In the interim Neville Chamberlain had risen to challenge Baldwin's leadership of the party, basically on the economic question of Free Trade or Protection. Churchill at this point nearly rose above them both—his was the voice most often heard in violent opposition to Labour, and many Conservatives found his adamant stand against Dominion status for India a rallying point—but he was seen by the party as a friend and supporter of Lloyd George, and another Liberal coalition was anathema. In January 1931, he made his bid for leadership by resigning from the "Shadow Cabinet," as the leaders of the Opposition were called, over the question of India. He called the proposed Government of India Act "a gigantic quilt of jumbled crochet work, a monstrous monument of shame built by pygmies." You couldn't fault his language. But both Baldwin and Chamberlain supported the Labour policy of giving that land its freedom, and though Winston's move sparked a violent debate among the faithful, in the end the Chamberlain-Baldwin axis held firm and he fell from grace into a political wilderness.

His career seemed to be over. When MacDonald (the Labour leader) and Baldwin formed a National coalition government later that year, he was not included. Nor was he asked back during the second National Government nor when Baldwin took sole charge in 1935 nor when Chamberlain assumed the leadership in 1937. As C. P. Snow later described him, he was seen in those years as a "great man *manqué*," one whose future had never been and was now past.

Churchill's importance to the development of radar is both general and particular: in general he was an important factor one way or another to everything that happened in England and indeed anywhere in the world during the nineteen thirties and forties; in particular he nearly destroyed radar before it was born through his friendship with F. A. Lindemann, professor of physics and Fellow of the Royal Society, friend and confidant of the Duc de Broglie, the Duke of Westminster, and the Earl of Birkenhead, and of lesser mortals down to but not below the levels of the lower aristocracy. Lindemann was a firm supporter of the Royal Air Force, who in 1941 was to prove to the consternation of the Air Staff that only sixteen percent of bombs dropped landed anywhere near the target, forcing a total revision of bomber tactics. And he was a man who in the early days—when radar was just being developed—was a catastrophe looking for a place to happen.

He nearly found it.

C. P. Snow, in the Godkin Lectures at Harvard University, 1960:

Nearly everyone I knew of my own age who was politically committed, that is, who had decided that fascism had at all costs to be stopped, wanted Churchill brought into the government (in the late 1930s).... We signed collective letters about Churchill; we used what influence we had.... We wanted a government which would resist [the Nazis].... But if Churchill *had* been brought back to office? We should have been morally better prepared for war when it came. We should have been better prepared in the amount of war material. But Lindemann would have come with him... different technical choices would have been made... and without getting radar in time we should not have stood a chance in the war that finally arrived.

They were to have met at a weekend at the Duke of Westminster's palatial Eaton Hall, where the Duchess had paired Lindemann, the former tennis champion of Sweden, with Clementine, Winston's wife, who was a keen and competitive player. The two of them got along famously, but Winston, busy with political affairs, never showed up. The Duke then arranged a quiet evening at home for them. (Lindemann to his father: "I refused an invitation to the Marquess of Headfort and another of the Duke of Westminster to Scotland to meet Winston Churchill [but] the Duke was very keen on my meeting Churchill and arranged a special meeting in Town last night for the purpose. It was interesting....")

It was a milestone in both their lives. Lindemann, later Lord Cherwell, was to remain for more than thirty years one of Churchill's most intimate and influential friends. In the words of Sir John Colville, Churchill's principal private secretary, "If he were writing fiction an author might reasonably be criticised for creating such [an] improbable character... as Professor F. A. Lindemann, commonly known as 'the Prof.'"

"He had a distinguished presence," according to Colville. "His deep-set eyes looked out beneath gently arching brows and a curving, well-developed forehead. His nose was of noble Roman shape, his moustache was flat and neatly trimmed, his mouth suggested refinement.... He wore beautifully cut clothes, out of date in design... spotlessly clean, almost clinically so." But according to C. P. Snow, "one would have taken him for a Central European business man—pallid, heavy featured... [with] a faint Teutonic undertone to his English, to his inaudible, constricted mumble."

In more than physical appearance, he was a man of contrasts. He presented different aspects of character, personality, and intelligence to different people. To Snow he was "a very odd and a very gifted man...the sort of man that makes a novelist's fingers itch....He was formidable, he was savage, he had a suspicious malevolent sadistic turn of what he would have called humour....If one was drawn to him at all, one wanted to alleviate it." To unidentified people quoted by Lord Birkenhead he was "an offensive man of alien extraction, professing an abstruse subject, spending much of his time ingratiating himself with exalted people." To Colville he "had an ear for music and [was] a skilled pianist, he was modest, he was a tennis player of professional quality, he could quote long passages from the Bible by heart and his knowledge of almost all periods of history was impressive." But to Snow, again, he had "no interests in literature nor in any other art." To Churchill he had "a beautiful brain," while Churchill's wife, Clementine, enthused, "What fun it was to see him!"

When he met Churchill he was Professor of Experimental Philosophy, in charge of physics at Oxford's Clarendon Laboratory. As a scientist he was "highly distinguished...on a par intellectually with Lord Rutherford," according to Colville. But according to Snow, "he could not compete with Rutherford....He was an amateur among professionals, which is how Rutherford always regarded him." He was merely "a gadgety scientist who soon got tired" (Snow); he was, on the contrary, "primarily interested in fundamental theories, much less in the [details] of the laboratory" (Lord Birkenhead); he was "a natural revolutionary with little respect for traditional thinking....He jumped to conclusions about fundamental assumptions without analysing the evidence in detail. His occupation with administration, University affairs and politics allowed him little time [for science]" (Professor Max Born).

It was "difficult to think of him undressed or in pyjamas"; for one person it was "impossible to imagine him in bed with a woman," while to another it was "remarkable what a sexual appetite for women he had." He was a venomous anti-Semite, who was the first to rescue Jewish scientists from Germany when Hitler took power. He was "always immaculately dressed," "never in style"; "no one seems to know to this day what his father's nationality was"; but "His father was born in the German Palatinate, his mother was American."

He was everything to all people, but to everyone he was brave. Turned down in his attempts to enlist in the Royal Flying Corps at the outbreak of war in 1914 because he was essentially blind in one eye, he joined the scientific staff of the Royal Aircraft Factory. There he pleaded that the scientists should learn to fly, though it was exceedingly dangerous in those early days when no one knew what was actually happening to make an airplane fly or fall from the sky:

My dear Papa,

We went to a lecture at the Aeronautical Society on Wednesday. The lecture was very good but the discussion was absolutely puerile. Everybody disagreed with the lecturer in the matter of instruments. They all said the pilot ought not to use them as they might go wrong. A lot of people get killed by not using them. Evans (for example) flew much too slowly and stalled on the turn. Having done this he seems to have lost his head and failed to recover himself. Our head pilot (also) got killed today. Nobody can make out quite what happened.

It was his job to "make out what happened," but he couldn't do that from the ground. In particular, the spin was the terror of the skies, death to any pilot who slipped into one. This could happen when the airplane stalled—when the speed dropped low enough so that the rush of air over the wings was no longer sufficient to keep the airplane flying. When this happened, the nose would drop and, if the wings were not kept absolutely level, one of them would dip and the airplane would fall over on that side. The other wing would flip over and the plane would plummet earthwards, spinning faster and faster as it fell uncontrollably.

This sequence of events might happen as the pilot was slowing down in preparation for landing, or during maneuvers even at fairly high speeds when the wings were tilted in a tight turn and thus lost their lift. Once in a spin, no one had any idea how to get out of it; it was generally regarded as fatal.

Lindemann came up with a theoretical description of the forces which must be operating during the spin, and so figured out how to counteract them. The instinctive response of pulling back on the stick was no good, he said; one should concentrate on first straightening out the aircraft by use of the rudder pedals, *keeping the nose pointed down toward the earth*, until the spin stopped and flying speed picked up. Then one might pull back on the stick.

And the wings might be ripped off by the pressure buildup. It was obvious that he couldn't ask any pilot to put his aircraft intentionally into a spin in order to test his theory, so he decided that he should learn to fly. The Superintendent turned down his request, on the grounds that it would "weaken the position" of the professional pilots "who were the recognized experts in flying," but he allowed him to send a written request to the War Office, which surprisingly gave permission—if, of course, the stringent physical examination could be passed.

His colleague at the time, Warren Farren, wrote:

> The sight in one of his eyes was so poor that it was hardly an exaggeration to say that from the point of view of a test it was blind. It certainly hindered his flying afterwards, particularly in landing, since his stereoscopic vision was seriously impaired. When he came out from the test, naturally triumphant, I asked him how he had done it, and he said: "I went in and [the doctor] sat me in the chair, and whilst we were chatting I used my good eye to memorize the letters on the side of the card I could see. When he started the test I knew that he would either leave it that way round or turn it over. [He turned it] so I shut my bad eye and read the new side without any difficulty. He then turned it back the other way round and I looked at it with my bad eye and read out the letters from memory."

And so he learned to fly. And then he took off in a B.E. 2 fighter, an open-cockpit biplane, built of the thinnest and lightest of wood struts and covered with fabric, capable of seventy miles an hour downhill in a light wind. And he climbed her up 14,000 feet, and then he pulled the throttle back, chopping power, and he pulled the nose up into the sun and held her there while the flying speed ebbed away and she began to shake and shudder complainingly, warningly, until finally she just couldn't fly anymore and her nose dropped and she fell out of the air. As she did, Lindemann coolly pressed the spade-handled joystick over to the right; and the right wing dipped and the left wing rose and, falling, she fell over on her back and out of control. As the nose pointed itself down at the green fields below, the wings spun around and around, faster and faster; and she was well and truly in a spin.

Watching from the grass of Farnborough aerodrome, Mervyn O'Gorman, the superintendent who had reluctantly and "with immense misgivings" allowed the test, winced, turned away, and then turned back to look again.

Faster and faster she spun; from where Lindemann sat in the cockpit the ground was now a blurring kaleidoscope rushing up at him. He held her steadily spinning as the blood was forced by centrifugal forces up into his head, hazing over the instruments with a blood-red gauze covering his eyes. He held her until there could be no doubt that he was in a totally uncontrollable spin.

And then he controlled her. Pushing hard on the stick to keep the nose from rising—the craft had a life instinct of her own; she kept trying to raise her head and pull away from the onrushing ground—he applied pressure on the left rudder pedal. Nothing happened. He sat there and held the pressure; he did nothing else as he hurtled through the empty air, down, down, faster, faster. He sat there in the open cockpit with the wind screaming crazily in his ears, and he held the left rudder pressure as his theory said he must; and finally the spinning began to slow, the blurred earth took shape again, and the trembling craft began to straighten, and finally held steady—steady, but still diving straight toward the ground at more than a hundred miles an hour.

Now he began to pull back on the stick, cautiously at first, still cautiously as the earth roared closer and closer, ever cautiously so that the flimsy wings might not buckle and tear off. He never lost his nerve, he never relaxed nor increased the pressure on the control stick, and finally the nose slipped up and the deadly earth slipped down under and with a great surge the B.E. 2 roared over the heads of those watching on the ground and soared away back up into the sky.

Back up 14,000 feet again she went, and as O'Gorman's "heart sank into his heels," Lindemann pitched her into another spin; the first had gone clockwise, this one turned counterclockwise—just to make sure that those watching were convinced.

And so they were. Before Lindemann's flight that day, "an airplane seen to be spinning was regarded as doomed"; after that day the spin became a standard method of escaping from an uneven combat, a frightening but routine maneuver taught to every fledgling pilot.

In his duties Lindemann had often to visit one airfield or another, and he always flew himself. The ground staff, who didn't quite trust him because of his Germanic accent, kept an unofficial but firm limit on the amount of fuel they would put into his plane, never enough to allow him to fly across the Channel. He was never aware of this, so it was lucky he didn't try. When he was due to fly somewhere he would show up at the aerodrome dressed as always in bowler hat and long coat with velvet collar, and carrying, of course, a rolled umbrella. He would take off the coat, roll it carefully, and store it under the seat with his bowler and umbrella. He would then put on his flying coveralls and helmet, climb into the open cockpit, and take off. Upon arrival at the other aerodrome he would taxi across the grass to his assigned space and cut the engine. Then anyone watching would see a long spell of time go by with nothing visible but the top of his head scrunched down in the cockpit. He was, in fact, taking off his flying coveralls and pushing them under the seat, and taking out his coat, hat, and umbrella. Eventually he would step down from the cockpit immaculately dressed in his long coat with the velvet collar, carrying his rolled umbrella, with the bowler hat sitting carefully on the top of his head. (He had a dread of ever appearing badly dressed or out of place; of course in this situation the bowler, coat, and umbrella were precisely that. People used to stare in wonderment at this apparition descending from the primitive biplane.)

He was extremely rich. After the war he drove away from Farnborough in his chauffeured Rolls limousine, accompanied by his personal servant, to seek an academic position. Before the war, when they were both studying in Germany, he had made fast friends with Henry Tizard, who in the thirties would chair the committee set up to investigate aerial defense and who would serve as midwife to radar. They studied and worked together, played tennis and boxed together (although the latter only once, since Tizard was a better boxer and Lindemann an extremely poor loser), and Lindemann stood godfather to Tizard's son. Tizard had become a chemistry don at Oxford, and through his recommendation, Lindemann was now offered the Chair of Experimental Philosophy (physics). He took over the moribund Clarendon Laboratory and began to build it up. In this he was somewhat less than successful. One of the standard undergraduate jokes in Oxford during the twenties was the following. Question: "To whom does the Cavendish (the physics lab at Cambridge) owe its preeminent position?" Answer: "To Rutherford (head of the Cavendish, who brought together a world-class group of physicists) and to Lindemann (who didn't provide any competition)."

He dropped out of the first line of scientific research, surprising those who had predicted a fine career on the basis of his student days. He never published anything of lasting importance after that. Not many scientists do, of course, but it was ridiculous of some of his friends (Churchill among them) to put him in the same class as Rutherford. Instead, he cultivated the aristocracy. (Question: "Why is Professor Lindemann like a Channel steamer?" Answer: "Because he runs from Peer to Peer."). In this he came dangerously close to that most frowned-upon practice, sucking up to one's betters. But he was not merely sucking up. The world of politics and power was the world to which he belonged by instinct and by background. Though it was a world separated as if by an iron curtain from the equally insular world of the university, he moved easily into it with the assurance bred by second-generation money. Lord Birkenhead:

> His (aristocratic) hosts were not slow to perceive what an excellent guest he was. With his Rolls-Royce, chauffeur and valet, he lived in a manner they understood, which came as a surprise as they did not expect a scientist to be an *homme du monde*. They found his manners perfect, his presence sympathetic and his conversation stimulating. It was a relief to them that he knew many of their friends, so that there was no embarrassing need to grope for

topics of conversation from which he was not excluded.... He was an exceptional tennis player.... He loved feminine society and was adept at talking to women, for whom he reserved a fund of up-to-date, somewhat malicious gossip, punctuated by giggles. The men were also impressed, over the port which he did not drink, by his views on politics. He appeared usually to be fresh from contact with some Cabinet Minister whose opinions on the vital questions of the day he relayed to his friends....

His friendship with Churchill is a bit harder to understand. Winston did everything to excess, particularly eating, drinking, and smoking, while "the Prof" was a "cranky vegetarian" who neither smoked nor drank. But there were compensations. Lindemann admired Churchill's position of power in the days when they met, before The Fall, and Churchill admired Lindemann's scientific knowledge, which he squandered as if he were throwing pennies to a crowd. Winston would write to him asking for the best dimensions to build his swimming pool or the best mixture of mortar for his continual wall building, and Lindemann never failed him. In 1931 Churchill was hit by a taxi in New York and nearly killed. From the hospital he wired across the Atlantic, asking the Prof to calculate in mathematical terms the effect of the trauma he had suffered; he wanted to impress the readers of the *Daily Mail* back in England, for whom he was writing a report on his New York adventures. Lindemann immediately wired back:

> Just received wire delighted good news stop Collision equivalent falling thirty feet on to pavement equal six thousand footpounds energy equivalent stopping ten pound brick dropped six hundred feet or two charges buck shot point blank range stop Shock presumably proportional rate energy transferred stop Rate inversely proportional thickness cushion surrounding skeleton and give of frame stop If assume average one inch your body transferred during impact at rate eight thousand horsepower stop Congratulations on preparing suitable cushion.

Churchill loved it. But any first-year student in physics could have done as well, except perhaps for the final line. It was this sort of showboating, or "shooting a line," as the British say, that at the same time endeared him to the aristocracy and disendeared him from those who should have been his scientific friends.

There was more in the Churchill-Lindemann friendship. Both men were exceedingly brave physically, scorning those who were not. They were both aviators. (Winston had learned to fly when, as First Lord of the Admiralty, he founded the Royal Naval Air Service in 1912—Britain's first air force. He had over a hundred hours of instruction and yet never soloed, in days when a man generally took a plane up alone after three or four hours of instruction. The reason was partly his ineptitude, but more importantly the timidity of his naval instructors, none of whom had the courage to take responsibility for killing a First Lord of the Admiralty—and in those early days, a fatal crash while learning was a not unusual end to a man's career. Winston finally gave up flying after a series of pleas from Clementine.)

And there was more. Despite Lindemann's "cranky vegetarianism" and principles of neither smoking nor drinking, he never imposed such principles on his guests. Instead he lavishly supplied cigars and cigarettes, red meat and port, brandy and whiskey for all. And there was yet more: both men were intensely patriotic. In Winston's days as Chancellor of the Exchequer immediately after the Great War of 1914 he cut finances for all the fighting forces ruthlessly, thunderously decrying their need: "Has the English Channel dried up? Are we no longer an Island?" But as the 1920s swung into the 1930s, his cry was answered by the unbelievably rapid progress of aeronautical research, as airplanes flew faster and higher and farther: 100 miles an hour, 200, 300! Three hundred bloody miles an hour, flying so high they were invisible from the ground, covering the long miles from Paris or the coast of Germany without need to refuel—and suddenly the vision of tons of bombs plummeting from airplanes unseen and unstoppable provided the answer to his question: "No, we are *not* an Island any longer!"

In 1932, arguing in support of the disarmament conferences to which Churchill was opposed, Stanley Baldwin had made the blunt statement that "the bomber will always get through." For the next several years that was the dominant view, and there were continuing arguments in Parliament as to what the country should do to prepare a suitable military stand against aggression, given that there could be no defense against the bombing of English cities. On the eighth of August, 1934, under the heading "Science and Air Bombing," the *Times* published a letter from the Prof:

> Sir,
>
> In the debate in the House of Commons on Monday on the proposed expansion of our Air Forces, it seemed to be taken for granted on all sides that there is, and can be, no defence against bombing aeroplanes and that we must rely entirely upon counter-attack and reprisals. That there is at present no means of preventing hostile bombers from depositing their loads of explosives, incendiary ma-

terials, gases, or bacteria upon their objectives I believe to be true; that no method can be devised to safeguard great centers of population from such a fate appears to me to be profoundly improbable.

If no protective contrivance can be found and we are reduced to a policy of reprisals, the temptation to be "quicker on the draw" will be tremendous. It seems not too much to say that bombing aeroplanes in the hands of gangster Governments might jeopardize the whole future of our Western civilization.

To adopt a defeatist attitude in the face of such a threat is inexcusable until it has definitely been shown that all the resources of science and invention have been exhausted. The problem is far too important and too urgent to be left to the casual endeavors of individuals or departments. The whole weight and influence of the Government should be thrown into the scale to endeavour to find a solution. All decent men and all honourable Governments are equally concerned to obtain security against attacks from the air and to achieve it no effort and no sacrifice is too great.

The only response he had came from Winston, who enthusiastically supported his proposal. Unable to get in to see any members of the Cabinet, the two of them traveled by Channel steamer and train the next month to Aix-les-Bains, where Stanley Baldwin (then Lord President, sharing powers with Prime Minister Ramsay MacDonald in the coalition government) had retreated for his annual vacation. This vacation was a sacred ritual to Baldwin: he did no business, saw no one, even refused to read the newspapers; he rested, read good books, breathed slowly and deeply the fine French air. To have two such energetic, nearly brawling, interruptors was a disagreeable strain to him. In vain did he explain that he was there to rest and would not listen; in vain did they continue to talk, gesticulate, and insist. Finally they left in disgust (on both sides), and Baldwin tried once again to concentrate on the fine healthy air.

But Churchill and Lindemann did not stop; they were like twin locomotives with full heads of steam and no brakes, and the parallel iron rails of their logic and fears carried them straight ahead. They wrote, they badgered, they visited, and a few months later their efforts were crowned with success—of a sort. Lindemann was invited to attend a meeting of Sir Robert Brooke-Popham's committee which was studying how best to defend London from air attack. It was success of a very limited sort, since that committee had already accepted the premise that the bombers could not be stopped; their mission was to determine how London could best withstand the siege and how to counterattack with their own bombers to bring the aggressor to a standstill. Lindemann wrote next to Lord Londonderry, Secretary of State for Air, a personal friend, insisting that "an antidote has always been found hitherto for every offensive weapon and I see no reason to suppose that aircraft are the only exception." He wanted a small but powerful committee to be formed that would not accept the bomber as invincible but would "find a method of preventing bombers...from reaching their objectives."

He was not aware that the Tizard Committee had already been formed, on Wimperis's advice, to do precisely this. But his attitude was precisely the same as that of Wimperis, Tizard, and Watson Watt; his objective—finding a scientific method of defeating the bomber—was precisely their objective; so why was he a catastrophe looking for a place to happen, and why did he very nearly find that place to happen when he was, on Churchill's insistence, at last invited to join the Tizard Committee?

Chapter Six
The Battle of Barking Creek

> Indeed, it has been singularly unfortunate....
> —*The Last Journal of Robert F. Scott*

It happened because although radar was ready in time to win the Battle of Britain, it was *just* ready. It was not ready at the time of the Munich crisis in 1938, nor was it even ready when the war began on September 3, 1939. One of the reasons the Allies lost France in the spring of 1940 was that Stuffy Dowding, who in his previous job as Air Officer for Research and Development had provided the first funds for radar and had godfathered its entire development, was by then head of RAF Fighter Command. He refused to permit any of his Spitfires to leave England because there was no radar system in France, and he knew that without it his precious Spits would be destroyed by the Luftwaffe. The system worked to perfection in the summer of 1940, and the Luftwaffe was driven from the British skies, but a delay in its development of just a few months would have been fatal.

Consider, for example, what happened in the skies over London during the first great aerial battle of the Second World War, on September 6, 1939, just three days after war was declared—the greatest *unknown* battle of the war, The Battle of Barking Creek.

At 11:15 a.m. on the morning of Sunday, September 3, 1939, Prime Minister Neville Chamberlain went on the radio to announce that his world had collapsed. "I am speaking to you from the Cabinet Room at 10 Downing Street," he began. "This morning the British Ambassador in Berlin handed the German Government a final Note stating that, unless we heard from them by 11 o'clock that they were prepared at once to withdraw their troops from Poland, a state of war would exist between us.

"I have to tell you now that no such undertaking has been received, and that consequently this country is at war with Germany."

At 11:25 a.m. the air raid sirens sounded over London.

Ten feet beneath the ground at Bentley Priory (a site that goes back in history to the year 61 A.D., when Queen Boadicea, after struggling furiously but vainly against the invading Romans, took poison and was buried there), the current headquarters of Fighter Command was in operation. The same Squadron Leader who had marked the *Graf Zeppelin* on her first reconnaissance along the coast the previous May, and who was eventually to become Air Marshal Sir Walter Pretty, was on duty in the Filter Room watching a cardboard marker being pushed across a huge map laid on a table. Experienced as both a pilot and an electronics specialist, he realized clearly the implications of the plot being traced across the table by the marker. As the WAAF pushed it along with the type of long-handled rake used at the gaming tables in Monte Carlo, Pretty's hand strayed to the phone by his side. The marker's path reproduced the trace of a blip on one of the radar screens looking out past the southeast coast of England, and at 11:20 a.m. it had progressed far enough across the Channel to set the system into motion. Squadron Leader Pretty picked up the phone to call his Controller, and within minutes the sirens began to wail.

In the Central War Room in London, Air Vice Marshal John Slessor (who the previous day with Cyril Newall, Chief of the Air Staff, had gone to the House of Commons and summoned the Secretary of State to make sure that there was no chance of the Government's backing down on its promise of war if Poland was invaded) was discussing plans with the Minister for Coordination of Defence. "My word," Lord Chatfield said, glancing at his watch as the sirens went off, "these chaps don't waste much time!" They looked at each other: German efficiency. Christ. In the Air Ministry, Air Vice Marshal Sholto Douglas, Assistant Chief of the Air Staff, filed down to the shelters, as did millions of other people in London, where "we all sat around looking rather stupid and waiting for the Germans to knock hell out of us." At Biggin Hill, the most famous fighter base of them all, no one knew what to do; they weren't ready yet for war. Some of the pilots ran for their planes, thinking perhaps they should take off even though they had received no orders—they hadn't been told what to do when they heard the sirens. They slipped into their parachutes and walked nervously around their Hurricanes, keeping an eye on the skies to the east, not wanting to be caught on the ground by the German raid but unsure of just what action they should take. In London the streets went curiously silent—buses and automobiles and horse-drawn carts left alone and unattended, sidewalks empty, stores quiet and lifeless, as Bobbies wheeled down the streets on their bicycles ushering the few lost dawdlers into the shelters. In the Biggin Hill dispersal hut assigned to No. 32 Squadron, the telephone finally rang. Nearly two minutes had passed since the sirens had begun their wail. The airman on duty yanked it from its cradle. "Thirty-two Operations," he yelped. "Yes, sir,...yes, sir,...yes, sir,.... What height, sir?" And then he yelled out the window to the pilots who had been stalking the ground around their machines and who were now still and silent, turned attentively toward him, "Scramble! Blue section, scramble! Patrol Gravesend, 5000 feet."

"Let's go, chaps! This is it!" someone shouted, and three minutes later two Hurricanes were tearing across the aerodrome and pulling themselves into the air.

"Hullo, Sapper Control," the radiotelephone crackled almost immediately in the Biggin Hill control room. "Jacko Blue Leader calling. All aircraft airborne."

"Hullo, Blue Leader," the Controller answered. "Vector 045 degrees, angels five. Unidentified aircraft."

"Hullo, Sapper Control. Blue Leader answering. Message received. Listening out."

But no bombs fell that day on England. The radar blip was an airplane, all right, but it carried nothing more dangerous than Captain de Brantes, a happy Frenchman who served as assistant French military attaché to his embassy in London, flying back to England from a carousing weekend in Paris. Incredibly unaware of all that had been happening that weekend, he had also neglected to file a flight plan for his private airplane, so that when his blip was picked up on the radar screen it was listed as unidentified, and the air raid sirens went off and the fighters scrambled.

Well, no harm done, as they say. It was even a bit of good, proving in an unauthorized and unplanned test that the radar system worked. It threw the censors into a bit of a flap: they weren't prepared for events to happen so quickly, and the afternoon papers were going to bed before anyone was sure what had caused the radar blip. They decided to play safe and not allow any stories of the "raid" to be published. But their censorship system wasn't yet airtight, and the *New York Times* published a photo of a whistle-blowing Bobby careening down the streets of London on his bicycle with a sign hung from his neck proclaiming "Air Raid! Take Cover." But no accompanying story was published, and in the English papers there was no mention of the incident at all.

That night the alarms went off again, and the situation proved more serious. The censors had had time during the previous afternoon to decide that they couldn't keep air raids secret forever and that honesty was the better part of discretion; the *London Times* the next morning announced that "Air raid warnings were sounded in the early hours of the morning over a wide area embracing London and the Midlands...."

The alarms went off at 2:30 a.m. and the all-clear was sounded at 4:12 a.m.

Again no bombs were dropped. On September 5 the *Times* announced that the air raid warnings were "due to the passage of unidentified aircraft. Fighter aircraft went up and satisfactory identification was established...," all of which was nonsense. It was impossible for fighters to *find* any aircraft at night, let alone identify them. What had happened was far more serious than a repetition of the September 3 scare: what had happened was nothing at all.

There had been no aircraft, unidentified or not, German or not, penetrating English air space that night. Yet just before 2:30 a.m., the radar station at Ventnor had picked up a plot in mid-Channel, heading in toward the coast. At Tangmere aerodrome, No. 1 Squadron was maintaining an all-night vigil, with one section at cockpit readiness, when the telephone rang. Flight Lieut. Johnny Walker led Pilot Officer Richey and Sergeant Soper off to intercept. They searched for an hour; when they returned other sections were scrambled. None of them found anything. The radar system had simply misfired, had reported a thundercloud or a flock of pigeons or God knows what else; what it had *not* reported was an airplane, for there were none flying in the vicinity that night. The story of "satisfactory identification" was given out to placate the population which had been wakened and sent to the shelters in the middle of the night; what kept the RAF awake the following nights was wondering what had happened. They couldn't be sure there had not been a flight of enemy aircraft. But then why had no bombs been dropped? They sat around trying to convince themselves that there *had* been something flying up there, and secretly each of them was wondering if the radar system had let them down, and what they would do if they couldn't trust the one system all their defenses depended on.

Their answer came two days later.

On Wednesday afternoon, September 6, all England was talking about the gigantic air battle seen that morning over London and the southwest counties. The talk was lively and excited as stories were compared and swapped, and accounts of a tremendous victory for the RAF were cheered and drunk to. All that afternoon army vans scoured the countryside looking for the wreckage of the German bombers.

By evening it was clear that there were no wrecks, except for two Hurricanes. Nothing else was clear. The Government tried to stifle the stories. On September 7 the *Times* reported:

> The Ministry of Information issued the following announcement to the Press yesterday afternoon.

"We are officially informed that enemy aircraft were reported near the East Coast of England early this morning. So far as is known, they did not penetrate our defences at any point, and no damage has been reported."

The Press Association issued the following statement late last night.

"The Press Association is officially informed that the enemy air reconnaissance off the East Coast referred to in this morning's Ministry of Information bulletin led to the dispatch of fighter aircraft.

"Contact was not made with the enemy, who turned back before reaching the coast. On returning, some of our aircraft were mistaken for enemy aircraft, which caused certain coastal batteries to open fire. This accounts for the rumours of a heavy aerial engagement."

In fact there *was* a heavy aerial engagement, but it had not included the Luftwaffe. Again there had been a radar warning of incoming aircraft, and again it had been a friendly plane coming in from France with no flight plan filed, and again a flight of Hurricanes had been scrambled. But from here on the story got worse and worse.

A new blip on the radar screen was observed: more enemy aircraft. The Bentley Priory Controller scrambled a whole squadron to deal with it. But no sooner had they become airborne and were climbing to fighting altitude when another blip appeared on the screen. More squadrons were scrambled.

King George VI was visiting Bentley Priory on that day. Air Marshal Dowding, in his capacity as head of Fighter Command, was showing him how the system worked. As more and more squadrons were sent up he began to look distinctly worried. By the time every single squadron east of London had been scrambled, the system was overloaded: the controllers had too many aircraft aloft, and neither they nor the radio telecommunications link to the aircraft nor the telephone lines from the radar stations could handle the traffic. Dowding was beginning to sweat; the King, slipping protectively into the inscrutable armor of royalty, was looking excessively calm.

Over East Anglia, No. 54 Squadron, which had reequipped from biplanes to Spitfires just six months before, was searching the skies. Alan Deere, later one of the Battle of Britain aces, had had trouble starting his aircraft and was now trying to catch up with his mates, but each time he changed his course to intercept them they were given new instructions. As he later said, they were "receiving so many vectors that it was impossible to follow them. When I did eventually join up the squadron was near Chatham where the anti-aircraft guns heralded our presence by some lusty salvoes, at which we hastily altered course despite the controller still insisting that we investigate the area."

The battle raged for an hour, from London down through Kent and out nearly to the coast. Antiaircraft bursts peppered the skies and fighter contrails were visible swooping down and around and back up again. "Sailor" Malan, a South African who was to become one of the RAF's greatest fighter pilots, led a squadron of Spitfires south from London. He never saw the enemy, but his rear section of three Spits became separated from the rest of the squadron during the frantic vectoring, and as they came out from behind a towering cloud they saw a formation of Messerschmitts below. Immediately shouting the "Tallyho!" they dove in to the attack, shot down two on the first pass, and kept on going.

After an hour of fighting on emergency boost, the fighters began to run low on fuel. Although the controllers were still trying to direct them onto new plots, the pilots had no choice but to return to their aerodromes and "pancake." With no reserves left on the fields ringing London, the controllers were terrified of further raids which might sweep in to catch all their forces on the ground: on this first day of action it looked like they would lose the war. But luckily, as the fighters returned to base, the enemy plots faded away too, and in another thirty minutes the skies were clear and the radar scopes were empty. The WAAF's, emotionally and physically exhausted, took off their headphone sets and leaned wearily against the gigantic map tables, the controllers sat quietly by their silent telephones, the King of England scratched his beard, and Dowding began to breathe again.

Back at the aerodromes, the squadron intelligence officers interviewed the returning pilots and sifted the combat reports, trying to get an estimate of how many Messerschmitts had been destroyed. The first reports were, of course, excited and confusing, but they set to work to straighten them out.

As the hours passed, they got only more and more confused. Only two groups seemed to have made actual contact with the enemy, although the antiaircraft organization was sending in report after report of sightings upon which they had fired. But for the RAF, only "Sailor" Malan's rear formation had attacked anyone, and only one squadron of Hurricanes had been attacked by anyone.

The intelligence officers began to feel a bit uneasy. One of the downed Hurricane pilots was taken by van to the airbase at Martlesham Heath, where he reported that he *thought* he had seen RAF roundels on the wings of the plane that had shot him down. Were the Luftwaffe disguising their fighters? Air Vice Marshal C. H. N. Bilney drove out to take a look at the lad's wrecked airplane. He saw that although there were a number of machine-gun bullet holes in the radiator and wings, there were no large holes from exploded cannon shells; and while the RAF fighters used only 0.303 caliber machine guns, the Germans used both these and twenty-millimeter cannon. "Knowing that German bullets had steel cores as opposed to the lead cores of ours," he said, "I got a piece of wood, put some glue on one end and fished around in the wing for bullet fragments. I soon found quite a lot of lead...."

It took a few days to work out what had happened. The entire fighter defense of Great Britain was primed to go off when the watching radar towers gave warning. Again, as in the early morning hours of September 4, they had given a false warning, resulting in a section of Hurricanes being scrambled. And then there was another, more serious malfunction.

The radar transmitting antennae radiated their signals both forward and backward simultaneously, and an operator couldn't tell from which signal—forward or backward—an echo was obtained. That is, an aircraft directly east or directly west of a radar station would give exactly the same blip on the scope. This was obviously an impossible situation, and the solution at that time was to "blind" the antennae to their rear by electronically screening the backward emissions. Therefore anything they saw would be in front of them.

Fighter Command thought the system worked. It didn't. At least, as they now discovered, it didn't *always* work.

The electronic screening had failed. When the first section of Hurricanes rose into the air from their aerodrome *behind* the coastal radar towers, they were picked up on the scopes and plotted as *incoming* unidentified aircraft in *front* of the towers—as enemy aircraft coming in across the sea.

A squadron of Hurricanes was scrambled to meet this new "threat." And this squadron, taking off from an aerodrome behind the radar towers, was also picked up on radar as another enemy formation in *front* of them, a bigger one this time. Two squadrons of Spitfires were scrambled to deal with them, and *voilà!*: escalation ad infinitum.

None of the fighter pilots were yet experienced in war, and the most inexperienced fliers of all naturally were stationed in the rear sections. When the rear section of No. 74 Squadron broke cloud and saw a group of Hurricanes below, they misidentified them as Messerschmitts and dove into them at well over 300 miles an hour—at which speed identification was hardly likely. The Hurricanes reported being jumped by Messerschmitts because it never occurred to them that their attackers might be Spits—except for the one puzzled pilot who *thought* he had seen RAF roundels on the attackers' wings; the others saw nothing but dark shapes flashing out of the sun before they broke formation and tried to spin away.

The Battle of Barking Creek, as it came to be known among the pilots who participated, was a sheep in wolf's clothing, a blessing in disguise. It revealed a disastrous failing in the radar system, but luckily there were no serious German attacks on England until the following summer, by which time the system had been completely revamped and put into consistent, reliable working order. Ever since that first day when Skip Wilkins had shown Watson Watt that the death ray was impossible but that radio fingers of detection could be built to reach through the skies and find enemy bombers in time to warn the waiting fighters, Britain's scientists and engineers had raced crazily against time and the fluctuating moods and countermoods of Hitler and their own politicians, teetering and tripping through the thirties into the first year of war and on into 1940, dancing and skipping crazily on the very edge of time. And all the while Churchill was nipping angrily at their heels and attempting to inflict on them the one scientist he trusted—the Prof, Frederick Alexander Lindemann—who in the end very nearly pushed them over the slippery, precipitous, calamitous edge.

Part Two
Toil and Trouble

...fire burn and cauldron bubble...

Chapter Seven

The Opposition: 'Round and 'Round the Mulberry Bush

> If you can't provide us with a Bren Gun,
> The Home Guard might as well go home!
> —*Noel Coward*

Not from any wickedness nor disloyalty did Lindemann act the role of *agens disruptus;* on the contrary, though "he regarded most men (with the exception of a few peers and scientists) as nothing more than so many furry little animals," he had a great, even passionate, love of his country. The problem was basically his egomania: he could never believe that he was wrong. Add to this his natural insensitivity to the feelings of others (e.g., of a scientist who expressed a view contrary to his own: "He's nothing but a dirty little Jew"); his sardonic sense of humor (of another scientist, he said: "I should like to castrate him—not that it would make any difference"); his contempt for others more successful ("The Blenheim dinner and dance was most amusing. They had got H. G. Wells, of all people, and the Duchess made him dance, a most comic business. He is very second rate as regards brains and was told off well by Fitswilliam who is not considered clever at all"); and his sense of inferiority, which drove him to pursue not merely success but success combined with the failure of others ("It wouldn't be any use getting an award if one didn't think of all the people who were miserable because they hadn't managed it"). All these combined with his love of gadgetry and his driving ambition to be in charge of all things, with the result that he was continually suggesting new schemes and new ideas or different techniques; thus he interrupted, disrupted, and antagonized the people upon whom Churchill inflicted him. This was hardly conducive to the sense of cooperation and tight organization that was necessary if the defense against aerial attack was to be ready in time for the war that was surely on its way.

First of all, Lindemann misunderstood the problem. "I am not making a great point of day bombers," he wrote to Lord Londonderry, explaining his concerns when he was urging the formation of a committee to study the problems of aerial defense, "as I take it the Air Force can deal with these." But it was precisely the bombing by day that was the problem, for though night bombers were even harder to intercept, the amount of damage they could do was limited by their own inability to see in the dark. Later in the war he would advocate concentrating British bombers on the destruction of German cities rather than on a defense against submarines; disastrous results would surely have ensued if his advice had been followed. (For a while it was, in fact, and we nearly lost the war as a result.)

Further, he pushed with all the rude insistence at his command for methods of aerial detection and destruction which were either in such primitive stages that they could not possibly be ready in time for the war (such as infrared, which was finally used in Vietnam more than twenty years later) or would never work at all (such as aerial mines and entangling wires). In the end, he disrupted the committee completely—and that is what saved it.

The committee which came together in 1935 (unknown to Lindemann) to investigate scientific means of defense against aerial bombing consisted of Wimperis, on whose advice it had been formed, A. V. Hill, and P. M. S. Blackett, and A. P. Rowe as secretary. Hill was the physiologist and First World War antiaircraft expert whose advice Wimperis had sought about the death ray. Blackett, professor of physics at Birkbeck College, London, had served in the navy during the Great War and had afterwards been a colleague of Rutherford's. Both Hill and Blackett were later to win Nobel Prizes in their respective fields. Henry Tizard served as chairman.

Tizard was then fifty years old. He had been Lindemann's friend from their early school days in Germany, and their subsequent careers had been separate but closely parallel. A lecturer in chemistry at Oxford before the Great War, Tizard had tried to enlist in the Royal Navy in 1914. But unlike his father, a career navy officer, he—like Lindemann—had poor eyesight and was turned down. (His father, he later said, was stunned; dizzy with disappointment, he went to a friend in the Admiralty to ask, "What would *you* do with a boy who can't get into the Navy?") Henry enlisted in the Royal Garrison Artillery and then managed, despite his poor eyesight, to transfer to the Royal Flying Corps, although, of course, in a nonflying position. In fact, he was an assistant equipment officer at the start, but soon raised himself to the level of scientific officer, and then immediately sought permission to undergo flight training, which was refused. They couldn't spare the time to train someone who wasn't fit for combat; training days were extremely scarce and valuable, since in those early days they could fly only when the weather was perfect—and in England days free of high wind, cloud, or rain were (and are) none too plentiful. Eventually, after insisting (like Lindemann) that his work as aerial scientist would be more useful if he knew by firsthand experience what he was dealing with, Tizard received permission to learn to fly—but only on those days when the weather conditions were too rough to allow the instructors to take up the regular students. So he went up in those flimsy string-and-canvas kites on days when, as they said, even the birds were walking. Like Lindemann, he took bravery for granted.

And like Lindemann, he had his own famous flight. He was readying a Sopwith Camel for a flight test when word came that a formation of German Gotha bombers had been sighted headed for London. Though he was not a combat pilot, he took off and chased after them, catching them as they were leaving London and heading for home. He had picked up enough altitude on his chase to come in on them in a tearing dive from the rear, picking out the last bomber in the formation and zooming right in through the defensive machine-gun fire until the black crosses on the wings filled his gunsight. He pressed down his thumb on the trigger button—and nothing happened. His guns were jammed.

He skidded out of the line of fire and flew alongside the formation, trying to clear the guns, but he couldn't. He was the only British fighter up there in the air with a formation of German bombers, and he couldn't attack them.

So he did what probably no other pilot in the RFC would have thought to do. He flew alongside the formation on its way home, taking notes of their cruising speed and favored altitude; he "attacked" from various quarters and noted their response; he gained the first information the RFC had on the actual performance and tactical characteristics of these raiders. Finally he flew in close alongside, waved goodbye, and went home.

He was "English of the English," according to C. P. Snow. "His whole appearance, build, and manner were something one does not often see outside England." It's an amusing description, especially when Snow adds that "he looked like a highly intelligent and sensitive frog." Shorter and lighter than Lindemann, he was still of average size and a good athlete; his special love was boxing (at which he infuriated Lindemann with his prowess), but he also enjoyed tennis and all sorts of outdoor activities. He had light-blue eyes behind his thick spectacles, thinning red hair, and the typically English mustache, which he wore full and thick on his upper lip, as did Lindemann, a compromise between the aristocratic pencil line of David Niven or Errol Flynn and the bushy, curling RAF handlebar.

Before the war he had gone to Westminster and Oxford, taking a First in science, and ended up teaching chemistry there. After the war he returned to Oxford, but found his experiences in the military had spoiled him for the quiet, contemplative life of a university don. The excitement of the "real" world was made more irresistible by the honest appraisal of his own abilities: "I knew (by then) I should never be any *real* good," he said in his unpublished autobiographical notes. "Younger men were coming on of greater ability in that respect." In this self-honesty he was a distinct contrast to Lindemann: while the Prof stayed on at Oxford and sublimated his vanities in socializing and politics, Tizard resigned and sought appointment in public service. He soon found it, accepting the position of permanent undersecretary to the Department of Scientific and Industrial Research. In the English system this position is the real and staying power of such a department, subservient to the politically appointed Minister in charge but continuous in tenure and more in touch with the day-to-day operations.

In 1929 he became Rector of Imperial College, London, but continued to serve on various official and unofficial bodies for the Air Ministry. In 1933 he was appointed to the chair of the Aeronautical Research Committee, so that by the time he was appointed to head the Committee for the Scientific Study of Air Defence (CSSAD, or Tizard Committee as it came to be known), he was already on close terms with the applied scientists who worked for the Air Ministry, with the military people, and with their political bosses.

This was an important consideration in picking a committee chairman who would be able to get things done, for in England the Old Boys' network is just as useful and effective for those on the inside as it is impregnable and infuriating for those on the outside. C. P. Snow has imagined for us just how things would have been, involving

> ...a great deal of that apparently casual to-ing and fro-ing by which high English business gets done. As soon as the Tizard committee thought there was something in radar, one can take it that Tizard would lunch with [Sir Maurice] Hankey at the Athenaeum; Hankey, the secretary of the Cabinet, would find it convenient to have a cup of tea with [Lord] Swinton [secretary of state for air] and Baldwin. If the Establishment had not trusted Tizard as one of their own,

there might have been a waste of months or years. In fact, everything went through with the smoothness, the lack of friction, and the effortless speed which can only happen in England when the Establishment is behind one.

Underlying all this happy confluence of purposeful activity was the necessary oneness of the Tizard Committee. They met for the first time on the twenty-eighth of January, 1935, and immediately accepted that Watson Watt's and Wilkins's idea of a radio-warning scheme was the only method of aerial defense likely to become workable within the few years they thought they might have before the beginning of hostilities. Following the successful demonstration at Daventry in March of that year, they never wavered in their wholehearted support, and were soon asking the Air Ministry for millions of pounds to follow up the initial 10,000-pound appropriation.

Into this happy, united group Lindemann was dropped.

By 1934 Churchill had been outside the inner workings of the government (though still a member of Parliament) for the previous five years. Mistrusted by both the Liberal opposition and the Labour and Conservative coalition government for his violent opposition to the Indian Dominion scheme, he was now doubling his unpopularity by speaking out in those tremulous bass chords of the dangers arising in Nazi Germany. The country did not want to hear; they turned away from him. They did not much care for their *own* Jews; they did not want to be troubled by the problems of Germany's. And like that of all the western nations, the British economy was a troubled and disconsolate one, recovering but slowly from the Depression. The Tyneside trinity of coal, steel, and ships was crippled and inactive; the industrial Midlands was rife with discontent; the miners were as badly off as ever. In South Wales, up to forty-six percent of the adult male work force was unemployed, while in comparatively well-off Scotland it never dropped below ten percent. In 1934 unemployment reached levels of sixty-eight percent in particularly hard-hit towns such as Jarrow in Tyneside and Merthyr Tydfil in Wales. And here was Winston Churchill roaring out that they needed to spend millions on defense rather than on the poor, that they must build bombers rather than public housing and hospitals, that they must buy bullets rather than bread.

Winston Churchill? The same Winston Churchill who in 1921 had pulled the army out of Iraq, turning the policing of that vast tribal territory over to the RAF because they could do it more cheaply by bombing native villages and killing women and children than the army could by standing patrols and showing the flag? The same Liberal Member of

Parliament who had so vociferously supported Lloyd George's "people's budget" in 1909? The same Conservative Chancellor of the Exchequer who in the later 1920s had instituted the Fighting Services Economy Committee and had ruthlessly cut military appropriations with the resounding cry, "Has the English Channel dried up? Are we no longer an Island?" Was this the same man now presenting figures and postulating terrors born of a borning Luftwaffe in a Germany all knew to be a defeated wasteland?

None other. "Good old Winston," they chuckled, and turned to the sporting news.

But he came back at them, again and again. Day after day he rose to speak in Parliament with page after page of figures concerning German aircraft production (expressly proscribed by the Treaty of Versailles), German submarine production (expressly proscribed by the Treaty of....), German conscription (expressly proscribed....), German this and German that, all creating a spectre of a warlike Phoenix rising menacingly from the ashes of a dead treaty. For the Treaty of Versailles had been grossly unfair, as Churchill himself had argued at the time, and its unfairness now spawned a grotesque, misshapen exaggeration of unfairness incarnate in the silly, posturing, impossible-to-take-seriously caricature of a bourgeois dictator, Adolf Hitler.

Churchill took him seriously. With ponderous seriousness he read to the Parliament his lists of German production statistics. Where did he get them? No one knew, though Baldwin railed at his security people. Some said from Brendan Bracken, some said from Lindemann; a recent study attributes them to disenchanted members of the secret Z organization which grew out of MI6, who wanted to discredit the "appeasement" Parliament and strengthen the "resisters." Wherever the statistics came from, they were correct. By the middle of the 1930s Churchill was the leader of a small but dedicated group of resisters, and though he was dismissed by the majority of the British people, they would remember him later when all their other leaders stood discredited by the savage slash of the Nazi war machine through Europe. In 1935 it was only these later-to-be-discredited leaders who took him seriously: he was a serious enough embarrassment to their lullabye of appeasement with which they sought to calm the ranting dictator and soothe him back to sleep.

Not knowing anything about either radar or the Tizard Committee, Lindemann and Churchill were fighting for the establishment of a committee to embody their own ideas, an independent group of scientists who would find some way of providing England with an adequate aerial defense. To quiet Winston down, Prime Minister Ramsay MacDonald agreed to form a committee. But when MacDonald talked to his Air Minister about it, he was informed that such a committee already existed: the Tizard Committee. Londonderry added that Lindemann had already written to him with his ideas, and MacDonald suggested then that Lindemann simply be informed about the existence of the committee. Londonderry therefore replied to Lindemann's letter, and suggested that he "get in touch with" Tizard and his committee.

Lindemann erupted. This was not what he had in mind, not at all. Tizard had been his friend twenty-five years ago, and Lindemann was not the sort of man to ever serve under a former friend, especially one who had beaten him at boxing and had then got him his job. (When Lindemann had left the Royal Air Force at the end of World War I and had gone looking for an academic post, it was Tizard, through his connections at Oxford, who had put his name forward for the vacant position in physics there and had been instrumental in securing it for him.) Now Lindemann complained to Churchill that the Tizard Committee would be ineffective: it was not chaired by a Cabinet minister and therefore didn't have enough clout, and it had been selected by the same Air Ministry that had told Baldwin "the bomber will always get through," so how could they be expected to now prove otherwise? And that was Lindemann's whole point: to find a way of *stopping* the bomber.

Churchill roared back to the attack in a private session with MacDonald at 10 Downing Street, and MacDonald agreed to terminate the Tizard Committee after its next meeting. But luckily, before then the secretary of the Cabinet, Maurice Hankey, returned from a visit to Australia and was horrified by the idea. (Lord Londonderry, the Air Minister, had been just as upset but had much less influence.) The two of them had been much impressed both with the rapidity with which the young committee had progressed and with its ideas on radar; to give it up because of the rantings of a man like Churchill was unthinkable. Baldwin, officially Lord President (an officially honorary office) but in actuality coholder of the premiership with MacDonald, agreed with him that Churchill must at all costs be shut up so that the real work of the government could proceed without interruption and continual altercation. It was Hankey who came up with the compromise: keep the Tizard Committee, but make it officially a subcommittee of a new committee, chaired by a cabinet minister as Winston wanted, who would oversee it but keep out of its hair. And so with smiles all around, on April 30, 1935, the Committee on Air Defense Research was set up under the chairmanship of Philip Cunliffe-Lister, the Colonial Secretary, with the political supervision of the Tizard Committee its raison d'être. (Actually, this new committee was itself a subcommittee of the Committee of Imperial Defense; thus do big fleas have little fleas, et cetera and so forth.)

As a further flourish, Winston was invited to sit on the committee (which became known as the Swinton Committee when Cunliffe-Lister became Lord Swinton in November 1935). This was seen as a political tour de force which would both end the complaints of the resisters that Churchill was excluded from foreign affairs and hopefully help to quiet his public outcries about the state of the aerial defenses. At the same time those in the know realized that the Swinton Committee was only a palliative and that the Tizard Committee would continue to do the real work. But when invited to join the committee by Baldwin (who had just switched positions with MacDonald), Winston immediately growled that he "was a critic of our air preparations and must reserve my freedom of action." Baldwin acquiesced, assuring him that his "invitation was not intended as a muzzle, but as a gesture of friendliness to an old colleague." Churchill then made it a condition that Professor Lindemann should be appointed a member of the Technical Subcommittee (the Tizard Committee), because he (Churchill) depended on his aid. Baldwin saw nothing wrong with this condition, and again acquiesced.

Lindemann dropped in on the next meeting of the Tizard Committee like a time bomb. The other members of the committee all knew him, of course, and were concerned lest his prickly manner disrupt the friendly cooperation that had marked their proceedings so far. They immediately found that the time bomb had been set to go off in about thirty seconds.

They told the Prof about the concept of radar. There was a moment's quiet, while he stared at them incredulously. Yes, that was all very well; it was a pretty concept, and that first test at Daventry had showed them there was an effect in which airplanes interfered with radio transmissions; but what of it? He could have told them that much without a test. The important consideration was not merely to know *if* airplanes were flying about, but precisely *where* they were.

Of course. No one disagreed. Further work was in hand.

Oh, no, that wasn't good enough. He wasn't going to take that kind of an answer. Could they guarantee a working radar system within the very few years they had before another war might break out? No, of course they couldn't. Then how dared they put all their eggs into this one feeble basket? No, they were quite wrong to do so; other schemes must be looked into at once.

Tizard looked at Blackett and Blackett looked at Hill. "What other schemes?" they asked bleakly.

He told them. By the first of July of that year, 1935, he'd sent them all a long memorandum "on some of the questions anent air defence, which I have been advocating for the past year to a number of different people. The thing I am keenest on is to get ahead with the small aerial mines. I am sure with goodwill they could be made to work."

The "small aerial mines" were a fetish of his that were to bother the RAF throughout the war, and which were never made to work. The concept was that as oceanic mines are sown in waters known to be used by the enemy, to float quietly and wait for a ship to come along and then to explode, so aerial mines would be dropped ahead of enemy bomber formations. Of course, such mines could not float in the air, so they would be dropped by parachute just ahead of the bombers, to sink slowly through their formations and disrupt them. Later other schemes were amalgamated with this one, in which long steel wires would be dropped by parachute, to enmesh and destroy the propellers of the bombers; even heavy steel nets were at one time envisaged, in a gigantic aerial tuna-fishing expedition.

None of these schemes ever worked. The whole concept of the sea mine rests on its invisibility; they would never be effective if the captain of the ship saw the mines being dropped ahead of him: he would simply turn away. But aerial mines had, of necessity, to be dropped immediately in front of the bombers. It was hopeless from the beginning.

Hopeless, but time-consuming. Instead of concentrating all their efforts on radar, the committee had to sit and listen to such ideas—and the ideas came one after the other. The aerial mine was by no means the only proposal Lindemann brought to the committee's attention.

Reginald V. Jones, a student of Lindemann's at Oxford, had become involved in purely scientific experiments with infrared radiation a few years previously. This is the type of radiation emitted by hot objects, and was a source of varied experimental and theoretical work at the time. As Jones later wrote in *The Wizard War,* his ideas "quickly brought me into conflict with Lindemann, who had novel ideas on how infra-red detectors should be made, but after some time I found that he had been leading me up a garden path because he had made some erroneous assumptions he had not troubled to check. When I told him so, he accused me of a defeatist attitude." By the time Jones received his doctorate, his differences with Lindemann over research work had reached the point where it seemed that he could no longer continue to work in Lindemann's laboratory, but when he was offered a position somewhere else, Lindemann apologized and told him he was welcome to work there as long as he pleased.

A few years later, in 1935, Jones got involved with a visiting American's scheme to detect airplanes by the infrared radiation emitted from their hot engines. Some weeks after Lindemann had joined the Tizard Committee, he dropped by Jones's room to ask after his work, and Jones told him about the infrared scheme. "Unwittingly I had presented him with an argument he could use against the Tizard Committee," Jones wrote. Lindemann's immediate reaction was, first, that this should be done for England rather than America, and second, that he had already invented it himself, back in 1916, but that no one then had been interested. (He had undoubtedly thought of the idea in 1916; he was not a liar. But every scientist thinks of many ideas; the only ones that count are those that are followed through to some demonstrable conclusion, and this he had never done with infrared. He had, in fact, forgotten all about it until Jones told him of the American scheme.)

Jones suggested that it might be possible to oscillate the detector so that any source of heat would be alternately focused and unfocused; this would give an alternating signal which could more easily be detected against the normal background of steady heat sources coming from reflected sunlight, etc. Getting more technical, Jones said that "*if* [such a detector] could be made, its rhythmic fluctuations could be used to generate an alternating current which could be amplified electronically, rather than detected by a galvanometer." This would give the possibility of making "a pattern of lights which would indicate the direction of the source."

Lindemann presented the concept of infrared to the committee, which turned it down. He brought it up again, and again. The committee pointed out that experiments had already been done some eight years previously, with no useful results. Lindemann countered, truthfully enough, that because one person fails does not mean the concept is at fault. He did *not* reply, even more truthfully, that he was infatuated with the concept because he had convinced himself that it was his own idea.

The committee pointed out that infrared radiation is strongly absorbed by water vapor and therefore could no more penetrate clouds than could visible light: it would be useless whenever there were clouds that a bomber could pop into or behind. Lindemann replied that there were many clear days, and that bombers were more likely to fly in good weather than bad. They pointed out that it would be a simple matter to screen the heat emitted from an engine by simply placing extra cowling around it. He replied that the exhaust gases, which cannot be screened, are also hot and must also emit infrared.

On May 16 the committee decided that "the detection of heat radiation... offers no prospect of success." But Lindemann continued to argue, answering every objection with a hypothetical circumstance which was clever enough that it could not be waved away, yet hypothetical enough that it was not convincing. Finally, angry at having so much time taken up by this scheme, the committee agreed to go ahead and authorize a test, if only to shut him up. The trials were to be made by the National Physical Laboratory, which then contacted Jones to ask him for his infrared radiation detector.

"What detector?" he asked.

The Opposition: 'Round and 'Round the Mulberry Bush

The gentleman from the NPL was nonplussed. Wimperis had told him that Lindemann had told *him* that Jones "had an infrared detector which flashed lights whenever an aircraft flew in front of it"—the hypothetical detector that Jones had said it *might* be possible to make, which Lindemann had evidently turned into a "fictitious reality" in his efforts to convince the committee. It was this nonexistent detector that the committee had been convinced to test. Nevertheless, Jones good-naturedly agreed to work with the NPL in putting together a much simpler kind of detector, one capable of resolving the question of whether or not such a system might ultimately work. In November 1935, the tests were carried out at Farnborough, with a parked airplane with engine running as the target. The results were positive in that infrared was readily detected from the engine, but, as they already knew, the engine could easily be screened; further testing showed that the exhaust gases did *not* emit enough radiation to be detected.

Lindemann was furious. He was positive something was wrong. The gases simply *must* emit radiation; *all* hot objects do. It turned out that indeed they do—but unfortunately at precisely the wavelengths which are strongly absorbed by carbon dioxide and water vapor. They are so strongly absorbed by these normal constituents of the atmosphere that they were undetectable, even in clear air. (Today our instruments are so sensitive that they can detect these radiations, but at that time it was hopeless.)

Despite these results, Lindemann persisted so strongly that the project was not dropped. With great sighs of irritation and a longing look at the money wasted, the committee authorized continued work by Jones in order to avoid the waste of their even more precious commodity, time, which they saw being used up by the continual wrangling with Lindemann as fast as toilet paper spinning off the roll. And so the infrared work went on, but nothing useful ever came of it during the ensuing war.

It would be a great oversimplification—indeed a falsehood—to say that Lindemann argued against radar because it wasn't his idea. The converse is instead true: he would argue *for* an idea because it *was* his, or because he saw it as his own. And though none of the ideas he proposed during these early years ever were of the slightest use, he would not have been such a detriment to the committee if it had not been for the unfortunate clash of personalities. After all, a devil's advocate is generally a very good thing, particularly in scientific research, particularly when the doors of the absolute unknown are being knocked against. But Lindemann's attitude so grated on the other members that even when he tried to do something useful for radar he precipitated conflict and angry tempers, something which a committee like Tizard's, which was trying to find the quickest way out of impending catastrophe while at the same time trying to impress the military to accept their judgment, could not possibly afford.

Lindemann was impatient. He decided that the committee was dragging its feet even on their primary inspiration, radar. He decided to do something about it. He talked tête-à-tête with Watson Watt and suggested that they go together to complain to Churchill; for some unfathomable reason Watson Watt agreed, although it is impossible today to understand what additional support he could have wished. He was already being given everything he asked for, and research was progressing as fast as his men could push it, which as we shall see was really very fast indeed. At any rate, when Lindemann offered to introduce Watson Watt to Churchill so that he could present his arguments personally to the great man, Watson Watt agreed. Perhaps he simply wanted to consolidate a personal relationship with Winston, who to a man in the technical branch, outside the corridors of power, must have seemed very much a man to know. Watson Watt was not an unambitious man and, as Hill later remarked, though he held him in the highest regard, it was true that absolutely no one ever appreciated him as much as he (Watson Watt) did himself.

And so they met, and Watson Watt complained that though he was well supported by the committee, he was not well *enough* supported. Winston immediately raised the matter at the next meeting of the CID subcommittee. And the Tizard Committee raised the roof.

They could not tolerate one of their members "running to teacher" behind their backs and bringing political pressure to bear on scientific decisions, particularly when that member was the Prof who sat at their table with "arrogant behaviour and saracastic tongue," and then "slipped away into his Rolls-Royce with relief, gliding away from discord to more harmonious company." Though they themselves were supporting radar wholeheartedly, they were not prepared to allow their decisions to be hustled for Churchill's political program—even though they entirely agreed with his program of preparedness. Tizard wrote to Swinton on June 12 that he was forced "to ask you either to remove Lindemann from the committee or accept my resignation...because his querulousness when anybody differs from him, his inability to accept the views of the committee as a whole, and his consequent insistence on talking about matters which we think are relatively unimportant, and hence preventing us from getting on with more important matters, make him an impossible colleague." Swinton tried to placate both men, but it was too late.

It was now July 1936, and a progress report was due. The committee took the opportunity to state clearly their conclusions: a small but permanent radar team had been formed and the first experiments had been satisfactory; there was every indication that a fully workable system could be prepared within a very few years, and there was no indication that any other system of aerial defense was practicable even in principle. They therefore recommended that full priority be given to the development of radar.

Lindemann disagreed. He wrote instead a "long and aggressive" letter rejecting the report and insisting that his idea of aerial mines be given "the highest priority."

The sound now heard softly reverberating through the rather smallish room in which the committee sat was the crunching collapse of the camel's back. This was truly the last straw. Blackett and Hill resigned, stating that they could not work effectively with Lindemann on the committee. Tizard backed them up, tendering his own previously threatened resignation. This was the culmination of the feud that had been brewing between the two former friends for years. It was not of Tizard's doing, but he was powerless to stop it. Though it *was* of Lindemann's doing, it had not been of his volition: he himself blamed Tizard, not knowing the roots lay in his own behavior, which was so insulting to others that no friend could have protected him from its fruits. The beginnings of the feud may be traced back a good ten years, when Tizard used his influence to have Lindemann appointed to the Council of the Department of Scientific and Industrial Research. Curious how one's good deeds backfire: when in the following year the other members refused to reelect Lindemann because of his insufferable behavior, despite Tizard's continued support, he seems to have blamed Tizard rather than himself. And now finally, in the Tizard Committee in July of 1936, the bitterness that had grown for ten years exploded and splattered over everyone in the room.

When Swinton received the resignations he sought to make peace, but it was impossible. Lindemann was "provocative, arrogant, and destructive"; he wasted their time by "continual and violent advocacy of a fantastic scheme for dropping bombs hanging by wire in the path of attacking aircraft"; it was "no longer possible to work with him." As evidence of his unsuitabililty, Hill had prepared a list of "ten major scientific crimes" Lindemann had perpetrated. Hill and Blackett were by now immovable and Tizard was, though saddened, nearly as angry; he and Lindemann had become "mortal enemies," he confessed, and one of them would have to go.

Churchill suggested, requested, and finally demanded "an impartial hearing to decide the relative merits of the case"; he proposed even that Lindemann be taken from the Tizard Committee and placed instead on the Swinton Committee, where he could effectively act as judge of the Tizard Committee's decisions. After considering the situation for nearly two months, Swinton made his decision in a manner which took a good deal of political courage, considering Lindemann's political protector was a man not hesitant about his martial advocacies and Tizard, Blackett, and Hill were alone in the political wilderness. Swinton accepted all their resignations and dissolved the committee, immediately reconstituting it with Tizard, Blackett, and Hill—but without Lindemann.

In his biography of Lord Swinton, F. A. Cross writes: "While Churchill and Lindemann were both undoubtedly prepared to recognize the potential importance of radar, they were for some time considerably more sceptical about its development than the other members of their respective committees. And, in particular, they disliked the priority that work on radar was being given in the allocation of limited resources." Three and a half years later, Churchill was to raise his voice in rich thunder over the air waves, blanketing England and rousing the people to a magnificent determination with those glorious words that will resound forever:

> We shall fight on the beaches, we shall fight on the landing grounds, we shall fight in the fields and in the streets, we shall fight in the hills; we shall *never* surrender!

Glorious words indeed. But the truth at that moment in history was that the English had just returned from Dunkirk by the scruffs of their necks, and they had left nearly all their weapons behind to rust on the sands of that beach. In all England there was not one tank or antitank gun, no cannon, very little ammunition. There were not enough rifles to supply the troops. Churchill renamed the Land Defence Volunteers, bravely calling them the Home Guard; but he had no guns to give them, and they practiced with pikes and clubs and wooden sticks, giving rise to Noel Coward's half-humorous lament quoted in the heading of this chapter. Winston's words were brave indeed, inspiring certainly—but no matter how much they wanted to fight the Germans on the beaches and the landing grounds, in the fields and in the streets, they had nothing to fight the Germans *with*. If the Luftwaffe were to defeat the Royal Air Force in the Battle of Britain, if the German Navy were to successfully land the Wehrmacht on the shores of southern England, not all of Winston's words nor all of his courage nor ten times his charisma and fighting spirit could suffice to repel them. The war would have been lost, and with it our western civilization would have "sunk into another Dark Age, made more protracted and more bitter still by the lights of perverted science." If Churchill had had his way in July of 1936, if Lindemann had been permitted to continue to "harass and restrict the work" of the Tizard Committee, the completion of radar would undoubtedly have been delayed. In September of 1939, when war broke out, it was still not ready—as we saw in the Battle of Barking Creek. If Churchill had succeeded in imposing his own particular scientific champion upon the nation, if the delays pursuant to Lindemann's combative behavior had been accepted, if the situation had been the same six months later, Churchill's "Finest Hour"—the Battle of Britain—would have been lost.

Such are the humors of history.

Chapter Eight
Technical Years: 1935–1940

> We few, we valiant band of brothers....
> —*William Shakespeare, Henry V*

Before delving into the personality-bedeviled political ruckus of these years, let us briefly review the technical considerations.

The concept of radar is most easily understood by analogy: the one most common to our experience, if not the most pertinent, is the case of sound echoes. If you were to stand on the edge of a cliff and give a sudden, sharp shout, the sound waves would travel out over the valley and if they hit a mountain on the other side, they would bounce back. On a dark night when you couldn't see a thing much beyond your fingertips, the echo would be the only indication you might have that something was out there on the other side of the valley.

Three more points about radar are illustrated here. First, you could tell how far away the reflecting object was by the time it took the echo to return. At sea level sound travels at about 750 miles an hour, which is just about one fifth of a mile per second. A mountain ten miles away would return the echo in just over a minute and a half, as the sound waves traveled out from your voice (the transmitter) and back again to your ears (the detector). A similar timing mechanism is a well-known part of folklore, although it doesn't depend on the echo effect. When a bolt of lighting strikes through the air it produces a rap of thunder; the lightning flash travels with the speed of light and so one sees it virtually instantaneously, while the sound of the thunder propagates much more slowly, with the speed of sound. By counting the seconds between the sight and the sound, one can tell how far away the lightning was: every five-second delay means one mile in distance.

The second concept is the diminution with distance. A mountain ten miles away would actually be undetectable by the shouting mechanism because the sound waves from your voice would spread out and be so faint at the time of impact on the mountain that only a small noise would be reflected; this too would spread out and be lost on its way back, and the echo would be too faint to be heard. If the echo isn't heard within a few seconds, it won't be heard at all because the distance is too great for the method to be effective.

The third concept lies in the requirement that the shout you give must be sharp and sudden. If you stand there howling continuously, any echo will be lost in the transmitting noise.

It was, in fact, these three points that destroyed the possibility of using sound locators to detect aircraft. The advantage of a sound locator is that the airplanes provide their own sound as an essential part of their operations, just as lightning does with thunder. But flying speeds in excess of 300 miles an hour are quite comparable to the speed of sound (which, in fact, is lessened substantially at the high altitudes at which bombers can fly), and so the first point, the comparative slowness of sound waves, meant that by the time an airplane a hundred miles away was located it would no longer be there, having traveled forty or fifty miles in the time it took the sound waves to reach the detector.

The second and third points combine: there is no way to keep the sound waves from spreading out and being rapidly lost with increasing distance, nor to keep other unwanted sound waves from spreading out from their sources and impinging on the detector. The combination means that the effective distance of a sound locator is negligible, and since airplanes go so fast, they do have to be detected a good distance away if the alarm is to be useful.

Radio waves have obvious advantages over sound waves. They travel at the speed of light, so the presence of a reflecting object is seen instantaneously, for all practical purposes; further, one can pick a particular wavelength not shared by any other process, so there is no interference from other "noise." Although our ears cannot hear radio waves and our eyes cannot see them, we can easily make radio receivers which can detect them and transform them into visible or audible signals.

The first problem arose from the first advantage: the speed of the radio beam. Imagine an airplane 100 miles away. The detecting radio beam, traveling at the speed of light, 186,000 miles per second, will reach out to it and bounce back again in 0.00108 seconds. Therefore the transmitting beam must be cut off within a thousandth of a second so that the small amount of reflected beam will not be swamped by the output of the transmitter. The solution to the problem was already technically known, for among the first radio studies of the atmosphere was the work by Gregory Breit and M. Tuve. In 1925 in Washington, D.C., they used *pulsed* radio transmissions, sending out small chunks of radio waves lasting only one-thousandth of a second in duration, enabling them to receive the reflected pulses during the quiescent interval before the next one was sent out.

The choice of wavelength was obvious to the British from the start, since the strength of the reflected beam is a strong function of the relative size of the object to be detected and the wavelength, reaching a maximum when the wavelength and the object are about the same dimensions. Taking the size of an average bomber, a wavelength of twenty-five meters was chosen, and within a week of the Daventry experiment in February 1935, Stuffy Dowding had allocated an initial year's support of 12,300 pounds sterling and Skip Wilkins and L. H. Bainbridge Bell were at work under Watson Watt's direction at his radio laboratory at Slough.

They started with an available pulse transmitter that had been used for ionospheric studies, using a frequency of six megahertz. (The relation between wavelength and frequency is a simple reciprocal one related to the speed of light: the wavelength λ is equal to the speed of light c divided by the frequency v. Both terms are frequently used in radio terminology. Six mega hertz is a frequency of 6 million cycles per second, corresponding to a wavelength of fifty meters, close enough to their preferred choice of twenty-five meters.) Each pulse would last about 200 microseconds (0.000200 second) and was to repeat roughly a hundred times every second. Obviously, the larger the duration of the pulse, the worse the precision of detection, for any echo would be a result smeared over from the beginning, middle, and end of the transmission. An infinitesimally short pulse would be ideal, and one object of the first experiments was to attempt to reduce the pulse width to fifteen microseconds; they also wanted to improve the power of the device so that it could reach out further than the distance to the ionosphere for which it was originally designed.

Not only must the equipment perform its task as required, but the information received from it must be translated into a visual or audible signal that a human being can comprehend. Luckily they had already at their fingertips the ideal device for displaying the echo pulse in visual and immediately useful form, dating from Watson Watt's researches into the causes of "atmospherics," which are natural electromagnetic waves that interfere with radio reception, causing static.

In 1916, in his first work at the Radio Research Laboratory at Slough, Watson Watt was trying to determine the conditions which caused these atmospherics. The first problem was to locate the source. For this purpose he had a pair of directional antennae, each of which provided a maximum signal when the incoming waves were arriving in its vertical plane, and no signal at all when they arrived at right angles to it. This effect is similar to the use of today's television antennae: to get the best signal from a distant station the antenna has to be rotated so that it points at the station; if the wind blows it around, one loses the signal and gets a bad picture until it is realigned. In Watson Watt's case he had two antennae looking out at right angles to each other, their signals feeding a "search coil" which had to be tuned by hand to find the atmospheric disturbance. It would have worked fine except that the "statics" lasted only a fraction of a second; the search coil couldn't be turned fast enough to locate them accurately.

The answer was the cathode-ray tube. This was basically the device used by Roentgen when he discovered x-rays. It had been developed during the ensuing twenty years so that it now provided a visual response to electric signals, as illustrated in the figure.

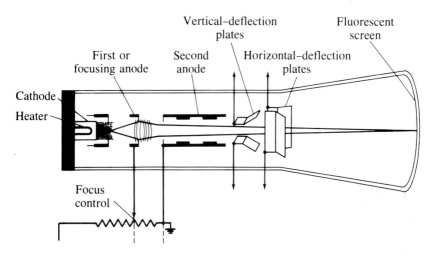

The cathode is simply a metal filament, heated to the point where electrons "boil" off. They are magnetically or electrically focused and collimated so they emerge from the barrel in a thin beam. If nothing else were to happen to them, they would continue in a straight line and strike the fluorescent screen at the end of the tube, causing a bright point of light there. But on the way, they pass through two electrostatic "traps," each consisting of a pair of metal plates which produce a field when a voltage is applied.

The first trap deflects the beam vertically. When a signal, in the form of a voltage, is applied to the pair of plates, the electron beam passing between them will be attracted to the more positive plate; the amount it is deflected is proportional to the voltage, so the greater the strength of the incoming signal, the greater the deflection. The second pair of plates deflects the beam in the horizontal direction, and works exactly the same way. In Watson Watt's early atmospheric work, the signal from one antenna would deflect the beam vertically at the same time that the signal from the other deflected it horizontally; the result of these two simultaneous effects would be to draw the beam out from its central position in a line which would indicate the direction of arrival of the "atmospheric" which had supplied the signal. The electrons are so small they can react instantaneously to the incoming signal, while the fluorescent screen can be made of chemical phosphors which continue to glow for several seconds after the electrons strike them; thus the indicating line remains visible after the brief incoming signal has vanished.

In 1935 the cathode-ray tube was modified for the purpose of radar. The horizontal deflection was not to be affected by any incoming signal; instead it would give a time base. The incoming signal would affect the vertical deflection only.

The way it worked was like this: A pulse would be transmitted out, and unavoidably some of that electric energy would be "seen" by the receiving device, giving a large vertical blip to the electron beam, which would then be seen on the screen as a vertical line. The horizontal deflection plate would be triggered by the outgoing pulse, and a steady signal would be fed into it so that the beam would sweep at a constant speed across the screen, tracing out a visual baseline. If an echo was received, it would provide an additional vertical blip somewhere further out on the steadily generated baseline, and since the horizontal sweep of the beam was constant and at a known speed, the time between the transmitted pulse and the echo could be read from a calibrated scale inscribed on the screen. Finally, since the transmitted and echoing radio waves traveled at the known speed of light, the elapsed time could be converted into a distance, and the scale on the tube could be inscribed in miles instead of microseconds. An ideal system would look like this:

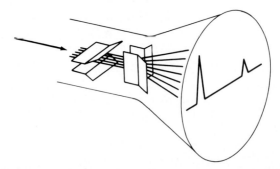

In this example the transmitted pulse is the large vertical line on the left, and the echo is seen as a smaller blip thirty miles away. Of course, nothing is ideal, and a real system would look like this:

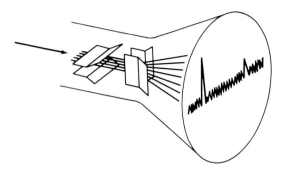

The jagged lines are caused by "noise," which is always present. It is a continuing job of engineers working on any system to reduce this as much as possible. It can't be eliminated totally because some of it comes from basic causes such as the thermal motion of the electrons, which is a vibration set up by the operating temperature. (The form of energy that manifests itself to us as heat is energy picked up by atoms and/or electrons, causing them to vibrate and giving rise to static, or "noise." This cannot be reduced to zero because quantum mechanics tells us that even at an operating temperature of absolute zero——−273°C or −459°F, the ultimate limit of "coldness"—there remains a permanent "ground-state" energy which will keep the electrons in motion.) This thermal motion could be reduced by refrigerating the apparatus, but to achieve any significant reduction would necessitate a cooling system, involving bathing it all in a liquefied gas such as nitrogen or helium, which would be horribly expensive and inconvenient. There was nothing to do but to accept the thermal noise and learn to live with it.

The vertical signals are spread out from simple lines to wider areas because the transmitted signal is not infinitesimally short. The wider the transmitted pulse, the wider the echo and the greater the resulting error in distance reading. As the echo signal moves further to the right, away from the source, it gets smaller and smaller because of the inverse-square law relating intensity to distance. It's easy to see that the sensitivity with which a signal can be picked up depends on the strength of the outgoing pulse, the distance of the echo object, and the noise of the system, for as the pulse gets smaller and smaller it eventually will disappear into the noise, as an ant settles in among blades of grass. The accuracy of the distance determination depends on how narrow the pulse can be made. The initial work at Slough concentrated on improving these parameters.

This initial system could not tell where the bombers might be: the position of the blip on the cathode screen simply said that there was something out there at a specific distance, but gave no information regarding direction. The system Watson Watt had used in his earlier work on atmospherics, pointing a pair of antennae at the emitting object to get a maximum signal, is not useful for covering a wide range of territory, and obviously the first step was to ensure that no enemy bombers got through unseen. In his first memorandum to Wimperis, Watson Watt had written of "hanging an electromagnetic curtain" along the shores of England, through which nothing could penetrate without creating a disturbance which would be seen on their equipment. In the first experimental setup they decided to "floodlight" a broad area with their transmitted signal; anything within a wide area would send back an echo pulse. The problem of finding the precise direction—and altitude—of the bombers was postponed. In fact, when at this initial stage it became necessary to find a name for the program (radar was the name that was to come from America several years later), Watson Watt decided that the one thing the system could not do was find direction, and so he named it RDF, Radio Direction Finding, in order to confuse the spies they were all afraid infested England at that time.

For this same purpose of security from inquiring eyes, the experimental group left the Radio Research Station at Slough and moved to a small, isolated spit of land jutting out into the North Sea about sixty-five miles northeast of London. Just seaward of where the river Ore curls around the village of Orford, the sedimentation patterns of sand and debris spilling down along the coast have built up a jutting point of land, a ness, which sits barely above sea level, treeless and buttressed by seawalls and dikes. It's an accidental, useless peninsula of sand where the only things that grow are a few yellow flowers called sea poppies by the locals, and a sprinkling of pea bushes originally washed ashore nearly a hundred years ago from a merchantman that sank in the vicious North Sea currents within view of the lighthouse on the ness.

In 1915 the Armament Experimental Flight was established there at Orfordness, and in 1921 it moved away amid cries of rejoicing and good riddance, leaving the barren marshland to rot or be blown away by the constant easterly winds. In 1924 it moved again to the nearby aerodrome at Martlesham Heath, and Orfordness was reopened as a supplementary bombing range and testing station. A wireless experimental station was also set up there, and it was into these wooden buildings that the Radio Direction Finding Unit moved on May 13, 1935.

There are no roads to Orfordness; the only way to get there is by ferry from Orford quay across the turbulent river, the scene of many drownings when the east winds blow. The thirteenth of May was an unusually sunny day, but by the fourteenth the wind and rains had come again and the transport and unloading of materials took place under what they were to learn were standard conditions of "torrential rain, hail, and thunderstorms." One large wooden shack left over and unused since 1918 was refurbished for use as their transmitter station, others were set up to house the receivers, and a half-dozen towers seventy feet high were constructed to hold the antennae. Watson Watt, still head of the Radio Research Lab at Slough, came down on weekends to prod them along, and within two weeks the construction was completed and the system was being tested.

Not fast enough for Watson Watt. On the sixteenth of May, while his scientists were still dragging equipment across the rain-swept Ore River and up through the sandy mud to the wooden shacks, Watson Watt promised the Tizard Committee that not only could he detect the presence of airplanes out to ranges of 100 miles, but that he would soon be able to locate them accurately enough to direct fighters on an interception course, and to aim antiaircraft guns to shoot them down without ever seeing them. In his autobiography he makes the curious statement that "On 15 June we tracked an unexpected flying boat out to 15 miles and back, continuously; on that and the next day we failed lamentably to show the Tizard Committee any echo from aircraft." The two statements seem contradictory. What actually happened was that on his weekend visit of June 8 he announced that he was planning to bring the Tizard Committee for a visit to show them how they were getting on, and when they showed up on the following Saturday it was immediately apparent that he had promised them a demonstration—although nobody at Orfordness had yet seen an echo from an airplane. Members of the staff took him aside to explain that the system didn't yet work, it had only just been set up, and they hadn't even tried to observe an aircraft in flight.

This was typical of how Watson Watt's impatience and his vision sometimes combined to produce statements of fact out of wishful flights of fancy. On this occasion he was upset at being let down by his staff, although clearly he had promised what was beyond their capacity to deliver. Nevertheless, the committee was there, and so a test flight was set up. A Valentia aircraft from Martlesham Heath flew over the ness at 15,000 feet and they turned the transmitter on. But there were thunderstorms in the area and static was intense, giving rise to large amplitude noise along the baseline of the cathode tube, hiding any possible return signal. Then, according to Wilkins, while "Watson Watt was observing the cathode ray tube [he] claimed to see a glimpse of an echo at 27 kilometers [17 miles]." No one else saw anything. That is all that happened on the day that Watson Watt claimed to have "tracked an unexpected flying boat out to 15 miles and back, continuously." They tried again the next day but the storms were still bad, and this time nobody claimed to see anything on the tube.

But five days later Watson Watt, Wilkins, and E. G. Bowen did get the system to work properly and observed real echoes from an aircraft. One month later the antenna masts had reached a height of 200 feet, and a successful demonstration was arranged for A. P. Rowe, a Bristol 120 aircraft being clearly seen nearly forty miles away. And a week later, on July 24, while watching the echoes from a test airplane, they spotted an additional echo having an "unusual amplitude variation." Wilkins suggested that this was the result of interferences from three echoes close together, and when the test airplane returned, the pilot told them that a formation of three Hawker Hart bombers had flown by. This was the first indication that they would be able to judge the size of enemy formations—although Watson Watt had already promised the committee he would somehow be able to do this.

By the end of the month they had progressed far enough to be bothered by interferences from radiations emanating from commercial broadcasting stations. One of the advantages previously mentioned of radio waves detected by receivers over light waves detected by eye is that the receivers can be built to be sensitive to only a narrow range of wavelengths so that any ambient noise at a different wavelength is invisible to them, just as if it didn't exist. But the wavelength chosen for the first work was too close to those used commercially, and so they decided to go from six megahertz up to 11.5, corresponding to a wavelength of twenty-six meters. During these discussions they realized that a potential enemy could jam the system on purpose by broadcasting on their transmission wavelength, so they began to work on several wavelengths, with the option of switching from one to the other if interferences occurred.

While these engineering tactics were being worked out, the overall strategy was also being designed. It was decided to line the entire shoreline with a chain of stations whose emissions would overlap, thus creating the "electromagnetic curtain" Watson Watt had visualized. Each of these Chain Home (CH) stations, as they were called, would be linked by telephone lines to a central system, which would coordinate the signals and present to RAF Fighter Command a visual plot of incoming air raids. By September the scheme was approved by the Air Defence Research Committee (in whose thinking we can still detect the sensibilities of a country at peace, not quite believing that the war would really come: one of the criteria for choosing the various sites was that the necessary construction must not "gravely interfere with grouse shooting.") By December 60,000 pounds had been allocated for the first five stations along the Thames Estuary, although not a single prototype station was yet operating.

We live in a three-dimensional world (neglecting the dimension of time), which means that to locate any object, three separate measurements must be given. To locate an aircraft, its distance from some point must be specified, along with the direction in which it lies and its height above the surface of the earth.

Distance is the easiest thing radar can measure. The elapsed time between a transmitted pulse and its echo gives the distance accurately if the duration of the pulse can be made short enough, and by that summer and fall of 1935 the pulses had been cut to just about ten microseconds, corresponding to a precision of about one mile. Direction and height were tougher problems, but one clever idea apiece from Wilkins and Watson Watt showed the way.

If a radio beam is considered as an analogue of an optical searchlight, direction finding seems easy enough. A thin beam of light is sent out by a searchlight, so that a visual return is seen only when the searchlight is pointed directly at the target; finding the target is essentially synonymous with determining its direction. A thin radio beam would work the same way, an echo being received only when the beam was pointed right at the airplane; but inescapable laws of wave diffraction allow a narrow beam to be produced only if the dimensions of the transmitting antenna are large relative to the wavelength. The wavelengths of visible light are on the order of 0.00000001 centimeters, so there is no problem in making a searchlight with dimensions greater than this; but at radio wavelengths of fifty meters, or even 26 meters, the problem is obvious. Even if antennae hundreds of feet long could be made, they could not be rotated quickly enough to follow an airplane in flight.

The height problem, at first thought to be equally difficult, was more easily solved to a first approximation, although precise height determination remained a problem throughout the Battle of Britain. Wilkins realized that if a receiving antenna is situated on a tower at a given height above a flat stretch of ground, part of the echo signal from the target will come directly to it through free space and part will come bouncing off the ground, so that the actual signal it receives will be different from what it would have received if the antenna were an infinite distance from the ground and received only the free-space signal. The ratio of the two (free space to actual) is related to the sine of the angle of elevation of the target, but is not single-valued—one such measurement does not uniquely determine the angle of elevation. Wilkins therefore devised a system using two aerials spaced at different heights on one tower, and the ratios of the signal strengths received provided the true angle of elevation of the target. By September of 1935 the towers were able to estimate to within 1200 feet the height of an aircraft flying at 7000 feet. This was exciting success, but a Spitfire pilot who came through a cloud expecting to find the enemy 500 feet below and who instead found them 700 feet above would in the next few seconds be one dead Spitfire pilot.

Because of the way the signal bounced off the ground, and because the ground would not be perfectly flat in all directions, the error in height determination was strongly coupled to uncertainties in direction of the target, and so it was imperative that this problem be licked. Later in the war it would be possible to use much shorter wavelengths and therefore to use directional antennae, but in 1935 this was impossible because the power needed is inversely proportional to wavelength and they simply couldn't generate enough power at the lower wavelengths. This time it was Watson Watt who provided the solution, in which a crossed dipole arrangement was used. The method was "so simple that it was remarkable no one had thought of it earlier," according to Wilkins. The outgoing pulse was radiated in a broad band which floodlit the entire region around the station, and two receiving antennae were set up at right angles to each other in the horizontal plane. If a target returned a blip from directly in front of one antenna, that antenna would receive a maximum signal and the second antenna would receive a minimum, whereas if the target was situated halfway between the two, their blips would be of equal size. Wherever the target was, a comparison of the amplitudes of the returned signals—the size of the blips—would determine the directional angle.

This is the type of set that was operative when war broke out. It had the disadvantage of mirror symmetry; that is, since the outgoing pulse was radiated in all directions, any returned blip which appeared to come from a target directly in front of the station could just as easily have come from one directly behind. (The sine of any given angle and the sine of that angle plus 180 degrees are exactly equal and could not be differentiated by the equipment.) To resolve the difficulty, a reflector was placed behind the transmitter; the reflector would increase an echo from in front and decrease one from behind. In practice it didn't always work, as was to be proved in the Battle of Barking Creek. (Incidentally, Watson Watt never accepted any deficiency in the system. Like Marie Curie's devotion to radioactivity—which she claimed would cure cancer and would never cause it, an idea she clung to until the day she died of leukemia, probably caused by her exposure to radioactivity—Watson Watt's love affair with radar precluded his acknowledgment of any possible weakness. He insisted that there *were* enemy airplanes in the air over 'Barking Creek'; there *must* have been.)

The RAF summer air exercises of 1934 were the last to be run without radar, and they were disastrous, with more than fifty percent of the nighttime bombers reaching their targets unmolested, and not many fewer by day. The nighttime attacks were unrealistic in that England's cities were not blacked out, so in effect the attackers were invisible while the defense points stood out like glistening jewels on a velvet curtain. But the daytime exercises were slanted in the opposite direction. Nearly half the bombers got through, though they flew at speeds under seventy-five miles per hour (sometimes running into head winds which brought their ground speed down to sixty miles per hour), and everyone knew that the next generation of bombers, scheduled to enter service within a very few years, would be cruising at over 200 miles an hour. Presumably the fighters would also be faster, but the lead times and reaction times would not be increasing. The Air Ministry had no alternative but to nod their heads wearily and acknowledge that yes, the bomber would always get through.

By 1936 radar (in the form of RDF) was ready to be tested. It was used by the defense forces in the summer air exercises that year—and it failed terribly. It gave warning of bomber formations as far away as seventy-five miles, giving a rough indication of how big the formation was and where it was and at what height it was flying—but when the defending fighters got there they found nothing but empty air.

It wasn't accurate enough, it wasn't organized enough, and the communications weren't reliable enough. Watson Watt claimed that his direction finding was accurate to two degrees. To visualize what this means, imagine the radar tower at the center of a circle. Divide the circle into 360 slices of pie. Each slice is then quite thin, and the angle at the tip of each slice, pointing into the center, is one degree. So an accuracy of two degrees is quite good; if Watson Watt said the bombers were at an angle of ninety-three degrees—practically due east—and they were actually at ninety-one degrees, his information was remarkably good. But if the circle is seventy-five miles in radius, corresponding to the distance at which the bombers were being detected, simple trigonometry tells us that the bombers might be as far as 2.5 miles away from the point to which the fighters were directed—and of course in *either* direction, left or right. The width of the transmitted pulse now added nearly another three miles to the error along the radius vector, spreading out the area in which the bombers might be to $3 \times 2.5 \times 2$, or fifteen square miles. Finally, the error in height determination was strongly dependent on the distance, the equation that related them being $h = 107 d\theta + 0.88 d^2$, where h is the height, d is the distance, and θ is the measured angle of elevation. Even if there was no error in the angle measurement, an error of three miles in distance would give a corresponding error of about 1000 feet in altitude; the error in angle measurement would roughly double this. So the bombers were somewhere in the box defined by the accuracy of these measurements, and the volume of that box was too damned big for them to be found with any consistency. Add in the difficulty of radio communication with the defending fighter aircraft (which often rendered messages from the ground totally unintelligible) and the difficulty of organizing the information that came in from the various radar stations (sorting out the data to determine if two stations were seeing the same bombers or different ones, and cross-checking vectors to minimize errors), and the result was a fiasco which could easily have signaled the end of radar.

Military men and organizations are not famous for their friendly regard of outside interference in general and of scientists—eggheads, boffins, queer, absent-minded creatures—in particular. Generals and air marshals are naturally reluctant to admit that someone not brought up through the military ranks could tell them how to run their own business better than they can themselves. But on July 14 of that year 1936 the RAF was reorganized and Sir Hugh Caswall Tremenheere Dowding (a.k.a. Stuffy), Air Chief Marshal of the Royal Air Force, became the first Air-Officer-Commanding, RAF Fighter Command. As Member of the Air Staff for Research and Development a year and a half before, he had authorized the first allocation of funds for radar, he had followed and nursed its development, and he believed in his heart that with it he could stop the bombers. He was also on the verge of certifiable insanity, and during the Battle of Britain he would plunge over that edge.

In the summer of 1936 he was still on this side of the abyss. As Commander-in-Chief of Fighter Command, he moved into a lovely old home named Montrose in Stanmore, a northwest suburb of London, just down the road from the nearly 2000-year-old monastery of Bentley Priory, which would be his headquarters. He had brought with him his sister Hilda to keep house, since his wife had died in 1920. His marriage had been extraordinarily happy, but he no longer talked of Clarice. His energies were concentrated from the time he woke in the morning till nearly the moment he went to sleep at night on his new command, on the challenges he faced, on the dangers looming over the horizon.

It was only at the very end of the day that he would climb the thin wooden stairs to his immense bedroom and lock the door. The windows would be open on these warm summer evenings, and he would look out over the low hills to the southeast where he could see the warm yellow glow of London diffusing through the night air. He would go over his day, which had been spent in trying to prepare a suitable defense for that city and for the country it served like a central beating heart. And then finally he would sit at his desk and take from a locked drawer a file envelope. From it he would lift a number of newspaper cuttings, some fresh and some yellowed with the years, all dealing with people who had come back from the grave.

He could not quite face the fact that his wife was gone.

Dead, yes.

But gone?

* * *

By the end of 1935 Watson Watt's small team of scientists had outgrown their initial 12,300-pound allocation and were beginning to swell in numbers and equipment, well on their way to the million-pound annual budget they would reach before the war's end. The wooden shacks and towers at Orfordness could no longer contain them, and Watson Watt's weekend visits were no longer a sufficient expenditure of time. It is at this point that his contribution really becomes clear. Credit must go to him not for suggesting the original idea, for as we have seen it was actually Wilkins who did that. Nor can the measure of his contribution be contained in the ideas he put forward to make the early system work, for valuable as those were, others contributed more than equally, and by the time the system was functioning so effectively that it became the single most important weapon of the war, he had really been left behind in a scientific sense. But now, before a plethora of scientific ideas swamped and buried him, before the potential of the system became apparent to every scientist so that they all wanted in, now when the attainable goals seemed so far away that they constituted little more than a dream whose fabric tore easily at the first poke of inquisitively doubting fingers—now Watson Watt stood upright and planted his feet firmly and assured everyone that it would work, it would function, and it would be ready in time. Full and total credit must be given him for having faith in his proposals and guts in his belly, for leaving his lifetime's work and staking his reputation on the success of this scheme when it was based on nothing more substantial than a few quick calculations and a rather marginal experiment. With no outward sign of the doubts he must have felt, he left his position as head of the Radio Research Laboratory, left the NPL completely, and transferred to the Air Ministry to direct the new effort.

Even before his move, it was clear that the work itself must move to bigger quarters. The previous summer he and Wilkins had inspected a possible new site at Aldeburgh and rejected it. Wilkins then remembered coming across a perfect place during an earlier exploratory journey around the neighborhood: about ten miles below Orfordness, along the wind-swept smuggler's coastline, just past the huddled few houses and single pub that are called the village of Shingle Street, where the Suffolk sands are cut through by the river Deben, standing in lonely, lovely majesty in the midst of nearly 200 acres on the head of a seventy-foot "mountain" overlooking the sea stood Bawdsey Manor, the ancestral home of Sir Cuthbert Quilter. The central enclave was surrounded by thick woods, intergrowths of a wide variety of trees and shrubs; it was said to contain the largest bougainvillea in England. It was isolated by the North Sea to the east, the river Deben to the south, and these acres of forestry to the north and west. In the midst of formal gardens, a cricket pitch, and a tennis court stood the Manor, a melange of styles built over several centuries; with its central gothic hall attended by Elizabethan and oriental towers, it appeared to Wilkins as awesome, lonely, and altogether perfect for their purposes. Watson Watt agreed as soon as he saw it, and Sir Cuthbert agreed to sell it to the Air Ministry for 24,000 pounds. It boggles the imagination to wonder what Hilton or Club Med would pay for it today.

By March, 1936, the first antenna had been installed atop a 250-foot mast at Bawdsey; Canewdon, Dover, amd Bawdsey were to be the first three of the Chain Home stations. All three stations were to be ready for the summer air exercises, but Canewdon and Dover were not operational when the games began in September, so Bawdsey went it alone.

Alone and untested. The equipment was ready barely in time, and they went into the exercises without the rigorous checkout such equipment demands. In the receiving room, which was set up in the Manor stables, Watson Watt, Wilkins, and two other scientists sat, stood, walked around, and cursed under their breath as not one echo showed up on the cathode screens. Wilkins checked through all the equipment but couldn't find what was wrong. Eventually, bewildered and exasperated, he stepped outside for a break—and saw several workmen casually sitting on the aerials.

Even after the workers were forcibly removed, however, nothing worked right. There were interferences, either from commercial radio or from atmospherics, which gave spurious echoes; there were no echoes where they should have been observed; and finally there were no bombers where the echoes said they should have been, when the defenders finally got there. It was all very discouraging. Even Watson Watt, who never acknowledged anything bad about his baby, was forced to record in his history that "we did not greatly distinguish ourselves."

The next exercises were scheduled for April, 1937, and with Dowding's strong backing it was agreed that the Chain Home system would get another chance. They changed the frequency from thirteen to twenty-two megahertz after experiments showed that this wouldn't reduce the sensitivity much and that it would eliminate any possible radio interference. They brought in a whole new crew to work on the problem of transmitter instability. They spent several weeks deciding on a new type of coaxial cable, only to find that the weight of the cables on the 240-foot wooden towers would be several tons, unquestionably more than the flimsy construction could bear; luckily they found that the Post Office was using a new, thinner type of cable, which they modified for their own use.

The work proceeded well and by April they had not only Bawdsey, Canewdon, and Dover, but also Great Bromley operating with at least partial efficiency. An aircraft coming in over the North Sea would be observed by at least two of the stations, and if the same story was obtained, everything would be fine—but this rarely happened. Watson Watt, in his subsequent telling of the story, pretended complete success by the simple expedient of concentrating his attention on what they did well and ignoring what they did not: "Most of the aircraft operating in the exercise area were reported by RDF with good accuracy at 80 miles...." But in truth the system just wasn't working well enough, and if one station had a small error, say in calibration, the plots of the two stations would be different, and it wasn't possible to know which one was wrong. During the air exercises they did manage to give warning of aircraft as far away as eighty miles with good accuracy as to range, but distinctly less so as to direction and height. "Systematic and haphazard errors of the three stations made the task of assessing aircraft tracks extremely difficult," A. P. Rowe reported. "Some aircraft approaching the coast were even recorded as flying parallel with the coast.... The results seemed nearer a failure than a success." Nevertheless, Dowding had faith enough to recommend to the Air Staff that the Chain Home station network be completed along the entire coast of England facing the continent, from the south to the northeast.

Most of the system was completed by the time of the Munich crisis in September 1938, although direction finding and height estimation were still not sufficiently advanced to be of practical use in intercepting enemy raids. The annual exercises were held in August that year, with five stations fully operational on one frequency (22.64 megahertz). Problems with analyzing the data became apparent, now that the stations were providing enough of it for the simulation of a full-scale air raid. It also became apparent that the system couldn't see aircraft below several thousand feet, where reflections from the ground swamped those from the airplanes. Still, the results were sufficiently accurate that Dowding irrevocably placed his whole hope for the future on the foundation laid by a radar defense. This was a final commitment; there would be no turning back, no thoughts of alternative schemes; there was now only a desperate sense of trying to get the damned system working accurately and reliably enough to save the country from the devastation that was clearly on its way.

In September, during the Munich crisis, the first five stations went operational on a twenty-four-hour basis. By April there was an entire line of operational stations stretching from the Isle of Wight in the south to the Firth of Tay in the north. They were operational, but not yet dependable. It was just a few weeks later that the *Graf Zeppelin* came poking its nose along the eastern coastline and was detected—the first time; the second time no one knew it was there.

Aside from reliability, the major remaining unsolved problem was identification of blips as friendly or enemy. By the summer of 1937 the station at Dover had seen blips over Belgium, and plotting their location on a map, the operators saw that they corresponded to the known locations of military airfields. So clearly they would be able to see bombers taking off from the continent, and as they followed them on their cathode screens across the Channel there would be no mistaking who they were. Similarly, the RAF fighters rising from their own fields would be obvious, as long as the two plots were separate. But once the battle was joined and the plots merged, it would be impossible to tell who was who. To avoid sending new fighters to intercept the first fighter groups as they returned, there would have to be some method of distinguishing friendly aircraft from hostile bombers.

The principle of the method had been discussed at the very first moment radar was conceived. Wilkins had pointed out to Watson Watt in their original conversation that a suitably designed dipole could be placed in our own aircraft so that the reflected echo from it would give a signal distinctively different from that of the airframe alone. The idea was simple, but to get a signal sufficiently distinct to be recognizable proved difficult, and it wasn't until what was to be the final Air Defence Exercises in August, 1939, that the problem was solved by fitting the RAF fighters with powered devices that were activated by the transmitted pulse; these devices in turn sent out a distinct signal which accompanied their echoes and identified them as friendly aircraft. The device, known as IFF (Identification-Friend-or-Foe) was approved for installation in all British airplanes.

The system worked. Height estimation was still a problem, but by August both direction and distance were consistently given with good accuracy. Dowding pronounced himself satisfied. At the invitation of the prime minister he made a radio broadcast, assuring the English audience that their island could and would be protected by Fighter Command.

Three weeks later they were at war. And three days later they waded into the Battle of Barking Creek.

Obviously the system didn't yet work. From September through October and November and on into the winter they worked at a feverish pace, hoping to iron out the last bugs before the bombers came. Miraculously, it seemed, the Phoney War, the Bore War, stretched through those fall, winter, and spring days, and when the Luftwaffe finally showed up in the summer of 1940, the system did work. Wherever and whenever the Heinkels and Dorniers and Messerschmitts came slipping out of the clouds that bright and terrible summer, they found always another small band of Spitfires and Hurricanes waiting for them.

Chapter Nine
The Spectre and the Terror

> Everybody was too frightened
> of Winston Churchill.
> —*H. C. T. Dowding,*
> *Head of Fighter Command*

In 1940 Hugh Dowding, AOC Fighter Command, was fifty-eight years old. Just under six feet tall, he was slim and correct at all times. He did not wear his top tunic button undone, as did the Spitfire and Hurricane pilots, his "chicks"; on the contrary, his formality extended to putting his cap firmly on his head whenever he left his own room, even to walk to the next office in his headquarters. It says a lot about the man to realize that he was the only one of his staff to do so; it's impossible to imagine such self-determination allowed around MacArthur or Patton, for example. His hair was gray, his eyes blue; he wore a neat triangular mustache. He was unfailingly courteous, quiet, and determined; "a grey, austere spectre of a man."

His career began, as so many do, rather by accident. His father had founded St. Ninian's, a preparatory school in Dumfriesshire, Scotland (although the family was not Scots). After a few early years at St. Ninian's he went to Winchester, his father's old school. He embarked on his military career there, he later claimed, simply because the military students were the only ones who weren't required to learn Greek. He went to the Royal Military Academy at Woolwich and intended to make his way in the Royal Engineers, but through laziness failed to get high-enough marks. Instead he joined the Garrison Artillery in August of 1900 at the age of eighteen. He served in Gibraltar, Ceylon, and Hong Kong, leading the rather pleasant life of a British officer, learning to ride and play polo. In India he was not permitted by his commanding officer to apply to the Staff College, because he "didn't take life as a soldier seriously enough."

Six years later he did make it to Camberley (the Staff College), where he was "irked by the lip-service that the Staff paid to freedom of thought, contrasted with an actual tendency to repress all but conventional ideas." A rebellious spirit like that should have ended up with a different nickname than "Stuffy," but Stuffy it became. He was different; he loved sports and physical testing but not the rowdy games the other officers delighted in. He was only drunk twice in his life, he later said, and he didn't enjoy either time. It was impossible to think of him playing soccer in the mess hall with an empty Scotch bottle as the ball, or tearing the pants off a fellow officer in drunken delight. And so it became "Stuffy," but bestowed with all the connotations of a British Army nickname: someone thought to be really stuffy would not end up with a nickname at all.

In 1913 he decided to transfer to the Royal Flying Corps, which had just been organized. The rules were that you first had to learn to fly at your own expense, and were reimbursed if you were subsequently accepted. He took an hour and a half's instruction on credit, the arrangement being that he would pay when he got his refund; he soloed, got his Royal Aero Club ticket, his acceptance into the RFC, and his refund.

But then his father found out. In vain did Hugh argue that the RFC was the quickest route to promotion and a military career of distinction; his father replied that it was simply too dangerous, he forbade it, and Hugh had to resign his transfer and go back to the Garrison Artillery. He was at the time thirty-two years old.

According to the military regulations of that day he remained in the RFC reserve, and when war broke out later that year he was remobilized back into the Flying Corps; the army did not think to ask his father's permission. He was posted to Dover as commandant of a temporary camp, a way station for squadrons en route to France. He made such a nuisance of himself—trying in desperate haste to get sent to France, because like everyone else he thought the war would be over in a few weeks—that Hugh Trenchard, his immediate superior and later boss of the RAF, sent him over as an observer. ("It was by way of being a fearful insult to send a qualified pilot as an observer," Dowding later said, "but I was well enough content.") At first the airplane crews of the opposing sides would wave to each other as they flew by on their way to the front lines to observe ground movements and back again to report (there being no radio), but soon they began to throw rocks and then to shoot at each other. Dowding was one of the first. ("I had a Mauser pistol...but I never hit anything.")

In 1915 he was sent as flight commander to the Wireless Squadron, which was set up to investigate the possible use of radio telephony as a means of communicating with airplanes from the ground. They determined that the system would work; Dowding himself was the first person in England, perhaps the first anywhere in the world, to sit in an airplane several thousand feet in the air and hear someone talking to him from the ground. But the War Office informed the squadron that they had determined that "radio-telephonic communication between air and ground was not...practical," and that was the end of that.

Later in the summer he was posted as commander of No. 16 Squadron, part of the wing commanded by Trenchard, now a lieutenant colonel. Trenchard was a huge man, "straight as a ramrod, [who] covered the ground quickly with huge strides...forcing [others] to move in a quaint kind of turkey-trot at his side, trying to keep up with him," and he had earned the nickname "Boom," not for the effect of his bombs but for his voice. (Question: "My Lord, what's that awful sound?" Answer: "That's Trenchard, talking to headquarters in Paris." Question: "Why doesn't he use the bloody telephone?")

Dowding's fighter squadron was using Maurice Farman airplanes, and he was shipped the wrong replacement propellers. He complained to Trenchard, who impatiently told him to improvise. Dowding said it was impossible, the propellers simply didn't fit. Trenchard, irritated by his subordinate's "pernickety primness," told him brusquely that he was informed on good authority that they could be made to fit by simply drilling a larger central hole for the spinner. Dowding said that would weaken the wooden shaft. Trenchard told him to fit the damned propellers; he wanted no more argument.

Right. Dowding drilled the larger hole, fit the propeller on, took the Farman up himself, and tried to break it off. He landed, to his disgust, still in one piece, only to receive a telephone call from Trenchard with as close to an apology as Boom was able to make. Trenchard told him that further information indicated that he, Dowding, was right, the propeller probably would fall off in flight. Dowding, understandably miffed and less than diplomatic, expressed his regret that Trenchard had preferred "to take the word of some half-baked motor salesman against mine." There was a moment's silence, then Trenchard said that new propellers would be sent and hung up.

By 1916 Dowding was himself a lieutenant colonel, commanding No. 70 Squadron, which was equipped with the new Sopwith "fighters," as the scout planes were now called, a name more in line with their new function of protecting their own observer planes and attacking the enemy's. Trenchard was by then a major general, in command of the entire RFC in France and still Dowding's immediate superior. During the battle of the Somme, Dowding's men fought desperately, and after it was over he requested rest leave for them.

Denied.

The situation was well described in the classic film *Dawn Patrol*. Dowding continually badgered Trenchard, protesting against the practice of sending new replacements directly into battle and demanding that his squadron be pulled out of the line for rest; the strain was growing too much for his men, their life expectancy was measured in weeks for the veterans, days for the fledglings. Trenchard was not impressed. "He grew very angry," Dowding said, "though our casualty rate was 100 percent a month."

Trenchard was indeed angry. He complained to his staff that Dowding was no fighter, that he was afraid, that it was impossible to fight a war with "dismal Jimmies" as commanding officers, and he got rid of Dowding by having him promoted to colonel and sent home in charge of a training command, never again to see combat.

The slaughter went on, and though physically removed from it as commander of the Southern Training Brigade back in England, Dowding was still involved. Continual orders for more replacement pilots came in from Trenchard as his men were swatted down like flies over the trenches of France. When Dowding replied that it took time to train a pilot, Trenchard replied that it would have to take *less* time: cut the training period and send them across the Channel.

But Dowding knew what happened to new pilots thrown into action, even with a full period of training. To cut their training would be an act of murder. He would not do it. Trenchard's answer came booming back: then send over the instructors! Dowding of course refused, only to find an immediate order coming in from HQ, RFC, transferring all instructors to the front with the exception of one per training squadron. It was clear by this time that Trenchard was the power in the new service. So Dowding made one last effort, writing to Trenchard's senior personnel staff officer and asking him to intercede, to explain that they simply couldn't afford to lose their instructors. But that gentleman took Dowding's letter directly to Trenchard, and the ensuing boom was heard clear across the Channel, frightening the gulls from Dover to Brighton and unborn children in their mothers' wombs. "That finished me with Trenchard," Dowding realized.

In February of 1918, at the age of thirty-six, he suddenly married, to the surprise of all his friends who thought of him as a confirmed bachelor. One year later his only son, Derek, was born, and two years later his wife Clarice was dead.

And so was, apparently, his career. With the ending of the Great War, Dowding was one of the most senior officers in the newly formed Royal Air Force, but the new Chief of Air Staff was Boom Trenchard. And so, despite his seniority, it should not have come as a surprise when Dowding received a letter one day informing him that with the cessation of hostilities his services were no longer required and that he was dismissed from the RAF, with orders to return to the Garrison Artillery. This would have been the end of his military career, since the Garrison Artillery had by this time a number of officers who had grown more experienced in their own methods of warfare during the struggle, while Dowding had been away playing with airplanes. Luckily for him, though, despite Trenchard's animosity, there were a few highly placed RAF officers who took up the battle on his behalf, though he didn't make it easy. He had at the time a young officer named Sholto Douglas under his command, who was brought up on court-martial charges by the Air Ministry. Dowding thought Douglas was not at fault, and refused to proceed with the court-martial, despite a direct order from the Air Ministry who, he said, simply "were being stupid." His stubborn stand prevailed, and the future Marshal of the Royal Air Force, Lord Douglas of Kirtleside, had his career protected. Despite this perfect example of Dowding's intransigence, his friends eventually prevailed: he was granted a permanent RAF commission as Group Captain.

By 1936 he was the most senior officer in the RAF. Trenchard had long been retired and Edward Ellington was chief of the Air Staff. Dowding, as Member of the Air Staff for Research and Development, had just allocated the first funds for radar. In a general shift of appointments, he was made the first Air Officer Commanding-in-Chief of Fighter Command, newly organized out of the former Air Defence Great Britain, and was told by Ellington that upon his (Ellington's) retirement the next year, he (Dowding) would take his place as C.A.S., the top RAF position.

On February 3, 1937, Ellington sent him a handwritten note: "The S. of S. [secretary of state for air] has asked me to let you know...that he has decided that [Air Chief Marshal Cyril] Newall will succeed me as C.A.S....." Newall, junior in appointment to Dowding, was the man Trenchard had chosen to succeed Dowding when he had fired him from his fighter squadron command twenty years before.

Five days later, on February 8, 1937, Ellington wrote to spell out Dowding's future: although he was not to be C.A.S., he would be expected to continue to serve in his present rank "up to the age of 60," i.e., till 1942. Then, in July of the next year, Newall (now C.A.S.) wrote to inform him that his "services would not be required after the end of June, 1939." A month later official confirmation came from the Air Ministry: "I am commanded by the Air Council to inform you that they have recently had your case under review and have come with regret to the conclusion that they will be unable to offer you any further employment in the Royal Air Force after the end of June, 1939." Dowding's position as AOC, Fighter Command, was to be taken over by Air Vice Marshal Christopher Courtney.

In February, 1939, an article in the *Evening Standard* gave public notice of his retirement. The Air Ministry replied, however, that "no changes would be made during the present year." Dowding was informed of this decision by phone, and was later officially asked to defer his retirement—as if it had been his own idea—until March, 1940. A follow-up letter read: "I am commanded by the Air Council to inform you that, after further consideration, they have decided to continue your employment until the end of March, 1940."

On March 30, 1940, just one day before his scheduled retirement, with the nation at war and the Luftwaffe perched like an obscene buzzard on the windowsill, he was asked to defer his retirement again, this time until July 14, 1940. On July 5, with the English army back from Dunkirk huddled in retraining depots and with the Luftwaffe already beginning its attacks on convoys in home waters and on coastal towns and aerodromes, he was once more asked to defer retirement, this time until October, 1940. Some weeks later, during the height of the Battle of Britain, his employment was continued indefinitely.

Three weeks later, in November 1940, after he had won the Battle of Britain, the most important victory since Waterloo, he was fired and told to vacate his office within twenty-four hours.

The reason for this treatment is not simple, one-sided, obvious, nor unimportant. It is composed of a gallimaufry of interwoven and complex relationships and postures. To put it most simply, Dowding was the right man in the right place at the right time to rally a discarded system, invoke a despised and distrusted resource, and ultimately save the western world; but in the furtherance of his own personal career, he was always and most definitely the wrong man in the wrong place at particularly the wrong time.

He had alienated Lord Trenchard at the very beginning of his military career, and an overwhelming percentage of those who governed the fate of the Royal Air Force in the 1930s and on through the Second World War were protégés of Trenchard, handpicked and guided by him into positions of authority. Dowding had an indication of the prejudices ranged against him when he was nearly sacked and sent packing back to the Garrison Artillery after the 1918 war; and though he survived that attempt, he was never to outgrow or outdistance the enmity inculcated by Trenchard in his numerous seed.

Dowding demonstrated continually, in his own quiet, gentlemanly ways, an uncanny ability to rile people of all levels except the lowest: the workers always loved him, from the typists at Bentley Priory to his "chicks" in the Spitfire cockpits they were strangely dedicated to this "gray, austere spectre of a man" who never came around with fiery speeches like Patton, never waded through the surf with a corncob pipe in his mouth like MacArthur, never employed a private press corps like Montgomery, nor smiled like Eisenhower—in fact I have been unable to find a picture of him smiling anywhere, under any conditions. Nevertheless, he had an incredible level of support from those who served under him.

Higher up it was a different story. Air Vice Marshal (later Air Chief Marshal Sir) Trafford Leigh-Mallory, Dowding's Air Officer Commanding No. 12 Group, which was to be one of the two most important forces under his command in the Battle of Britain, said that he "would move heaven and earth to get Dowding sacked from his job." Churchill at one time stood sputtering with rage and glowering at the slim, unbending figure, surely in his soul remembering the words of Henry II: "Will no one rid me of this troublesome priest?" All around Dowding, at levels high and low, were professional air force officers who were building their careers on the sacred concept that "the bomber will always get through," and there he stood stubbornly proclaiming—quietly but defiantly—that he could stop the bombers, that they were all of them wrong. "He was a difficult man," reminisced Sir Frederick Pile, Commander in Chief of Anti-Aircraft Command during the Battle, "a self opinionated man, a most determined man, and a man who knew more than anybody about all aspects of aerial warfare." Stubborn and unyielding as the blocks of Scots granite out of which he had been carved, he was not an easy man to like; his policies had been "proven" wrong before he even embarked on them, and his contemporaries had been prejudiced against him by their godlike mentor. On top of all that, it couldn't have helped when at the height of the Battle of Britain he went calmly, austerely, and supremely self-confidently, crazy.

When he became Commander-in-Chief of Fighter Command in 1936, he was stepping into a professional grave; his official job was to defend the indefensible. It was just four years from the time of Baldwin's pronouncement which stated in public terms the consensus view of the Air Ministry: "The bomber will always get through." It was not merely the consensus view, it was all but unanimous. RAF appropriations centered almost exclusively on its bomber force which, acting as a deterrent in a plan of mutually assured destruction, was thought to be the only possible defense.

But Dowding was sitting on top of a revolution which he alone among the top RAF brass appreciated. The operational warplanes of the day were wood-and-canvas biplanes, held together by struts and wires, whose design characteristics were such that, aside from acrobatic ability, there was little performance difference between bombers and fighters. The fighters had two fixed machine guns firing through the propeller arc, while the bombers had at least two rotational guns that were fired by an air gunner. Both types had top speeds of nearly 200 miles an hour, bombers and fighters alike. In fact, the fastest plane the RAF had was the Hawker Hind bomber which was four miles an hour faster than the Demon fighter.

But in 1931 Britain won the Schneider Trophy, the prize for seaplane speed competition, with a sleek, streamlined monoplane designed by Supermarine's Reginald Mitchell. Dowding, at that time just taking up his post as Air Member for Research and Development, saw in the lovely, speeding single-winged watersprites something different, something useful, something deadly. As themselves they were useless—with their huge dangling floats, they couldn't possibly do acrobatics or any kind of stressful maneuvering, and they could take off and land only from sheltered harbors in low-wind conditions. But they were *fast*. Dowding suggested that instead of continuing to design planes to go faster and faster under useless conditions, the Air Ministry should immediately "cash in on the experience that had been gained in aircraft construction and engine progress so that we could order...the fastest [war] machines which it was possible to build." This resulted in Design Specification F.5/34 to which two firms replied, Hawker and Supermarine. Both designs were accepted, and five years later the RAF was reequipping with the Hawker Hurricane and the Supermarine Spitfire.

The revolution brought about by the monoplane, with the elimination of supporting struts and wires, was not limited to fighter aircraft. In the case of bombers, the increased aerodynamic efficiency and more powerful engines went naturally into longer range and heavier bomb loads. It was true that the Bristol Blenheims which replaced the Hawker Hinds also increased maximum speed by nearly 100 miles an hour, reaching 260 miles per hour with no bomb load, but the new fighters turned the aerodynamic revolution into avenues of speed and destruction that had been unthought of a few years previously. The Spitfires could now barrel along at the incredible rate of more than 360 miles an hour in level flight and could add a good hundred miles an hour more in a power dive, carrying buried in their single wings no less than eight machine guns whose combined rate of fire could shred a bomber's skin, set fire to its engines, and decimate its crew in one three second burst. With the Spitfire on the drawing boards in 1935, Dowding had a weapon which could claw the bombers from the sky, if only he could find them in that deep, wide, cloud-filled vastness.

And then at Daventry in March of that year, three men stood in a soggy, muddy field and watched a cathode-ray screen hiccup and blip as an antiquated bomber flew back and forth in the distance, interrupting and reflecting the invisible radio waves of the BBC transmitter, and Dowding began his crusade.

*   *   *

Even accepting the assumptions that the new high-speed, eight-gun, monoplane fighters would be ready, and that radar would be ready, and that both would perform as their designers claimed, there was still one ingredient missing. The incoming flood of information had to be collected, organized, merged with the visual reports of the Observer Corps, and fed to the proper fighter aerodromes. It had to be evaluated, so that one flight of Hurricanes should be scrambled from Hawkinge instead of two squadrons of Spits from Hornchurch, if that was the proper response. The chicks had to be sent off at precisely the proper moment, giving them enough time to intercept the bombers at the right place and the right altitude—neither too late to make the interception before the bombs were dropped, nor so early that precious time would be wasted waiting for the bombers, for fighting time in the air was measured not in hours but in minutes before fuel ran out. And it had to be remembered that fighter pilots were human; their own resources had to be conserved as carefully as their engines' fuel. They could not sit in those cramped cockpits for hours at a time and still be fresh when called into combat, so they had to be precisely moved from calls of "readiness" to "scramble." Finally, once in the air, the fighters would have to be directed by radio not to vague locations over Kent or the coast but to precise points in four-dimensional space-time as the bomber formations changed direction or as new raids came up on the screen.

It was a formidable logistics problem, involving new techniques and technological specifications never before developed. And in the end, the system succeeded brilliantly.

The first step was to provide a network of trunk telephone lines on a scale never before proposed; these would link each radar station to Bentley Priory. The initial information would be fed into a Filter Room which, as the name implies, would filter, sort, and organize the information, comparing each bit of data with similar bits from neighboring stations, filtering out duplications and contradictions, and finally estimating the position, speed, direction, altitude and size of any incoming formation. This would result in a plot, a red marker being placed on the large map which was laid on the central table in the hollow of the Filter Room, viewed by the controllers from a balcony running around two sides. The appropriate response to each plot would be decided upon, and a chart on the wall would indicate by flashing lights the state of each squadron at each aerodrome: thirty-minute readiness, five-minute readiness, waiting in cockpits, or scrambled. As the fighters left the ground, a black marker would trace their progress across the map, moved by WAAFs wielding casino croupiers in response to telephoned instructions.

This central Filter Room was duplicated in Operations Rooms at the

Toil and Trouble

Group Headquarters and at each of the Sector Stations. The Group controllers would initiate each reaction, with the Sector controllers translating their orders into the appropriate response. Above them all would be Dowding in the center of the cobweb at Bentley Priory, but in practice his method was to have the system working so well that he would rarely have to interfere.

At Bawdsey, Watson Watt had initiated an atmosphere that blended camaraderie, relaxation, and urgency, perfectly matching the situation and the personnel. The work to be done was not the kind of drudgery that could be accomplished by driving a large crew to greater and greater exertion; it was rather creative work which needed continually rested and fresh minds, stimulated by each other and driven by their own sense of professional pride, motivated by a full understanding of the ticking seconds of the world clock.

As in America a few years later the universities would be raided and stripped of nearly all their physicists to work on the atomic bomb at Los Alamos and Oak Ridge, now in England whispered recommendations through the Old Boy network brought the scholar-scientists to Bawdsey. From Oxford and Cambridge they came, from Manchester and Birmingham and Sheffield and London, and in the old manor house on the edge of the North Sea they established one of the most efficient scientific and engineering establishments ever known. It resembled outwardly the relaxed colleges they had come from rather than a military laboratory. Watson Watt kept them ignorant of and free from the constricting coils of the government machinery. They had no set hours, but rather worked when they pleased; he was clever enough to understand that this would ensure greater rather than fewer hours of work. At any time during the summer days, one might come across a group playing tennis or cricket, or swimming or fishing. In the autumn and spring the woods and neighboring coast would see small groups wandering around at random or hiking vigorously; in the winter there would even be snowballs flying through the air. But never was the manor dark and quiet.

At night there would be always lights on in the offices and laboratories which had been carved out of the old, curtained, high-ceilinged rooms, as the minds of the young men who had been walking, playing, swimming during the day came alive from their rest and returned to their labors. Even while they fished or threw snowballs, their minds were ticking over, reviewing problems, searching for alternative pathways to the final end result—a bouncing radio echo that could be snatched out of the ghostly ether and captured on the greenly glowing cathode screen, there to be measured and evaluated until the last tiny ounce of meaning was wrung from it and sent by radio to another young man sitting in the cramped cockpit of a Spitfire hurtling across a grass field and scrambling into the air which could no longer hide secrets behind its thick clouds or in its endless reaches. Long after midnight and on into the small, cold hours of morning, the lights would shine and the equipment would hum, the arguments would rage and the slide rules would be slammed down and picked up again, and again, and again and again.

They lived together in one section of the manor house, and they worked everywhere: in their laboratories and offices, in the wood-paneled dining room, on the cricket pitch and in the woods and on the sandy beach beneath the cliffs. They ducked their heads under the waves and came up with a new electronic circuit; they smacked a spinning cricket ball safely away from the wicket and redefined a geometric parameter; they sat quietly with a bottle of beer and a lure floating on rippling waters and they thought, and talked, and extended Skip Wilkins's inspiration of a single moment into a solid, workable, efficient system of aircraft detection.

A. P. Rowe took over as superintendent of Bawdsey in 1938, when Watson Watt became the first Director of Communications Development at the Air Ministry. Rowe introduced an organized pandemonium called the "Sunday Soviets," which owed nothing to communism beyond the concept that the worker and the boss should sit down as equals over a cup of tea or a bottle of beer and argue out the problems from their own particular points of view, with no dispensation given or taken in regard to rank, status, or Oxcam accent. Engineers from industrial firms now began to seep into the ranks of academic scientists, wrinkling their noses in disgust at the taped and wired equipment the physicists had lashed together to test their ideas, producing in their stead solid marvels of tubes and dials that could be mass-produced and worked not by geniuses but by ordinary men.

And women. It was Watson Watt who first proposed that the radar scopes, telephones, and filter rooms be staffed largely by women. There was sure to be a shortage of manpower in the event of war, and nothing in the work of radar was beyond a woman's capacity. He was met, naturally enough in those days, with smiles and supercilious murmurs of the danger to the radar stations from bombing and the necessity of minds and hands that would remain cool under attack. Luckily he persisted, and Dowding backed him up, and when the bombs fell and the pressure mounted there was found to be not the slightest difference between the men and women serving at the scopes and tables.

Alongside the creative activity, the more prosaic labors went on. The engineers at the General Post Office started their work of planning and installing the nationwide network of telephone cables. The Director of Signals at the Air Ministry began to train the tens of thousands of men and women who would maintain and operate the radar transmitters and receivers. The Director of Works and Buildings organized and built the necessary structures. And all this before a working radar system had ever been convincingly demonstrated.

Most important of all, liason with the working personnel of the RAF was established and strongly cemented. Due largely to Tizard's long involvement with the service and to Dowding's total support of the scientific concept, radar was immediately accepted and integrated into the everyday tactics of Fighter Command. The pilots accepted the need for obeying radioed instructions, the group commanders accepted the need for obeying and immediately acting upon the telephoned orders to scramble whatever components the controllers decided should be sent off. In contrast to almost every previous scientific innovation in the history of warfare, radar was welcomed by the warriors. Despite its shortcomings, despite its inefficiencies, despite the incorrect altitudes and misinterpretations, the glamorized and romantic "Brylcreem Boys" of Fighter Command took into their cockpits this magic eye that Aladdin was bringing them, and looked at what it showed them, and listened to what it told them.

For despite the system's problems, it was the only game in town.

When the war came in September of 1939 there were eighteen stations operational. Two more were functioning but were not yet hooked up to the central filter room and were therefore useless. Operations were conducted on a single wavelength in the frequency band from twenty-two to twenty-seven megahertz, which meant that the entire system was susceptible to jamming. The filter room at Bentley Priory was ten feet underground and safe from aerial attack, as were most but not all of the sector and group operations rooms, but the flimsy transmitting antennae, mounted on towers 250 feet high and of necessity exposed along the coast, were extremely vulnerable to bombing.

In addition, the entire system was inhabited by gremlins: wee, winking, giggling, malevolent creatures who blew fuses and bypassed connections and threw calibrations out of joint and sinusoidal waves out of phase, whose greatest triumph was the Battle of Barking Creek, and who continued their mischiefs throughout the autumn and winter of 1939.

By the greatest good fortune it didn't yet matter. Poland fell to the combined invasions of Germany from the west and Russia from the east; the two dictatorships met at a prearranged line and stood there facing each other distrustingly but peacefully, gorging themselves on the dead Poles rather than looking for further battle. In a few weeks it was all over; neither Britain nor France had been able to move quickly enough to provide the faintest aid. Their only hope had been that they could tie down a goodly portion of the German air and army forces which would be forced to defend Germany's western front, and that Poland could withstand the remainder. But Hitler gambled on a quick decision in the east, sent his Stukas and panzers crashing over the Polish border, taught the world a new German word—*Blitzkrieg*—and swept through Poland like a small boy through a pile of autumn leaves.

It hadn't really been much of a gamble. The French military strategy in the preceding years had been no secret; it was based on the Maginot Line, a fortified series of pillboxes and modern forts stretching along the Franco-German border from end to end, through which they grandly boasted the German armies could never penetrate. But, of course, it worked both ways. The Maginot Line was a defensive fortification, not an effective staging base for offensive operations. It was unacceptable to the German mind that the frugal French would have spent all those millions on a defensive line only to send their armies out beyond those well-prepared, expensive fortifications. As for the businesslike British, that "nation of shopkeepers" would send money and equipment to the French, but would never fight with their own men.

The Germans were right. The French and British made angry sounds but though the British pushed the French to actually do something, neither did anything. Truth to tell, they were frightened. For ten years all the prophets had been warning of the Armageddon that would be modern war, led by bombers of destruction. H. G. Wells, in his novel *Things to Come,* had warned of the awesome desolation that would be left by the incendiaries, high explosives, and gas; he even warned of entire cities wiped out by atomic bombs. In the same year that war broke out, a terrifying film based on his book was released, and in the weary lines of Raymond Massey's face the end of civilization was traced.

Though Dowding spoke on the radio with firmness of his ability to defend England from aerial attack, the Government was taking no chances. Women and children were evacuated from London to the outlying villages, and from England to the colonies. The upper and middle classes often had friends or relatives to go to, the others were sent to hastily organized receptions among well-intentioned strangers who opened their homes to those fleeing the holocaust. But there were problems. Harold Nicolson wrote: "Many of the children (most of them from the East End slums) are verminous and have disgusting habits. This horrifies the cottagers upon whom they have been billeted. The mothers refuse to help, grumble dreadfully, and are pathetically homesick and bored." By the end of September many of them began to drift back to London.

Meanwhile the western armies sat behind the Maginot Line and waited while the panzers swept through Poland, and by the end of September much of the Wehrmacht was on its way back again to take up positions along the western front.

Now, Hitler thought, they could have peace again. On the twenty-eighth of September he and Stalin issued a joint proclamation calling for an end to hostilities since neither of them had any "further territorial claims," Poland was now a dead issue, and the status quo was obviously stable.

There was strong impetus from many directions to accept this proposal. The communist parties in France and England echoed Stalin's line that the war was a capitalist affair in which the working people of the world had no interest. From America President Roosevelt warned that nothing on earth would induce him to send America's boys to fight in Europe. It was not only the professional pacifists who wanted to surrender Europe to Hitler; in every thoughtful mind there now rose the spectral memory of that holocaust that had ended only twenty years earlier. In the first week of the war Harold Nicolson recorded in his private diary that "there is a little timid selfish side of myself that tempts me by still murmurings to hope that we shall reach a form of appeasement." Sir Robert Vansittart, chief diplomatic advisor to the Foreign Secretary, advised that the British should not attempt to make war but should instead "hope" that Germany would not attack them. When the Russians moved into Poland it seemed that the Russo-German nonaggression pact might mean even more; their linked forces could together swamp the Allied armies. Lloyd George, the old tiger of the 1914 war, was "frankly terrified" and "did not see how we could win this war." He was arguing with his friends for a secret session of Parliament, to discuss how and under what terms they might sue for peace.

Churchill quickly established himself as the center of opposition to any such thinking. At the outbreak of hostilities it was reluctantly clear to the government that if any war cabinet was to be accepted by the people, he must be brought in. For years his had been the only political voice raised in protest and warning against Germany's warlike awakenings, and now that he was proved to be right, he was virtually the only politician trusted by the people. Chamberlain's prestige and authority had vanished out the window alongside his "peace with honor," and it was now to Churchill that the nation looked. Some suggested that he be brought into the Cabinet as minister without portfolio, simply as a gesture to the public, but more realistic minds warned that such a move would only give him an excuse to poke his nose into *everyone's* business, and so he was made First Lord of the Admiralty, his old position in the Great War, to give him something to do and keep him busy. He lost no time.

From Harold Nicolson's private diary, 26 September, 1939:

The Prime Minister [Chamberlain] gets up to make his statement [in Parliament]. He is dressed in deep mourning, relieved only by a white handkerchief and large gold watch-chain. One feels the confidence and spirits of the House dropping inch by inch. When he sits down there is scarcely any applause. During the whole speech Winston Churchill had sat hunched beside him looking like the Chinese God of plenty suffering from acute indigestion. He just sits there, lowering, hunched and circular, and then he gets up. He is greeted by a loud cheer from all the benches and he starts to tell us about the Naval position....

The effect of Winston's speech was infinitely greater than could be derived from any reading of the text. His delivery was really amazing and he sounded every note from deep preoccupation to flippancy, from resolution to sheer boyishness. One could feel the spirits of the House rising with every word. It was quite obvious afterwards that the Prime Minister's inadequacy and lack of inspiration had been demonstrated even to his warmest supporters. In those twenty minutes Churchill brought himself nearer the post of Prime Minister than he had ever been before. In the Lobbies afterwards even Chamberlainites were saying, "We have now found our leader."

In his long career Churchill was wrong about so many things, but all of them pale into insignificance before the fire of his one great insight into the psychopathic nature of Hitler. There was no peace to be made with such as he; there was, as Churchill warned, only a bloody struggle for survival—survival not merely for England but for the whole concept of democracy and civilization, survival for the "impulse of the ages," survival, or final defeat accompanied by an irrevocable slide into a despair and darkness unlit by the faintest flicker of humane or civilized reason.

While he was calling for peace, Hitler was meeting with his generals and planning his attack in the west. The Maginot Line was the greatest joke in history; it was enough to make a German laugh. It was designed to be absolutely impregnable along the entire length of the German-French border—which ended at the border with Belgium. And the Belgian border was almost entirely unfortified. Hitler had seen to that with angry speeches warning against any Belgian action that he might interpret as nonneutral. In vain did the English and French urge Belgium's King Leopold to allow their troops onto his soil in order to set up defenses; he clung to his neutrality like a virgin trusting in her holy cross for protection against vampires—and in this case the vampire was all too real.

The German plan was to attack without warning through Belgium and Holland, thus to outflank the Maginot Line which, though presumably impregnable where it existed, was immobile, sunk in stone, powerless to move to where it did not already exist. The date for the onslaught was set for November 12, then postponed day by day and week by week, fourteen times in all, while Hitler vacillated between conquering by waiting for England and France to collapse from the weight of their fears or by trampling them with his panzers. Finally the winter weather closed in, the fields of France turned to mud, the heavy clouds kept the Luftwaffe on the ground, and a final postponement of any action until spring was agreed upon by Hitler and his generals.

It was the grace of God that settled over Europe that winter. As we have seen, the radar system in September 1939 was operational but not effective. Aside from its inadequacies in regard to rearward echoes and precise determinations of position and height, it was totally 100 percent helpless against low-flying aircraft. The broad wavelength of about twelve meters spread out and bounced off the ground when directed low; the inevitable ground-return echoes swamped any aircraft returns and hid them in the noise. The problem was realized right at the beginning, and work was begun on a series of Chain Home stations having low-flying detection systems. The solution was to reduce the wavelength to 1.5 meters and use a searchlight-type beam which could be controlled and kept away from the ground, but the problems involved had not been solved at the time war broke out. By the middle of October, 1939, it was clear that there would be no solution in the near future, and so a line of trawlers was set up in the North Sea to provide visual sightings—obviously a short-range and inadequate solution, indicative of the desperate situation.

The radar solution, Chain Home Low (CHL), was not available in even its first rudimentary form until November, 1939, and the first chain of eleven operational stations didn't go on the air until the following February. None of these stations had even the slightest ability to measure height; the best that could be done was to give warning of aircraft that were unseen by the normal CH stations and therefore presumably below their height limit.

All that winter and on into the spring the scientists and engineers worked to improve the system at the same time that the factories were working overtime to produce the sets and the RAF was trying to learn how to use them. At the same time, Churchill was arguing for a naval offensive or an amphibious attack through the Mediterranean or Scandinavia, and others were urging peace. Cyril Joad, the Great War pacifist, was still around, loudly proclaiming that the average Briton would be less unhappy with a Nazi overlord than with the horrible results of bloody war. Philip Lothian, the ambassador to America, warned that the United States would never fight in Europe and that Hitler sincerely wanted peace with England. Joe Kennedy, our Jack's father, then American ambassador to the Court of St. James, warned everyone who would listen, from Roosevelt to the cockney chimney sweeps, that England was finished. "I always thought a dandelion was yellow," the British said, "until I met Joe Kennedy." He wasn't yellow; he was uninformed. He didn't know about radar, and without radar he was right: the British hadn't a chance.

A group in the War Cabinet who knew about the radar concept but didn't understand its significance, rallying loosely around Lord Halifax, was in secret consultation with Heinrich Bruning, former German chancellor then visiting the United States, trying to put together a peace that would involve the deposition of Hitler, together with his replacement by a group of seemingly democratically minded replacements. And there were faeries dancing at the bottom of all our gardens.

On April 4 Chamberlain made a speech in which he announced that Hitler had "missed the bus" by not attacking England and France immediately after the fall of Poland. At dawn on April 9 the Wehrmacht answered by marching across the borders of neutral Denmark, despite a nonaggression pact which had recently been signed. Simultaneously, helmeted figures arose from the holds of several merchantmen sitting in Oslo harbor, clambered down the sides of their ships, motored into shore, and occupied the fortress and then the town. Junkers 52 transports dropped parachutists over the Oslo airfield, and as soon as it was secured, further transports landed with more men and heavy guns; other landings took place at Stavanger, Bergen, and Trondheim.

Churchill immediately thundered out his response in the councils of war in London; this action gave him the opportunity to suggest just what he had wanted ever since he had become Lord of the Admiralty—not a land war on the continent but amphibious actions along the coast of Norway. If he had been given command of the RAF, he would have seen victory in the air; if the BEF, he would have seen their salvation waiting on the continent; as First Lord of the Admiralty he saw Britain's triumph to be on the high seas. He proclaimed that his navy would wipe out the silly Boche whom the madman had sent to what must be their destruction. On April 9 five British destroyers entered the west fiord at Narvik to prevent a German landing at that crucial port. Finding that five German destroyers and a number of merchantmen had preceded them, they waited through the night and then attacked the following dawn. Achieving surprise, they swept through the German forces. But then out of the mists appeared five more German warships. Two of the British ships were sunk, and the remainder limped away.

"They left their mark on the enemy and in our naval records," Churchill recorded in their honor; but the Germans occupied Narvik.

Guided by Churchill's warlike spirit, the British immediately moved in to help the Norwegian forces which resisted. The Royal Navy successfully landed ski troops at Namsos and Andalsnes, but their ski straps were sent separately on another ship which was sunk. They were helpless in the deep snow, and were evacuated two weeks later.

The RAF was useless at first, since the original German paratroop attacks had captured every airfield in Norway, but then they brought No. 263 Squadron with biplane Gladiator fighters over by aircraft carrier and flew them onto frozen Lake Lesjaskog. On the morning of April 25 they were spotted by German reconnaissance aircraft and were immediately attacked by low-flying Heinkel 111s. Before they ever had a chance to get their fighters into the air they had only five left. By the next day they had three, and on the day following they had none.

The fighting straggled on, and one month later No. 263 Squadron was back with more planes, accompanied by a squadron of Hurricanes. They fought for two weeks, and finally the British captured Narvik—at just about the time that the higher brass realized that the whole situation was hopeless. A week later the city was evacuated. The Gladiators and Hurricanes flew onto the aircraft carrier *Glorious,* which was intercepted on its way home by the battlecruiser *Scharnhorst.* The carrier went down with all planes and nearly all hands; two of the pilots out of the original two squadrons eventually made it back to England.

The Norwegian debacle was "a stunning blow to public confidence, the more so after Churchill and Chamberlain had spoken in optimistic terms in mid-April. The government yielded to demands for a debate on the failure of the expedition, which began on May 7. There have been few debates of greater importance in parliament's history."

Chamberlain began with a routine defense of the government's actions. He was interrupted with taunts about "missing the bus." Others rose to criticize him: "Everywhere the story is 'too late.'" The army was criticized for being unprepared, the navy for not gathering its forces and overwhelming the Germans. Leopold Amery climaxed the day when, to thunderous applause, he repeated Cromwell's famous speech to the Long Parliament: "You have sat here too long.... In the name of God, go!"

The next day a vote of censure was moved by Labour. Lloyd George said: "The Prime Minister...has appealed for sacrifice.... I say solemnly that the Prime Minister should give an example of sacrifice, because there is nothing which can contribute more to victory in this war than that he should sacrifice the seals of office."

The Government won the vote, but by too small a margin to long exist. Chamberlain spent May 9 trying to reconstitute his cabinet. On

May 10 the German armies invaded Holland and Belgium in a massive new offensive. Now even Chamberlain's friends turned away; the nation needed a Mars, not a Jupiter; it did not even matter that the fall of the Government had been precipitated by the Norwegian disgrace, which was largely attributable to Churchill—it did not matter at all, for he was the only Mars they had. That evening he was sent for by the King and asked to form a new government.

In the First World War, the Gallipoli expedition, whose terrible failure was not his fault, had brought Churchill down. In this new war the Norwegian expedition, whose terrible failure *was* his fault, brought down Chamberlain and installed Winston as the prime minister of England. The "has-been who never quite was," the great man manqué, the old, worn-out warhorse "without a seat, without a party, without an office, and without an appendix," entered his "Finest Hour."

He entered it in somewhat less than heroic fashion. The land forces which had to withstand the German onslaught were necessarily mostly French; a continental land army had never been a feature of English policy. Instead its strength was traditionally in the Royal Navy, which was fighting an heroic but largely unnoticed war against the German submarine forces. When asked by the French government for help, Churchill had little to offer except the Royal Air Force—and even there he had not much. Bomber Command had begun its expansion into modern bombers with the twin-engined Wellington, but even this was a slow, clumsy weapon not suitable for attacking an army's mobile targets. The RAF had sadly neglected what they called Army Cooperation Command, and had nothing to match Hitler's *Sturzkampflugzeugen*—the deadly Stuka dive bombers. Willingly they sent what they had, several squadrons of Fairey Battle single-engined bombers, which were useful only for exhibiting the bravery of young men ordered to die.

The Allied forces retreating through Belgium had failed to destroy the two bridges spanning the Maas River at Maastricht, and now the German army was pouring across in force. Twenty-one year-old Flight Lieut. "Judy" Garland of No. 12 Squadron was told to gather volunteers from his Battle squadron to destroy the bridges at 0900 hours on May 12. Every man in the squadron stepped forward. They took off early that morning and reached the target precisely on time, coming in at low level and then pulling up to begin their bombing run—"one of the greatest acts of cold heroism in history." One by one the three-man crews of the Battles tipped their aircraft over and slid down the still air into the muzzles of 300 antiaircraft cannon, and one by one they fell blazing out of the sky. The bridges were untouched, the panzers kept rolling. None of the Battles returned to base.

> Honor the charge they made!
> Honor the Light Brigade,
> Noble six hundred!

The only result was the RAF's first Victoria Crosses—awarded to Judy Garland and his gunner, Sergeant Grey, posthumously. (The other member of the crew, Leading Aircraftsman Reynolds, was ignored by the awards board; no one knows why.) Ten days later other squadrons of RAF Battles attacked the pontoon bridges at Sedan, and out of more than seventy aircraft, only thirty returned. Before the end of the month they had been wiped out as a bomber force; they had accomplished nothing.

In the meantime the Germans swept through Holland and Belgium, turning the corner of the land defense. They then burst through the center of the Maginot Line by attacking at the one point which had been left unfortified because the terrain was "impassable to tanks." The Germans brought their panzers through the impassable terrain, and the French were as startled as Macbeth seeing Birnam Wood moving to Dunsinane when they looked up and saw masses of Tiger tanks oozing steadily out of the Ardennes Forest, where no tanks ought to be. "Confusion reigned throughout the [Allied] armies," Churchill reported, and they fell back.

On Friday, September 8, 1939, five days after war had been declared, Flying Officer Paul Richey of No. 1 Squadron, Fighter Command, was being visited by his father, a colonel in the army. Richey had been sleeping, and was now sitting up in his bed at Tangmere aerodrome in southern England, chatting pleasantly, when his orderly rushed in and shouted, "No. 1 Squadron called to readiness, Sir!" Richey hugged his father, dressed hurriedly, and dashed out with the other pilots. They were told to rip the squadron emblems from their overalls and were ordered into the cockpits of their Hurricane fighters. Waving goodbye to his father, who was leaning over the fence watching, Richey followed the others into the air. They buzzed the aerodrome, swung out over Beachy Head, and headed across the Channel to France.

They were the first to go, torn out of Fighter Command over the protests of Dowding. There was no radar in France, he pointed out, and without it the fighters would be lost. The Air Ministry ignored him, and other fighter squadrons followed throughout the fall and winter. At each one Dowding protested, to no avail. His one victory was to allow only Hurricanes to go; he would guard the more valuable Spitfires with the prostrate form of his dead body, if need be.

Throughout that fall and winter the Hurricanes departed; the air fighting was sporadic and light, their losses were not heavy, and Dowding was increasingly regarded as an old woman, a librarian who resented and mourned each book checked out as one lost. But suddenly, with the blitzkrieg of May 10, the situation changed. The Hurricanes were in battle every day, and without radar they were helpless against the overwhelming numbers of the Luftwaffe. Their aerodromes were raided and they were caught on the ground and bombed to bits, or they were bounced out of the sun by hordes of Messerschmitts. Within four days of the German attack, there were only 206 British aircraft flying; the day before the attack there had been 474.

And now the German panzers came roaring out of the Ardennes into the heart of France, splitting the Allied armies in two, driving the British on the north into the sea and the French on the south into panic and rout. And leading the panzers were the low-flying Stukas and Heinkels, sowing their bombs here and there and everywhere, swooping down out of the vast skies with only a second's warning, machine-gunning and strafing and tearing the ground armies to tattered, helpless, huddling and ineffectual shards through which the tanks then rolled with impunity. Field Marshal Lord Gort, commanding the British forces, telegraphed home: "We have to support in the air not only the BEF but also our Allies who have suffered heavy air attacks....Our main defence in the air is fighters....I earnestly hope War Cabinet will decide to give additional air assistance...."

Dowding refused. *There is no radar in France,* he warned. The fighters would be lost.

Churchill flew to Paris to find out what was happening, to bolster the French morale, to check the panic, to ask what the British could do to help stem the flooding tide. Writing about it in *Their Finest Hour,* he said:

> The burden of all the French High Command's remarks was insistence on their inferiority in the air and earnest entreaties for more squadrons of the Royal Air Force, bomber as well as fighter, but chiefly the latter. This prayer for fighter support was destined to be repeated at every subsequent conference until France fell. In the course of his appeal, General Gamelin said that fighters were needed not only to give cover to the French Army, but also to stop the German tanks.... [But] it was vital that our metropolitan fighter air force should not be drawn out of Britain on any account. Our existence turned on this. Nevertheless, it was necessary to cut to the bone. In the morning, before I started, the Cabinet had given me authority to move four more squadrons of fighters to France. On our return to the Embassy and after talking it over...I decided to ask sanction for the despatch of six more. This would leave us with

only the twenty-five fighter squadrons at home, and that was the final limit.

In fact, that was less than *half* the limit. When Dowding had made his radio speech just before the war began, promising that Fighter Command could protect England from aerial devastation, the agreed-upon limit had been fifty-two fighter squadrons. He was already below that limit when the German offensive began in France, and without his acquiescence another two squadrons of Hurricanes had been sent across the Channel on May 13. He saw his forces trickling away down that yawing abscess of France, and he determined to stop it. He asked for permission to address the War Cabinet personally.

Ordinarily it was the Air Ministry which dealt with the Cabinet, but Dowding, who knew that the Prime Minister was determined to send more fighter squadrons across the Channel in response to the frantic appeals coming from France, was afraid that the representatives of the Air Ministry would not stand firm: "I felt that everybody was too frightened of Winston Churchill."

The War Cabinet met in an underground warren at the end of Prince George Street, one block from Downing Street, across from St. James's Park. The meeting room was nearly entirely occupied by a large central table around which chairs were placed. At first Dowding was asked to sit on a chair by the far wall, while other business was discussed; then he took a place at the table to discuss the French requests for more fighters. He complained later of receiving little support from the Air Ministry personnel. Newall (Chief of Air Staff) had been "rebuffed a little earlier by Churchill over something inconsequential, and he was silent." Sinclair (Secretary of State for Air) was attentively and "eagerly trying to guess what Churchill was going to say next."

What Churchill said next was that the French needed more fighters and that England was honor-bound to give them what help she could. Hunched forward, his head down like a bulldog but his eyes lifted and moving around the table, Churchill waited for any response. Dowding also waited, but there was none. Neither Sinclair nor Newall said a word against the glare of those passionate eyes, despite all their later claims. Finally, feeling deserted by his superiors in what should have been their fight rather than his, but seeing no alternative, he rose to his feet without a word. That morning before leaving Bentley Priory he had prepared a graph showing the day-by-day loss of Hurricanes in France. Now he took that graph and walked around the table to Churchill. He placed the graph precisely in front of him, and said: "If the present rate of wastage continues for another fortnight we shall not have a single Hurricane left in France or in this country." He carefully and clearly enunciated those last words.

Then he turned and walked back to his chair and sat down. There was not a single word spoken in that cavernous room, there was not a sound but the constant whirring of the air circulation system. Churchill sat glaring at the chart Dowding had placed in front of him.

General Sir Ian Jacob, Military Assistant Secretary, wrote: "The decision was one of the hardest to make in the whole war, and opinion swayed back and forth.... The Prime Minister was torn in two."

Professor A. J. P. Taylor wrote, in his *English History, 1914–1945:* "When argument failed, Dowding laid down his pencil on the cabinet table. This gentle gesture was a warning of immeasureable significance. The war cabinet cringed, and Dowding's pencil won the battle of Britain."

Taylor evidently got this scenario from his friend Lord Beaverbrook, who was present, but Dowding later said that the remark about the Cabinet cringing was "really very absurd," although he admitted to perhaps having "thrown down the pencil in exasperation."

Pencil or no pencil, cringing or not cringing, the Cabinet and Churchill gave in to that gray, austere spectre and his grim chart. No more fighters would be sent to France, come what might.

Field Marshal Archibald Wavell was an old friend of Dowding's, from their schoolboy days together at Winchester. When Dowding told him about that day's confrontation, Wavell replied that his days as Commander-in-Chief were numbered. Dowding already knew that: "I had opposed him [Churchill], and he had had to change his very stubborn mind in front of a large gathering of senior officers and officials over a very important issue. I've always felt that he didn't like it."

The Cabinet meeting broke up, Dowding left to return to Bentley Priory—and Churchill began to have second thoughts. He stood talking with some of his ministers, and the next minute he vetoed the decision. Ian Jacob later remarked that, as far as he could remember, never before or since had Winston Churchill changed his mind and reversed a formal decision.

This cabinet meeting had taken place the morning that Churchill left for France, and what occurred during and in that short discussion afterwards is the basis of Churchill's later brief statement in his history, *Their Finest Hour,* that "In the morning, before I started, the Cabinet had given me authority to move four more squadrons to France." Not a word does he say about the meeting itself, about Dowding's argument, about the decision *not* to send fighters to France, or about his personal reversal of that decision after Dowding had left and the Cabinet meeting was officially over. Oh, well. As Sir John Slessor has remarked in another context, "The enormous interest and value of Sir Winston's memoirs sometimes suffers from his occasional genius for self-deception." Robert Wright, in his biography of Dowding, writes:

"When the question was raised [after the war] about the curious distortion of the decisions reached about the fighters that is found recorded by Churchill in his own account of those times, Dowding commented: 'You couldn't very well expect him to admit that he came within a hair's breadth of wrecking Fighter Command before the Battle of Britain ever started.'"

When Dowding was informed about the further four squadrons that were being sent to France, after the Cabinet had agreed that no more would leave England, he sat down for a moment in despair. He couldn't ask again to appear before the Cabinet, he would be refused; a war can't be run if every general comes running to appeal every decision. And yet he knew that upon this decision rested the entire outcome of the war; he couldn't let it pass. So he took up his pencil (the one he had *not* made the Cabinet cringe with) and wrote a letter to the Chief of the Air Staff, in terms which he knew would make its transmission to the prime minister mandatory.

> Sir, I have the honour to refer to the very serious calls which have recently been made upon the Home Defence Fighter Units in an attempt to stem the German invasion on the continent.
>
> I hope and believe that our Armies may yet be victorious in France and Belgium, but we have to face the possibility that they may be defeated. In this case I presume that there is no one who will deny that England should fight on, even though the remainder of the Continent of Europe is dominated by the Germans.
>
> For this purpose it is necessary to retain some minimum fighter strength in this country and I must...remind the Air Council that the last estimate...as to the force necessary to defend this country was 52 squadrons, and my strength has now been reduced to the equivalent of 36 squadrons.
>
> Once a decision has been reached as to the limit on which the Air Council and the Cabinet are prepared to stake the existence of the country, it should be made clear to the Allied Commanders on the Continent that not a single aeroplane from Fighter Command beyond the limit will be sent across the Channel, no matter how desperate the situation may become....
>
> I must point out that within the last few days the equivalent of 10 squadrons have been sent to France, that the Hurricane Squadrons remaining in this country are seriously depleted, and that the more squadrons which are sent to France the higher will be the wastage and the more insistent the demands for reinforcements.

I must therefore request that as a matter of paramount urgency the Air Ministry will consider and decide what level of strength is to be left to the Fighter Command for the defences of this country, and will assure me that when this level has been reached, not one fighter will be sent across the Channel however urgent and insistent the appeals for help may be.

I believe that, if an adequate fighter force is kept in this country, if the fleet remains in being, and if Home Forces are suitably organized to resist invasion, we should be able to carry on the war single handed for some time, if not indefinitely. But, if the Home Defence Force is drained away in desperate attempts to remedy the situation in France, defeat in France will involve the final, complete and irremediable defeat of this country.

This letter, together with Einstein's letter to Roosevelt which began the development of the atomic bomb, must rank as perhaps the most important ever written in the entire history of warfare. It now hangs framed on the wall of the Royal Air Force College, Cranwell, and a copy hangs in Dowding's former office at Bentley Priory. But it is not even mentioned in Churchill's voluminous history of the war.

It pulled Newall to his feet; he sent it on to the other chiefs of staff with his strong recommendation, adding that "it can be said with absolute certainty that while the collapse of France would not necessarily mean the ultimate victory of Germany, the collapse of Great Britain would inevitably do so."

On May 19 the War Cabinet met again to consider the letter, and a final decision was made: "No more squadrons of fighters will leave the country whatever the need of France."

At the end of May the British Expeditionary Force collapsed onto the beaches at Dunkirk. There was no question now of not sending fighters to defend their embarkation, but at least these fighters remained based in England, flying across the Channel each day into the fight. The radar system was not yet good enough to warn them effectively or consistently of enemy aircraft in battle over France, but their bases were protected and the planes that were lost—and there were many of them—at least went down fighting. Even the Spitfires were thrown into this desperate battle, encountering the Luftwaffe for the first time, and by the end of the month over 338,000 men were brought home by destroyer and fishing boat, yacht and sailboat, tug and trawler.

"We must be very careful not to assign to this deliverance the attributes of a victory," Churchill warned. "Wars are not won by evacuations.... But," he went on, "even though large tracts of Europe and many old and famous States have fallen or may fall into the grip of the Gestapo and all the odious apparatus of Nazi rule, we shall not flag or fail. We shall go on to the end, we shall fight in France, we shall fight in the seas and oceans, we shall fight with growing confidence and growing strength in the air...."

On June 17, 1940, the last twelve Hurricanes of No. 1 Squadron, the first fighting force to fly to France, flew home again to Tangmere in England. Though some units on the ground continued to fight for another few days, the Battle of France was over. The Battle of Britain was about to begin.

The *Graf Zeppelin* as seen from the east coast of England during its 1939 mission to investigate the Bawdsey radar emissions. (Photo by Mr. G. Soar, published in Kinsey [*Bawdsey*])

The Manor at Bawdsey.

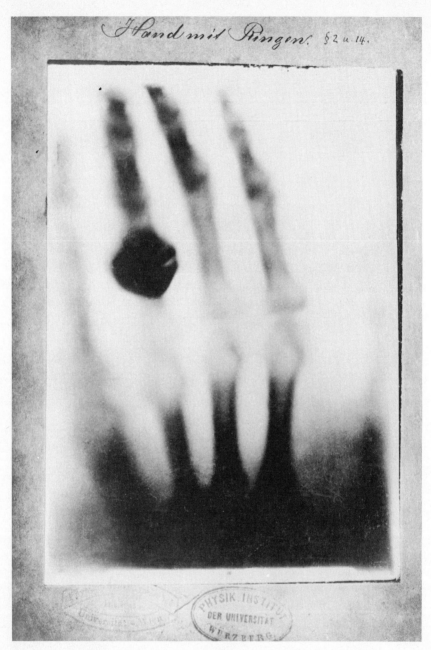

The first x-ray photograph: Frau Roentgen's hand, with ring. (Courtesy Kodak)

F. A. "Skip" Wilkins. (Courtesy Mrs. Wilkins)

Robert Watson-Watt. (Courtesy E. G. Bowen)

Professor Lindemann (Lord Cherwell) and Winston Churchill. (Courtesy Imperial War Museum)

Sir Henry Tizard. (Courtesy Imperial War Museum)

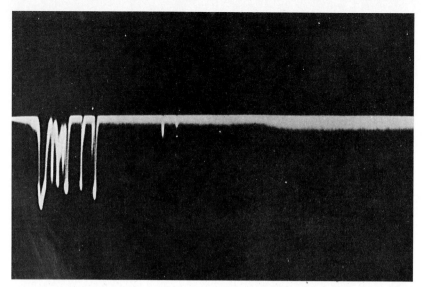

Chain Home Low radar display. The series of blips at the left are from the outgoing pulse. Two echoes are seen about one third of the way along the baseline, indicating aircraft at 50 and 55 miles. (Courtesy Imperial War Museum)

Chain Home radar towers in 1940. Horizontal platforms on the near towers bear the two sets of antennae which allow height determination. (Courtesy Imperial War Museum)

Operations Room at Bentley Priory. (Courtesy Imperial War Museum)

The map table at Bentley Priory.

Sir Trafford Leigh-Mallory. (Courtesy Imperial War Museum)

H. C. T. Dowding and Douglas Bader. (Courtesy Imperial War Museum)

Keith Park, in the airplane he used to visit his aerodromes. (Courtesy Imperial War Museum)

The Hawker Hurricane. (Courtesy British Aerospace, Military Aircraft Division)

The Supermarine Spitfire. This aircraft, which flew in the Battle of Britain, still flies today with the Battle of Britain Memorial Flight of the RAF. (Courtesy B of B Flight)

LEFT: E. G. "Taffy" Bowen. (Courtesy Mr. Bowen)

MIDDLE: Bentley Priory.

BOTTOM: The magnetron, showing the internal resonant cavities. (Printed by permission of AT&T Bell Laboratories)

The Bristol Beaufighter. This aircraft was fitted with a Mk IV airborne radar set, but—as in all photos released at the time—the aerials were deleted for security purposes. See Mosquito for details. (Courtesy British Aerospace, Bristol Division)

The DeHavilland Mosquito night fighter. Note the radar aerials extending from the nose and above and below the wing. (Courtesy British Aerospace)

TOP: AI radar scope in the cockpit of a Mosquito. To the left is the pilot's control stick. (Courtesy Imperial War Museum)

BOTTOM: The Mosquito in flight. The large aerial behind the cockpit is for the radio; the radar directional receiving antennae are visible on the wingtips. (Courtesy British Aerospace)

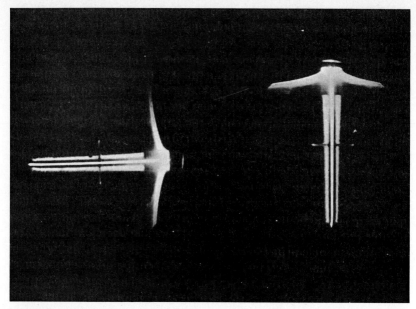

AI Mk IV radar display. The left-hand display shows a target aircraft above the fighter, the right-hand display shows him to the left. (Courtesy Imperial War Museum)

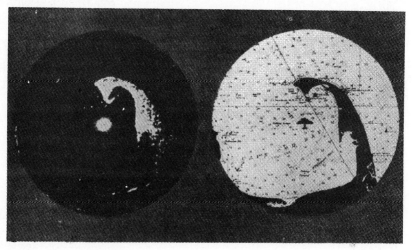

A radar picture of Cape Cod, contrasted with a map. (From *Radar System Engineering* by Louis Ridenour. Copyright © 1947 by McGraw-Hill, Inc. Used with permission)

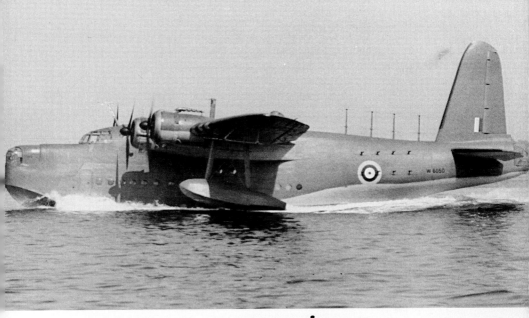

A Short Sunderland flying boat of Coastal Command, taking off on patrol. Note the side-scanning radar transmitting antennae on the top of the fuselage, and the receiving antennae below them. (Courtesy Shorts)

TOP: Boeing E-3A AWACS.
(Courtesy Boeing Aerospace)

ABOVE: Rear cabin of Boeing
E-3A, with NATO crew manning
the computer and radars.
(Courtesy Boeing Aerospace)

RIGHT: Pave Paws phased-array
radar, in Massachusetts.
(Courtesy U.S. Air Force)

Part Three
The Battle

The blood-dimmed tide is loosed...
W. B. Yeats, "The Second Coming"

Chapter Ten
In the Still of the Night

> *Owen Glendower:* I can call spirits from the vasty deep.
> *Hotspur:* Why so can I, or so can any man; But will they come when you do call for them?
>
> William Shakespeare, *Henry IV*

In the north of England early in the morning of February 3, 1940, the Hurricane pilots of B Flight, No. 43 Squadron, RAF Acklington, were cranking up their engines. As was usual on cold mornings, some of the starter batteries wouldn't work, and the engines had to be turned over by hand.

"Switches off?"

"Switches off."

A strong young man would reach up to his full height, stretching on tiptoe to grasp the top of the two-bladed propeller; he would lift his right leg off the ground, kick out with it, and then snap it down and bring his whole weight down, swinging the propeller with it. He would do this twice, pulling the propeller through to clear the frozen mush of oil in the lines and get the fuel moving through the carburetor. Then he would step back and take a deep breath or two, the exhaled air from his mouth condensing into white smoke in front of his cold face.

"Contact?"

"Contact."

In the cockpit, the pilot would flip closed the magneto switches. Outside, by the nose of the plane, the young man would grasp the propeller again, kick again, and swing it down. The engine would cough and sputter, black smoke would spurt out of the side exhausts, and if they were lucky, it would catch that first time and settle down into the lovely deep-throated Merlin rumble that has to be heard to be appreciated. If not, they would try again.

The Hurricanes, once started, taxied individually around the aerodrome to be parked a few yards in front of the wooden dispersal hut. There was no need to hide them, to spread them out or camouflage them; radar was on guard, and there could be no surprise attack—a luxury long forgotten by the squadrons then in France. The pilots turned off their warmed-up engines and left their parachutes sitting on their wings or tails or in their cockpits, according to individual taste. They trudged into the dispersal hut and congregated around the iron stove, warming their hands and faces. The corporal on duty picked up the hand telephone, cranked it to get through to sector headquarters at Usworth, and reported, "Blue and Green Sections, B Flight, 43 Squadron at readiness."

In the operations room at Usworth, a single bulb lit up on the readiness board, reflecting this information. The map on the table was empty, the WAAFs stood leaning against the wall, headphones on their ears, sipping coffee and chatting.

0903 hours.

In the radar station at Danby Beacon, the NCO on watch saw a blip appear on the cathode-ray screen. He blinked and looked again. He had never seen one before, except on exercises when he was told what to expect. But there it was: on the left-hand edge of the screen was the transmitted pulse; and there was a lot of noise stretching out to the right; and there—just on the edge—was most definitely a blip standing up and waving at him out of the noise.

He read the range directly off the calibrated scale: sixty miles. He twirled the goniometer dials and watched the echo fade and rise. Finally he zeroed in on it: it was directly out over the North Sea, heading straight in. He switched in and out of the vertically stacked array of antenna to measure the height: 1000 feet. By the time he had finished, the duty officer was already picking up the telephone and cranking away to the Filter Room at Bentley Priory. Within seconds the message was retransmitted to No. 13 Group headquarters at Newcastle and to sector headquarters at Usworth.

In the dispersal hut at Acklington, the telephone rang. Six months later the pilots, nerves jagged and raw, would flinch or jump at the sound; now, lounging around in wood-and-canvas sling chairs of the type popular in English parks and seasides, they didn't even bother to look up. "Bloody tea's going to be late again," one muttered.

The duty corporal listened to the phone for a moment and then shouted, "Blue section 43 scramble!"

With a rush of wind and a clatter of scattered deck chairs, three pilots dashed away. The duty corporal followed them to the door. "Angels one!" he shouted after them.

Peter Townsend led "Tiger" Folkes and Sergeant H. J. L. Hallowes off into the cold blue air. Townsend, a thin young man, rather tall for a fighter pilot and Hollywood-handsome, would later command a squadron and become an ace; later still he would serve as Equerry to King George VI, would have a world-famous but frustrated romance with Princess Margaret, and would write a wonderful book that reintroduced me to the Battle of Britain, *Duel of Eagles*. Sergeant Hallowes (non-University, nongentleman pilots were not commissioned as officers in those days) would also become an ace and would retire after the war as a wing commander (and a gentleman). Tiger Folkes one day would disappear without trace into the North Sea.

"Vector 190, bandit attacking ship off Whitby. Buster!"

In response to the "buster" command, they pushed their throttles to the wall and sped out to sea nearly due south from Acklington. At over 300 miles an hour they were soon out of sight of land. A few minutes later they spotted a Heinkel 111 twin-engined bomber just below the cloud base. It saw them as they banked and climbed toward it, but not in time. The bomber pulled up into the cloud cover at the same instant that the first bursts from the eight-gunned fighters ripped through its metal skin and glass canopy.

The Hurricanes followed it through the cloud and all came out on top together. Before they could fire again, they saw the smoke pouring out; the German pilot saw it too and knew he couldn't make it home across the deadly sea. He turned back toward England, belching smoke and losing height, with the Hurricanes trailing behind. He skimmed the cliffs at Whitby, just north of Robin Hood's Bay, slashed through telephone wires, skimmed over a barn, and crashed into the snow-covered ground.

It was the first German warplane to be shot down on English soil in the Second World War.

In the first half year of war not a single airplane of RAF Fighter Command was lost due to enemy action, though the Hurricanes detached and sent to France did see intermittent action and suffered some losses. There were training accidents, of course; flying these new Spits and Hurris was a dangerous business, even if no one was shooting at you. They landed on narrow wheels at nearly eighty miles an hour, and if the weather closed in and left you separated from the ground by a thick mass of dark clouds, there was no guidance system to bring you home to base and down safely onto the grass. There was just the one engine out in front holding you up in the air, and the sea was cold and endless when you flew convoy patrol. If the engine hit a snag, sputtered and quit, you couldn't possibly glide the long miles back to land; instead you fell out of the sky like a dead lump of steel, and within minutes you froze to death in those icy waters.

Still, loss of life in Fighter Command that half year was nothing like the nearly 500 people killed over a six-month period during 1984–1985 on the highways of Florida by drunk drivers. It was safer to fly a Spitfire over England then than it is to drive to 7-11 for a quart of milk in Miami today. In England in 1939, in Fighter Command, death was not yet an everyday part of life.

It soon would be.

In 1920, two years after her marriage and one year after the birth of her only son, Clarice Dowding died: one day she was a normal, healthy young wife and mother, the next day she was sick, and before her husband could realize how serious it was or adjust his thinking in any way, she was dead.

Somehow it all seemed terribly unreal.

No, it was real enough. He got up in the morning and she wasn't there, he went to bed at night and he was alone; it was real enough. And yet...he couldn't quite come to grips with it. With the fact of her *death*. With the suddenness of it, and with the finality. She was dead, she was gone. He couldn't reach out and touch her, talk to her, let his fingers and hers entwine, *be* with her. She had ceased to exist; how can that be understood?

Nearly twenty years later he was still lying in bed awake at night, trying to understand.

The belief that the dead have entered another level of existence is, of course, without the slightest shred of evidence, but it can easily and logically be argued that this lack is due to the intrinsic nature of the separateness of that existence. Science fiction writers have frequently drawn the analogy between birth and death: in each case an individual is ripped unwillingly from the existence he or she knows to go into separate, undiscovered countries, from which no traveler has ever returned, neither to the womb nor to life. A belief in such an existence, a life after death, is in fact almost mandatory, given the basic assumptions of most religions, and certainly of the Christian religion. The belief in life after death cannot be said to constitute insanity according to the chief criterion by which the charge is usually leveled, loss of touch with present reality, because such a belief in its essence deals with what happens *after* we leave this life, and has nothing to do with the reality we see in it. Nor can the believer be honestly described by that psychological euphemism for the insane, "abnormal," since the belief is more "normal" than not in our society. Many people, most without thinking about it very seriously, accept or assume or believe that a God exists who has plans for us not only in this life but also in the next.

This seems to be particularly true both among people who bear great responsibilities and among those who live dangerously. Stanley Baldwin in 1940 thought that he talked with God; he wrote to Halifax just after the fall of France that he heard the Lord quite clearly say to him, "Have you not thought there is a purpose in stripping you one by one of all the human props on which you depend, that you are being left alone in the world? You have now One upon Whom to lean, and I have chosen you as My instrument to work My will. Why then are you afraid?"

A fighter pilot wrote that he was outside working on his Hurricane when he suddenly grew thirsty. He went inside for a cup of tea, and while he was sipping it he heard a great crash. Running out, he saw that another Hurricane had come in to land trailing his antenna wire—and it had slashed across the Hurricane on which he had been working. If he had remained there another few minutes, he would have been decapitated. He took this incident to mean that Someone had plans for him, and would look after him until those plans were fulfilled.

(Not everyone, of course, looks on things in the same way. Later in the war and on the other side of the world Captain Gilven M. Slonim of the United States Navy left the radio room of the battleship *New Mexico* at the height of a Japanese aerial attack to present a message to Admiral Spruance. He intended to be back at his station in a moment, but the Admiral was in the head, so Slonim waited. While he was waiting, a Japanese bomb hit the ship and broke off one of the smokestacks, which fell directly on the radio room, crushing it. When Spruance finally emerged from his toilet, Captain Slonim presented his message and said, "Admiral, I owe my life to your constipation.")

Throughout the 1920s Dowding remained unconvinced but hopeful that his wife was still "alive" somewhere, still in some form of recognizable existence, still occupying some plane of consciousness. He wondered about the possibility of some connection across whatever plane or dimension might separate such an existence from his own, but he was an austere man with a scientific outlook, and he could not quite bring himself to attend a seance or consult a medium. On the few occasions when he did by chance come upon a medium, or talked to someone who had talked to one, he was convinced that they were fakes. But any phenomenon related to human suffering will spawn inevitably a number of charlatans who prey on the weak and unwary. Can this human frailty be taken as evidence that the phenomenon in question does not exist?

Certainly not. The existence of quack doctors with phony patent medicine cures does not negate the reality of pathogenic viruses and bacteria, though before the present century we could neither see them nor in any way detect their presence.

So perhaps there is life after death? Perhaps in the next century, or the one after that, communication across the Great Divide will be established? Who can say no?

Perhaps it is possible to talk to a wife long departed from this life? That was the question that occupied Dowding's mind during the long, sleepless nights, a question which he kept hidden in the soggy depths of despair and loneliness which sometimes overwhelmed him in the still of the night. He made no attempt to contact Clarice. He did not visit mediums or spiritualists; he held himself aloof and alone.

He simply thought about such things, and read about them. Not infrequently in those years—as today—there were stories in the newspapers about a dead child who had returned to its mother in the lonely night's quiet hours, a dead wife who had come back to her husband, a vision, a thought, a haunt. People had been seeing ghosts in England for many hundreds of years and would continue to do so. Most of these stories were and are obviously laughable, but every once in a while one of them would carry the small, silent germ of—if not truth, at least possibility. These stories Dowding would cut out carefully from the newspapers and magazines. He stored them in a file and kept them locked in his desk; and sometimes, in the chill still of the night, he would seek surcease of sorrow by leafing through them, searching in others' experiences for a hint of his lost Clarice.

June 18, 1940. Winston Churchill to the nation:

What General Weygand called the Battle of France is over. I expect that the Battle of Britain is about to begin. Upon this battle depends the survival of Christian civilisation. Upon it depends our own British life, and the long continuity of our institutions and our Empire. The whole fury and might of the enemy must very soon be turned on us. Hitler knows that he will have to break us in this island or lose the war. If we can stand up to him, all Europe may be free and the life of the world may move forward into broad, sunlit uplands. But if we fail, then the whole world, including the United States, including all that we have known and cared for, will sink into the abyss of a new Dark Age, made more sinister, and perhaps more protracted, by the lights of perverted science. Let us therefore brace ourselves to our duties, and so bear ourselves that, if the British Empire and its Commonwealth last for a thousand years, men will say, "This was their finest hour."

Twenty-nine of the Chain Home radar stations were now complete, and twenty-six of them had been linked by telephone into the command system and were fully operational. The entire coastline facing the enemy now installed in France was covered with their electronic curtain, from Lands End in the west to the Firth of Forth in the east. The IFF (Identification-Friend-or-Foe) apparatus was working and had been installed in virtually all aircraft of Fighter Command, although many bombers and Coastal Command flying boats had not yet been fitted. The first Chain Home Low station was operating, and seven others had been constructed and were being checked out. Height determination was still a problem, and would remain so throughout the battle, but the elimination of rearward echoes had been accomplished. Thousands of operators had been trained, and Dowding's idea of using fighter pilots themselves as controllers seemed to be working out: these men realized the strain of the men in the cockpits, they knew how to talk to them, and the pilots had confidence in the orders they received from such men. Throughout the winter, practice interceptions had taken place on every English bomber or transport or training craft that wandered through south England; without warning they would suddenly find a flight of Spitfires diving on them out of the clouds, roaring past, and disappearing again in a wild zoom.

The German attacks began on convoys coming through the Channel and on port towns. On the first day of July a dozen German bombers were shot down over Grimsby, Brighton, Dover, and the Channel waters. One Blenheim night fighter crashed when the pilot was blinded by searchlights; both crew members were killed. The next day the Luftwaffe lost a half dozen more; there were no British losses. But by the middle of the month, Fighter Command was beginning to suffer: five planes were lost on the fifteenth and eleven on the eighteenth. By the end of July it was beginning to rain burning Spitfires and Hurricanes, blazing Messerschmitts and disintegrating Heinkels; the green fields of England were being carpeted with burnt-out hulks carrying black crosses and bright RAF roundels—and shattered, bloody, and burnt bodies.

Day after day they came during the long, hot summer, from first dawn to last light, and hour by hour the dead and dying bodies tumbled out of the sky to lie crumpled and motionless on the ground. Dowding did not see the dead as Nelson had seen his; Dowding's airmen did not sprawl with shattered limbs around his feet as did the sailors on those ancient men-of-war. Dowding watched the war from the underground filter room at Bentley Priory and saw in the cold reality of day his losses

The Battle

only as numbers posted on the board: 43 Squadron returned to base, three aircraft missing; 85 Squadron back on the ground, two aircraft missing....

He didn't see them, but standing on the balcony of the filter room he heard by radio their voices as they attacked the overwhelming hordes of Huns; he didn't see them, but he heard their voices as they fell screaming to their death. And then at night when the day's fighting was over, when he was back in his bed at Montrose, when he closed his eyes to sleep—that was when he saw the bleeding, burnt bodies; and in the dead of night he heard the screaming again...and again....

"Yellow Two, break! There's a bandit on your tail!"

"Where? Where the fuck is he?"

"Break! Break!"

"Oh Christ! Leader, he's clobbering me! Get this bastard off me!"

"Yellow Leader, bandits at three o'clock high! Coming out of the sun!"

"Break! Everybody break!"

"Christ, I'm on fire! Oh, Jesus, save me!"

"Where's Yellow Two?"

"He's burning, below at five o'clock—"

"Yellow Two, get out of there! You're on fire, bail out!"

"Oh, Christ in heaven, get this son of a bitch off my tail! Somebody help me!"

"Bail out! Bail out, you're on fire!"

"I can't see! Leader! Johnny! Johnny, I'm burning! It's full of smoke in here! My hands—!"

"Tim, get the hell out of there! For Christ sake, bail out!"

"I can't see! I can't get the canopy open, it's stuck. Oh please, Johnny, help me—"

"Yellow Leader, watch out! Bandits coming in—Break!"

"Johnny! My face! Everything's on fire! Mother! Help me—"

And then nothing but screams, high and wailing, shrieks of terror, as they fell the long thousands of feet, spinning and burning and trapped in their falling, tumbling coffins.

Twenty years earlier, when Dowding had commanded a fighter squadron in France, they hadn't had radios. They died silently then. He had heard their screams in his imagination only, and he hadn't been able to imagine anything as horrible as he now knew them to be. Twenty thousand feet was such a long way to fall, spinning like a leaf, burning like a witch, trapped claustrophobically in a tiny, cramped cockpit like a tormented soul in hell.

All his life he had been associated with death, with the dying, and it was too much for a man to take. He spent each day in his office or in the filter room, watching the markers being pushed and pulled across the map table, listening to the radio reports, hearing his chicks die. At night he would drive around to the aerodromes to inspect the state of his defenses. He would try to get home by midnight, by one in the morning, by two. The house would be quiet and dark, the servants asleep up on the third floor. His sister would have been glad to wait up for him, to have a hot cup of tea or a bite ready, but he valued those last moments alone; he needed a few minutes of quiet darkness, to think, to plan, to remember.

The days were bad, and getting worse. Losses were high, and near the breaking point. Aircraft production was good, but he was losing pilots faster than they could be replaced. He remembered Trenchard's demand in the last war for more pilots, always more pilots! He was now in the same position. Should he order the instructors into combat? Cut the training period? Already he had begun scouring the other commands, stealing pilots away from Coastal Command and Bomber Command and even Army Cooperation, hustling them into Spitfires and Hurricanes as soon as he could, sending them up into the overwhelming sky and listening to them call out the "Tallyho!" listening to them scream for mercy, watching them fall back again from that merciless sky.

It was late one such night at the height of the battle when he returned home exhausted as usual. He left his chauffeured car at the door, mounted the two short steps, and took out his large brass key. In the blackout, he fumbled to find the lock, opened the door, and went inside. For a moment he simply leaned his back against the door and breathed deeply in the quiet dark. No lights had been left on, in order that there should be no telltale glimmer when he opened the door, but as his eyes grew accustomed to the dark, he saw a soft, warm glow coming from his den. Walking toward it, he saw that the door had been left barely ajar and there was a small fire burning in the fireplace.

This was unusual for summer, but in fact it was chilly in the dead of night and he was grateful for it. He went in and bent low over the flames, warming the cool skin of his face in the radiant heat. A small stool was kept there, and he pulled it over and sat down wearily on it, leaning forward with his hands clasped over his knees, staring into the flickering flames, feeling his soul drawn out into their mystery, resting and relaxing his tired body. There was so much to be done, and so little time. The radar system was working well as far as it went, but if the Hun should start serious night bombing, they'd be helpless: there were no radar sets small enough to carry aloft in the interceptors, and the ground-based units weren't accurate enough to bring the fighters close enough to see anything in the dark. Even in daylight the situation was serious; he hadn't enough Spitfires or pilots to fly them. That day three separate raids of twenty-plus had been reported, and in each case a single flight of three fighters had been ordered up to intercept them. The boys, his chicks, were magnificent, sailing in to attack against desperate odds, but how long could they prevail under such conditions?

He had no choice but to send up these "penny-packets" of three or four fighters against the German hordes, for in each case there had been further raids within the next half hour; if he committed whole squadrons instead of small flights he would have nothing left with which to intercept the next raids. He sighed and stared into the flames. It was strange, but in that fire he lost the visions of his chicks falling out of the hostile skies enveloped in brighter, darker, hotter fires. He sat and let the flickering forms rest his eyes; he sat and slowly the problems receded. They wouldn't go away, they'd be there again in the morning, but at least he'd have a few moments alone.

It was as he straightened up that he heard a soft movement behind him.

"I'm sorry, sir. I didn't see you come in."

He turned to see a young man in RAF uniform getting awkwardly and stiffly to his feet, standing at attention, obviously flustered. "I must have dozed off, sir. I'm terribly sorry. Have you been here long?"

"Quite all right," Dowding said. "No, I've just come in. Sit down, you look tired." He supposed Hilda, his sister, had let the lad in and then forgotten all about him. He smiled; that would be something to rag her about in the morning. The boy stood until Dowding took his own seat in the overstuffed chair to the right of the fireplace, then he sat down on the sofa.

"I'm sorry to bother you, sir," he began, then faltered. "It's terribly late, isn't it?"

"I suppose it is," Dowding admitted, "but we needn't let that worry us." He waited, but the boy didn't go on. He seemed bewildered, as if he didn't know where he was. He had lost himself in his sleep. Dowding envied him his youth: he wore the wings of a pilot, had probably been fighting that day, and would be off again tomorrow, and yet he could fall asleep on a strange sofa in his commanding officer's house. Dowding thought he would have given a good deal for one night's untroubled sleep, for the sleep of the young.

"Is there something I can do for you?" he asked.

"I—"the boy stopped. "I'm afraid I'm a bit confused," he said. "I don't really know why I'm here."

It was probably a mild case of shell shock, Dowding thought sadly. He would take the boy off operations for a few days, if he could. If he could spare him. Oh, dear Lord. "You must have wanted to see me," he prompted.

"I mean," the boy began again, and this time his voice began to rise in agitation, "I don't know where I am!"

Dowding leaned forward so that his face came out of the darkness, and was illuminated by the firelight. "Do you know who I am?" he asked gently. He was just a bit surprised that he wasn't surprised at all by the boy's puzzlement, nor by the incipient panic in his voice; he just wanted to calm him down, to help him.

The boy leaned forward too, squinting through the dark. "Why, you're Dowding," he said. "Excuse me, sir. Air Marshal Dowding, sir."

"It's quite all right then, isn't it?"

"But what am I doing here? Did you bring me here, sir?"

"Not to worry," Dowding assured him. "You've fallen asleep, you're exhausted. Just relax and it'll all come back. What's the last thing you remember?"

"We were scrambled this afternoon, just after tea. Paddy was leading, I was flying number three on his wing." He broke off, beginning to shake. "What's going on?"

"What do you mean?"

He sat there, his voice trembling. "Fire," he said. "Someone yelled to break, and there was fire everywhere... how did I get out?" He stared at his hands, holding them up shakily in front of his face. "My hands were on fire. Oh, God, my face—"

"It's all right, you don't have to worry," a soft, familiar voice said, and Dowding turned to see a woman in Red Cross uniform coming across the room. He couldn't see her face in the shadows, but he knew that he knew her, and that she would take care of the lad. "We lost you for a moment, but here you are safe and sound," she said. She took the boy's hand, but he pulled it away from her.

"It's burning," he said.

"But you can see it isn't," she replied.

"It *was!*"

"But it isn't now, is it? Everything's quite all right now."

The boy was shaking violently now, and beginning to call out spasmodically, uncontrollably. The Red Cross nurse turned to Dowding for help; as she did, the fire lit her face and he saw that it was his wife, Clarice. He stood up and went to the boy and put his hand on his shoulder and said, "This is my wife. You can go with her, you can trust her. She'll take care of you."

The boy snapped back to attention. "Yes, sir!"

Dowding smiled at him. "That's not just an order, you know. It's a blessing."

The boy hesitated for just one second, then he smiled back, and Dowding could see how he must have looked a few years ago when he was in school, a few weeks ago with his mother, a few hours ago with his mates, before he took off in his Hurricane.

Clarice simply looked for a long moment at Dowding. Then she took the lad by the hand and led him away into the darkness.

Dowding stared after them for some time. Then he put out the fire and, tottering from exhaustion, climbed the narrow wooden stairs to his bed.

This story was told me by someone who heard it from another who claimed to have heard it from Dowding in his later years. It may have been slightly twisted as stories are which are told and retold, but it carries the ring of truth as documented by Dowding in his own later writings on the subject. On that initial appearance, Clarice did not speak to him, but soon she did, and together they were involved in "rescuing" dead pilots. In *Lychgate* he describes the next time they met, at one of the seances to which he now went in search of her:

> I had been told that I might ask any questions; and so I asked about my own people, whether they were well and happy. After my mother had come and told me of my father, L.L. [his "guide"] said: "Here is a lady, very quiet, peaceful and dignified." I said, "Well, that's not my wife anyway; she was always full of laughter and fun and gaiety." Shouts of laughter from Clarice who always enjoyed dressing up and acting. She puts on her natural appearance. Astonishment on the part of L.L. "Why, she has been in my circle for a long time now. I had no idea who she was, or that she had anything to do with you." To Clarice: "What have you been doing in my circle all this time?" Clarice: "Oh, I just came to see if you were a proper person for Hugh to associate with."

He explained later that "Clarice is my wife who is a very active member of our group. She died in 1920...but nothing prevents her from manifesting as opportunity arises." In his book *Dark Star,* Dowding explained that we have, besides our physical bodies, another one: "The physical-etheric double...is an astral body which is not composed of physical matter at all, but of the 6-twist matter of the astral or emotional world." In *God's Magic* he describes why the dead pilots are sometimes confused:

> The etheric double is actually material, though invisible and impalpable. It is composed of a grade of physical matter finer than what we know as gaseous and intermediate between that state and the

ultimate physical atom. In a normal "natural death" the etheric double and higher bodies are slowly withdrawn from the physical (a process which can be actually seen by some clairvoyants), and then the etheric double is in turn discarded. But in cases of sudden or violent death it frequently happens that there is no time to discard the etheric double—the soul, in fact, has never for an instant lost consciousness—and this is the main reason accounting for the fact that those individuals do not know that they are dead.

Clarice and Hugh helped the poor dead pilots to this realization, and sent them off on their journey. Sometimes Dowding didn't remember what happened, and had to be told by Clarice when she visited him later, as in the case of four little boys who had died in a bombing raid, which he recounted in *Lychgate:*

This is from Clarice: "You were with me last night.... We had four little boys who had left their parents behind, and they were a little afraid. We couldn't seem to quieten one little boy. Then the Egyptian friend who has helped me so much said you might try. You stood there, oh, how I smiled, so shy and diffident, wondering what you could do. Then, to our surprise, you changed your etheric robe into an Air Force uniform. We none of us thought you could do that. You are full of surprises. As soon as the child saw your uniform he ran to you and said: 'My daddy is in the Air Force, too.' He was quite at ease after that and went with us quite happily."

At other times Dowding was quite aware of what was going on. The usual purpose was to make the dead realize that they were dead and that death was not the end:

...by trying to shake hands or to slap me on the back they would discover that we were intangible and the little shock of the discovery would bring realisation, but somehow or other they always eventually tumbled to what had happened, and then they could see the friends who had come to meet them and they would all go off happily together....

Now they are beginning to see us. They can't understand the new dimensions. With ten people in it the room ought to be crowded, but it doesn't seem to be. Five of them are sitting on the music seat, meant to accommodate two. Now the leader begins to talk to me. He says, "How is it that you are talking English?"

Self: "Because I am English."

Leader: "But we were shot down over the Ruhr."

Self: "You have been brought here that we may help you."

Leader: "But who are you?"

I go to the mantelpiece, take down a picture of myself in uniform, and hold it beside my face. "Do you know who I am now?"

Yes. They all recognized me now. One says, "I remember you when you came to inspect us at Biggin Hill."

Self: "Very well, then. Shake hands." (I hold out my hand.)

Leader: "I can't get hold of it. Why don't you grip my hand?"

Self: "All right, I will. Watch very carefully." (And I slowly close my hand *through* his without his feeling anything.)

Just then the dark lad comes up behind and gives me a terrific smack on the back. He utters a shrill Cockney yelp as his hand encounters no resistance.

Leader: "Look here, sir, are you trying to tell us that we are—that this is death?"

James explains that they were all blown to pieces instantaneously in their aeroplane three or four days previously, over the Ruhr. Now they can see all their other RAF friends who have come to meet them and they all go off together.

These instances Dowding wrote about did not merely illustrate a belief in life after death; they were in every sense visual and audial hallucinations over which he had no control, so real that he believed them completely and without reservation. They constituted a total break with reality, and a man who cannot see the silvery line between his fantasies and the real world, who has lost touch with reality, is insane in every possible sense of that much misused word.

These fantasies began during the Battle of Britain: "It is only by personal experience that complete conviction [of life after death] is possible. I had this personal experience in the Battle of Britain," Dowding wrote. They soon began to take other forms. One of his friends has said that one day Dowding claimed "to have discovered the secret of perpetual motion." When his friend replied somewhat skeptically, Dowding answered with "the utmost seriousness." Sir John Colville, Churchill's private secretary, notes that during the Battle of Britain, Dowding and a few other air commanders

> ...had their headquarters within easy range of Chequers and were regular weekend guests for dinner.... One night at Chequers [Dowd-

ing] told us that the battle was going well. The only things that worried him were his dreams. The previous night he had dreamed there was only one man in England capable of operating a Bofors gun and his name was William Shakespeare. We supposed we were intended to laugh, but I looked at Dowding's face and was sure he was speaking in deadly earnest.

The problem soon became rampant, extending not only to semireligious beliefs ("I say with absolute conviction that, but for God's intervention, the Battle of Britain would have been lost") but extending also into the scientific realm and the existence of mystical forces reminiscent of the death ray, although by now he was so sick of death and killing that the ray took the form of a healing force. He advised friends, and readers of the books he wrote on the subject, to take a gold ring and fasten it to a few inches of dental floss, and then to tie a string around an electric wallplug. The other end of the string was to be tied around the ankle, and the gold ring hung by the dental floss as a pendulum over the inert body for half an hour. The ring would supposedly begin to swing because "there is SOMETHING in the string... which is not an electric current nor electromagnetism... but which acts as a magnetic healing agent for certain diseases. Try it on a gouty knuckle."

He never lost his sense of humor. He liked to tell the story—as it was told him by a friend who overheard them—of two Americans in the audience at a spiritualism lecture given years later when he had become Lord Dowding. The speaker mentioned the books Lord Dowding had written about meeting with his dead wife and talking to dead fliers, and one of the Americans said to his neighbor: "Say, who is this Lord Dowding? Is he any relation to Sir Hugh Dowding who fought the Battle of Britain?" His companion replied, "Why, he's his father, I guess. You see, in England the son of a Lord is a Sir." The first man accepted this item of heraldic information without demur. Then he said, "Gee, I wonder what a smart guy like him would think of his old man going off the rails that way."

He knew what people thought of him, but he went his own way as always without regard to what others thought. Because he *knew*. ("The facts which I do know, I know with that complete certainty of personal conviction which nothing can shake.") He visited with people from Atlantis, who told him that their land was "the home of the Fourth Rootrace in which humanity reached a high degree of material civilization. But the priesthood used Black Magic and the evil caused volcanic catastrophes in 10,000 B.C. The survivors fled to Egypt and built the pyramids, which was a normal Atlantean structure." He saw fairies, which "are formless little globs of light, with the power to clothe their thought forms in etheric material (and therefore look exactly like our conceptions of them)." He *knew* these things to be true, because he saw them all.

If this degradation of his mind had occurred just a bit sooner, when he was the fountainhead of the effort to convince the RAF that the mysterious, invisible radar would prove to be their salvation against Germany, all would have been lost—for surely the radar would have been thrown out along with the etheric bodies and the pseudomagnetic healing force, amid mutters of "Poor old Stuffy, 'round the bend he's gone." England would have been stuck with the philosophy that the bomber would always get through, in which case it surely would have—for no defenses would have been prepared to stop it.

Instead, by the grace of God ("I say with absolute conviction that, but for God's intervention,...."), his mind lasted through the crucial stages *before* the war when radar was planned and worked out and installed in the RAF defense system. Dowding's genius as an officer, in fact, lay in the methodical way in which he perfected a system that worked without him: during the battle itself he was almost more of an onlooker than a participant. He had put the system together, he had trained his personnel, and now—despite the fact that the man in charge finally had gone absolutely bloody bonkers—all the components fit together and worked in unison, as more than one observer has remarked, "like a perfectly made clock."

Well, not exactly, I'm afraid. That last statement is true in a technical sense—the radar itself worked, and the control system worked—but there was also a human element involved, and the vagaries of that element nearly brought the whole system crashing down around Dowding's bewildered head while he stood above it all looking out over the map table clustered with markers of enemy raids, hearing the distant sound of trumpets amid the bursting bombs, seeing instead of hidden betrayal and conflict his own visions of life everlasting.

Chapter Eleven
The Human Element

> For I have sworn thee fair
> and thought thee bright,
> Who art as black as hell,
> as dark as night.
> —*William Shakespeare,*
> *Sonnet 147*

December 14, 1931.

Flying Officer Douglas Bader, upon graduation from Cranwell just one year before at the age of twenty, had been assigned as fighter pilot to 23 Squadron, Kenley. Like Dowding fifteen years earlier, Bader was a skillful pilot, a rugged sportsman (though he delighted in rugby football rather than skiing), and a nondrinker. He did not have Dowding's brain; instead he had charisma. Possession of the one without the other would one day prove troublesome.

Kenley was one of the classic prewar aerodromes: South of London, near Croydon, it was hidden pleasantly among the thick trees of the Surrey hills. Today it is a wasteland surrounded by suburbia, its runways slowly being encroached by the tall, waving grass, its red brick buildings untended and crumbling. But in the 1930s these buildings were austere yet comfortable, solidly built and adorned with perfectly tended gardens of roses and geraniums and thick, well-cut grass. The life of a pilot in the RAF in those halcyon days was that of a gentleman and sportsman: tea was brought with the wakeup call at 0700 every morning by one's personal orderly and left on a wooden chair brought over to the side of the bed. One's uniform had been pressed and one's shoes already shined and laid out. After descending the stairs to the dining room and eating a traditional English breakfast of porridge, fruit, kippers or eggs with fried bread and tomatoes and sausage and kidneys, washed down with mugs of hot coffee, one would head out to the flight line and begin the day's work, which was flying fighters; in 1931, that meant the Gloster Gamecock, an open-cockpit biplane with a huge, noisy engine out front, capable of 150 miles an hour and incredibly acrobatic.

Acrobatics, in fact, were what fighter pilots were made for, and what Bader was born for. 23 Squadron had won the RAF acrobatic displays the previous two years at the Hendon air shows, and in 1931—his first year with 23—he was picked as one of the two pilots from the squadron to compete again.

The Hendon air shows were the nation's outstanding annual aerial events in those prewar years. Hendon aerodrome was just north of London, on the site where the RAF museum stands today. In 1931 more than 175,000 people were sitting in the grandstands around the King and Queen or sprawled on beach chairs or blankets spread over the grass as Bader and his partner, with their aircraft actually tied together, performed a series of loops and rolls wound into a thrilling, intricate dance of aircraft from grass-top height up to a few thousand feet and then over the top and down again, whirling tightly around just in front of the Grandstand, coming down finally to land, perfectly synchronized, as the 1931 champions.

That had been in June. Now, on December 14, Bader spent the morning practicing acrobatics in the squadron's new fighter, the Bristol Bulldog. This was similar to the Gamecock, and in fact much more similar to the World War I fighters than to the Spitfires that were still not yet dreamed about; open cockpit, two wings connected by struts, a 500-horsepower engine roaring away and two machine guns, one on each side of the cockpit, fixed to fire through the propeller arc. Perhaps thirty miles an hour faster than the Gamecock, it was also sturdier and heavier; not quite as good for acrobatics, presumably it would be better for fighting.

After the morning's practice, the afternoons were free for sports, but that day there was no game organized. One of Bader's mates said he was flying off to an aerodrome near Reading to visit his brother's flying club for lunch; why didn't Bader come along?

Why not? It was the sort of thing one did in the RAF those days, so both he and another squadron member agreed to go. The three of them took off and flew their Bulldogs southeast to the Woodley aerodrome. As they came across it they put their noses down and roared over at ground level, "beating up" the grass and with their roaring engines pulling everyone out of the hangar building to watch. Then they circled, settled into a landing pattern, and set down as perfectly as befit members of the best fighter squadron in the RAF.

After lunch they were asked to perform. As the RAF champ, Bader was known to everyone who flew, and he never refused a request. He led the two others off in tight formation, held them low over the grass as they took off until their speed built up sufficiently, then pulled the nose up and brought them around for a low-level pass. At about the height of a tall man and at nearly 150 miles an hour, he turned upside down.

Now an airplane cannot really fly upside down. The way an airplane flies at all is to have a wing roughly flat on the underside and curved on the top, so that for the air rushing over it the path is longer than for the air that passes under. Normal, quiet, unmoving air is composed of molecules of oxygen and nitrogen that are always in random motion, and the repeated knocking of these moving molecules against any surface imparts a steady pressure, which is by no means negligible: it amounts to roughly fourteen and a half pounds every square inch. But when the air is moving, it has to spread itself out, its molecules are less densely congregated, and therefore the pressure the gas exerts is also spread out, so that every square inch of surface exposed to it feels a pressure *less* than fourteen and a half pounds. And since the air moving over the wing is moving faster than the air passing under it, there is less pressure pushing *down* on the wing from above than pushing *up* on it from below; the net effect is a positive pressure pushing up, proportional in its strength to the speed of the airplane through the air. If the wing is large enough and the speed is fast enough, this differential pressure is sufficient to hold the airplane up. (It is when the speed drops too low that the airplane stalls and falls out of the sky; this is done purposely to land a plane when the wheels are just a few inches above the ground.)

Obviously then, when an airplane is upside down the wings should work in the opposite direction, pushing the craft down into the ground rather than up into the sky. The trick to keeping an upside-down airplane flying is to have a lot of speed and to push the nose up away from the ground as you turn over on your back, so that the wing hits the onrushing air at an angle and the airplane literally bounces off this continual wall of air—it is not so much flying in the normal sense as skipping, the way a flat stone skips off the surface of water when thrown just right.

Obviously it's a very tricky maneuver. As the plane rolls over from its normal upright position, the nose must be gradually and continually adjusted to make just the right angle of attack against the air. As the plane turns onto its side, with its wings vertical to the ground, the nose has to be lifted by the rudder controls; as the plane continues onto its back, the adjustment passes to the ailerons. When you come diving onto an aerodrome and skim over the grass with your wheels practically skimming the blades, there is no margin for error. That's what makes it so exciting. That's why Bader loved to do it, and why everyone loved to watch it; that's why he was champion.

He came roaring in over the grass and began the roll, dropping his right pair of wings toward the ground. The wingtips skimmed over the grass and continued around, and there he was flying upside down, nose pushed high, the straps of his helmet hanging down and nearly bouncing on the ground as he skimmed past. Now the wing continued to revolve and he began to turn right side up, control of the uplifted nose passed from the ailerons to the rudder—

And he blew it. For one microsecond the rudder pressure wasn't quite enough, the nose dipped just slightly, so slightly you couldn't even tell from looking, but in the cockpit he felt it. Too late. He was only inches above the ground, and by the time he felt the nose begin to slip, the wingtips were already clipping the ground and immediately the whole craft was pulled out of its graceful flight and slammed into the speeding grass. It slid through the scorching tumbling grass and dirt and then suddenly the roar of the engine was gone and the speeding flash of godlike vision was gone and there was nothing but the distant sound of the other two airplanes pulling up and away in horror. On the ground was a cloud of smoke and dust, and hidden in that black eruption was the crumpled wreck of an airplane, and the crumpled wreck of a once-champion pilot.

He came to consciousness in a hospital bed, hearing a clatter of voices outside in the hall. As he came awake he heard soft footsteps hurrying across the linoleum and a nurse's voice hushing the people out there: "Please be quiet, can't you? There's a boy dying in there." As he faded back into sleep his anger rose and took over: he would *not* die.

They cut his legs off, but he lived. Both his legs gone, one just above the knee, one below, but he lived. They gave him a 100 percent disability pension of 200 pounds a year, and they removed him from the RAF roster. They fit him with mechanical legs and told him he might one day be able to walk with crutches. His anger broke out again, and he said he'd *never* use crutches, never use a cane. The doctor shook his head, but in the end Bader hobbled away from the hospital—halting, shuffling, but upright and without crutches or a cane.

He never played rugger again, but he played golf. And he flew. The first thing he did on leaving the hospital was to bring his log book up to date: "14.12.31. Bulldog K.1676. Pilot: Self. 1 hour 5 minutes. Cross-country, Reading. Crashed slow-rolling. Bad Show."

He taught himself to work the rudder pedals with his artificial stumps, and he flew. He applied for readmission to the RAF, but they shook their heads and said no. He applied again and again; he pestered them and wrote them letters, and they said no. He flew with his old friends from 23 Squadron, and they wrote letters saying that he could fly as well as they; he waited around offices, he badgered people on their way home and on their way to work, and finally the director of personnel of the RAF told him that if by chance there should ever be another war they'd find some way to use him.

That was on September 1, 1939.

A few days earlier Dowding had agreed to make a BBC radio broadcast to calm the fears of the jittery population. He gave his word that Fighter Command would be able to guarantee the safety of the skies; the bomber would not get through.

That promise was made on the basis of the radar defense screen which he thought was operational, of an enemy Luftwaffe whose airfields all lay to the east of England and beyond fighter range (which meant that their bombers would have to come alone and unescorted), and of a minimum Fighter Command strength of fifty-two operational squadrons of Spitfires and Hurricanes.

On the outbreak of war, at the battle of Barking Creek, the radar defense was proved to be not yet 100 percent efficient, but luckily there were no large-scale German attacks for nearly a year; the system's deficiencies were remedied in time. But when the Luftwaffe came, it came from airfields not 200 miles away in Germany but only thirty miles away in France: England was nearly ringed with them, from the northeast to nearly the southwest, and they were close enough for the deadly Messerschmitt 109 fighters to come along too, escorting and shepherding the bombers through the defending fighters to their targets. And the fifty-two squadrons never had existed: That number had indeed been reached at the outbreak of war, but included among its aircraft Blenheims, Gladiators, and Defiants. The Blenheim was a twin-engined bomber converted into a fighter by fitting a "blister" of four machine guns under its nose and assigning it to fighter squadrons; but calling something a fighter does not make it one, and the Blenheim was helpless against the Messerschmitts. The Gladiator was a biplane, an anachronism which didn't belong in the same skies with the Luftwaffe; although it created a legend over Malta later in the war, the legend was exactly that: it could no more fight the Messerschmitts than the Spitfire could match today's Migs. The Defiant was another anachronism, a single-engined monoplane fighter with no forward-firing machine guns but instead a power-operated turret bearing four guns. It was designed to fly alongside an enemy bomber and turn its turret guns around to blast it; it would have worked if there were no Messerschmitts around. But there *were* Messerschmitts around, and it did not work.

The Blenheims were restricted to convoy and night patrol duties and were replaced as soon as possible. The Gladiators were destroyed in Norway and never fought in the Battle of Britain. The Defiants had one day's success over Dunkirk when the Messerschmitts mistook them for a squadron of Hurricanes, dove on them from the rear, and ran right into the bullets from their turrets. But the Me 109s learned fast; in the next battle five of six Defiants were destroyed, and by the middle of the Battle of Britain they were no more.

In France, unprotected by a radar screen, the RAF had lost the staggering total of 959 aircraft, plus sixty-six more in Norway. Over 500 of that total were fighters. In July 1940, when the Battle of Britain began, Dowding had twenty-nine squadrons of Hurricanes left, plus nineteen squadrons of Spitfires. But two of the Hurricane squadrons had been virtually destroyed in France and Norway, while six others had been so mauled that they had to be reconstituted and rested and were not available as fighting units. Eleven of the Spitfire squadrons had been equipped with their fighters so recently that the pilots didn't yet know how to fly them in combat; it would take weeks or months to get them ready to fight. There were actually just twenty-three operational Hurricane and Spitfire squadrons instead of the fifty-two agreed on as a *minimum;* the twenty-three included about 500 fighters with which to face the German hordes of bombers more than 1500 strong, which could now come at them from the north, the east, and the south, escorted by more than a thousand sleek, deadly Messerschmitts.

Fighter Command was centered at Bentley Priory, from which its spiderweb spread out into four sectors. 13 Group guarded the far north and Scotland, under the command of Air Vice Marshal Richard Saul. This was the least likely area for aerial attack, both because of the lack of suitable targets and because of the flying distance from the French and German airfields. Mostly the squadrons of this group would be training, resting, and regrouping. South of them was the territory of Air Chief Marshal Sir Trafford Leigh-Mallory's 12 Group, covering Yorkshire and extending southwards roughly to Cheltenham in the west and Norwich in the east. 11 Group, commanded by Keith Park, would cover the "hot corner," the southeast of England which faced France across the Channel. Finally, to the west, 10 Group under Air Vice Marshal Sir Christopher Q. Brand would guard Park's flank.

Keith Park was a lean, tough, slim New Zealander who had been associated with Dowding since the First World War. He had come from his native island to fight with the ground forces at Gallipoli and the Somme, where he was wounded and hospitalized. His wounds were severe enough to prevent his acceptance into the Royal Flying Corps, so he managed to arrange to have his medical records "lost"; he became a fighter pilot and shot down twenty planes, ending up as New Zealand's top ace. After the war he stayed on in England, attending the staff college at Cranwell and then commanding a fighter station. When Dowding took over Fighter Command, Park was assigned as his Senior Air Staff Officer.

Trafford Leigh-Mallory was also a World War I pilot, but in Army Cooperation rather than fighters. Some of the more undesirable of his later characteristics have been ascribed to an inferiority complex deriving from this circumstance. He was an industrious, ambitious, and intelligent man, an honors-degree history scholar from Cambridge and an Olympic-class athlete; he and his even more famous mountain-climbing brother made a handsome splash among the smart set during the years between the wars. As he grew older, he put on a bit of weight, though never quite to the point of fat. His mustache, like his body and his mind, was heavier and thicker than Park's, but trimmed as carefully and neatly. In 1937 he was made commander of 12 Group, which at that time, facing Germany across the North Sea, occupied the hot spot. In 1940 the then-commander of 11 Group, Air Vice Marshal Ernest Leslie Gossage, was promoted to the Air Council. By then it was apparent that this Group, facing France across the Channel, would be the pivotal formation of Fighter Command. Leigh-Mallory consequently wanted and expected to be transferred to take over. Instead the post went to Park, and Leigh-Mallory remained where he was throughout the Battle of Britain. It would be nice to record that though in his darkest moments he sulked at what he considered an insult, he remained a loyal subordinate commander to his chief; it would be nice to say this, but it is impossible to do so. Given his ambitious insubordination (even treachery is not too strong a word), it would be nice to record his administrative downfall; again it is impossible to do so.

Keith Park, on the other hand, who pleased both his superiors and his subordinates, who was as popular among his own pilots as he was with his chief, Dowding, and who directed No. 11 Group with consummate perfection during the most difficult fighting of the war, was immediately fired when Dowding was ousted and Leigh-Mallory rose to power later in the year.

March, 1940. Dusk is falling over Duxford aerodrome, one of the main stations of No. 12 Group, a few miles south of Cambridge. Most of the pilots of 19 Squadron have already repaired to the Red Lion pub across the road. The hectic pace of the day's preparations for the coming battle is slowing down with the approach of night; the clanking sounds from the machine shops are becoming more sporadic, as is the spanking tread of marching feet. The dusk of an English spring is heavy, almost palpable, somehow nostalgic, bittersweet and sad, comforting and just faintly lonesome for young men barely out of their teens, away

The Battle

from their families for the first time in their lives. The Red Lion will do a brisk business tonight.

High over the nearly deserted airfield hums the smooth, deep rumble of a Merlin engine as one last Spitfire comes in from the day's flying. It approaches from the east, high in the sky, silhouetted darkly against the grey-bluish fading light. It begins to settle into the proper landing pattern, but then at the last moment the pilot seems to change his mind: its wings tip over and instead of entering the gentle gliding turn, the plane slips around, points its nose straight down, and comes roaring perpendicularly in a power dive straight for the edge of the field.

The station commander, "Pingo" Lester, is standing outside the control tower, enjoying the last peaceful moments of one of the last peaceful days left to this island. Now he tenses, leans forward, staring.

The Spit comes zooming down at well over 300 miles an hour, the thunder of its powerful Merlin bouncing off the buildings and shaking them, shaking the very ground. Down and down it comes, and then it levels off at the very last moment. Its propeller whipping the grass to froth, it comes charging like an ancient bat from the Stygian caves across the 'drome—

And then, as it skims past the bulging eyes of the station commander, it laconically rolls over, turns upside down, its nose held high just perfectly, exactly right, and with its jutting tail parting the grass inches above the deadly ground it sails across the 'drome and rolls upright again and zooms up in a power climb to the landing pattern, lowers its screaming roar to a muted rumble, and settles down and onto the grass in a perfect three-point landing.

The Spitfire taxies to its revetment; the engine is cut off and dies. As the pilot climbs heavily out of the cockpit, stumbling a bit to the ground with his parachute, the engine ticks away quietly, cooling down, settling itself for the night. The pilot comes limping jauntily across the tarmac, pulling off his flying helmet, waving a casual greeting to the station commander.

"Oh, Douglas," Pingo Lester sighs, "I *do* wish you wouldn't do that. You had *such* a nasty accident last time."

Across the Channel, Field Marshal Hermann Goering was purring like a fat cat as he stretched out his claws. He had been a bit worried just after Dunkirk, when it seemed that Hitler might turn on him. He had overstretched himself a bit there, promising to destroy from the air the British forces pent up on the beach, urging Hitler to hold back the panzers and let the Luftwaffe finish the job. It was an argument that pleased Hitler. He had continually to struggle with the old-line Wehrmacht generals to maintain his supremacy, but the Luftwaffe had been born under his overlordship and was his alone, as was Goering. He had smiled and nodded and turned the pocket of British soldiers over to the Luftwaffe for destruction.

But then out of every corner and crevice of England where the river waters leaked out to the sea came crawling an amazing armada: sailboats and luxury yachts, fishing smacks and trawlers, manned by professionals and amateurs, herded by destroyers and guarded by the once jealously horded Spitfires and the untested Defiants and the Hurricanes and the Blenheims, and even the American Lockheed Hudsons, a commercial craft transmogrified into a bomber. This ill-assorted medley of anything that would float came sailing across the Channel to pick up their soldiers and bring them home again, while up above the clouds everything the RAF could swing into the air guarded the beaches against the screaming, diving Stukas.

On the beaches the huddled masses of soldiers saw nothing of that hidden aerial battle. They saw only the Stukas and Heinkels that broke through the thin blue line above, and as they dug deeper into their shallow holes on the unprotected dunes of France, they cursed the bloody RAF for deserting them. But they came home; in overwhelming numbers they came home. As the massed Luftwaffe broke through only in isolated holes, the constant stream of navy ships and merchantmen and private boats brought them safely across the Channel to Ramsgate, to Dover, to all the Cinque Ports and beaches in between.

Winston Churchill, June 4, 1940:

> This was a great trial of strength between the British and German Air Forces. Can you conceive a greater objective for the Germans in the air than to make evacuation from these beaches impossible, and to sink all these ships which were displayed, almost to the extent of thousands? Could there have been an objective of greater military importance and significance for the whole purpose of the war than this? They tried hard, and they were beaten back; they were frustrated in their task. We got the Army away....

Goering very nearly was angry. But, helped by his intelligence officers, he kept his temper. The British Army had escaped, they admitted, but the Royal Air Force had not. They produced the numbers he wanted, the victory he demanded; and when they were finished he triumphantly announced to his Fuehrer that the Royal Air Force was decimated, was down to its last few Spitfires, and the British Isles lay open to conquest from the air.

This was one of the great differences between Goering and Dowding; the one believed everything his intelligence men told him—and made sure they knew in advance what he wanted to hear—while the other believed only what he saw with his own eyes. Dowding well knew from his own experiences in the previous war how three fighter pilots could dive on an enemy, see invisible bullets tear it to shreds, and each of them claim it as destroyed. Even if it *were* destroyed (and airplane after airplane came limping home with fuselage and wings and engines shot nearly to pieces), it was one plane destroyed, not three. He accepted as destroyed only those planes which he saw lying crumpled on the ground, and so throughout the battle he kept an honest and true evaluation of the situation.

Goering, even more than Dowding, was an experienced flyer; in the First World War he had been one of Germany's leading aces and had ended up commanding the famed Richthofen Squadron after the Red Baron was killed. But he was so proud of his Luftwaffe, so eager for success, that he encouraged reports optimistic to the point of insanity, and he believed them all.

He told Hitler that England was defenseless now, that if the British had any sense at all they would surrender—that if they did not he would destroy them from the air and would escort the Wehrmacht across the Channel to occupy their island. Hitler nodded in agreement. His war against Britain had been over the moment France surrendered, for alone Britain was helpless. Her world policy had been based on standing behind the great French land army while controlling the seas with her Royal Navy. Now that the French Army was no more, and the navy was afraid to move out of Scapa Flow because of the Luftwaffe bombers, the English were finished. He broadcast his willingness to make generous peace terms; he was ready to turn his attention eastwards to their common foe, the Slavic hordes of communism.

Winston Churchill, June 17, 1940:

> What has happened in France makes no difference to our actions and purpose. We have become the sole champions now in arms to defend the world cause. We shall do our best to be worthy of this high honour. We shall defend our island home, and with the British Empire we shall fight on unconquerable until the curse of Hitler is lifted from the brows of mankind....

From the German embassy in Washington came secret feelers; the British ambassador there reported to London that Hitler's peace offer was "unusually generous," that Hitler did not want the defeat of England. On the contrary, he was more than willing that Britain retain her position in world affairs. Virtually his only condition was that she should relinquish her concern in continental European affairs. On the continent Germany was master, elsewhere Germany and England together would reign. England would, of course, retain total independence.

Winston Churchill, June 18, 1940:

When final victory rewards our toils [the French and all the conquered people of Europe] shall share the gains—aye, and freedom shall be restored to *all*. We abate nothing of our just demands; not one jot or tittle do we recede.... Czechs, Poles, Norwegians, Dutch, Belgians have joined their causes to our own.... All these shall be restored.

Hitler turned angrily to Goering, and set the Luftwaffe loose.

In the last weeks of June they moved into position. *Luftflotte 2,* under the command of Albert Kesselring—who with the Stukas of *Luftflotte 1* had led the blitzkrieg over Poland and who then with his new command had spearheaded the conquest of France—took over airfields in Holland, Belgium, and northeast France. Hugo Sperrle, who had led the Legion Condor in Spain in 1936, moved his *Luftflotte 3* into northwest France. *Luftflotte 5,* commanded by Hans-Jürgen Stumpff, occupied the Scandinavian airfields.

Hitler was disappointed that the English had not yet replied officially to his surrender offers; their only answer had been Churchill's belligerent speeches in the House. Goering convinced him that the English could be guided to the conference table by a taste of what his Luftwaffe could give them, and so the muzzles on the Luftwaffe were loosened just a bit.

The Ultra radio interception and decoding unit at Bletchley Park picked up a signal appointing Kommodore Johannes Fink, leader of *Kampfgeschwader 2* of the *II Fliegerkorps,* as *Kanalkampführer*. This stumped them for a moment, since *Kanal* translates most usually as "sewer"; were they planning to attack through the London sewers? But it stumped them only for a moment, as they realized it can also translate as "canal" or "channel"; not through the sewers but through the English Channel was Fink going to strike, hitting convoys and towns along the coast as leader of the Channel battle. He was given two Stuka groups and two Me 109 fighter squadrons in addition to the Dornier 17 bombers that were normally part of his command—for a total of nearly 150 bombers and 200 fighters.

In England the war minister, Anthony Eden, had announced a call for civilians to form into Local Defence Volunteers. Churchill quickly changed their name to the Home Guard, and by the droves they swarmed out of villages and farms and cities to sign up. They were old soldiers from the last war, engineers or doctors or anyone else in restricted occupations not subject to the draft, the maimed and the crippled and the ill. They learned to drill and obey orders, they marched with broomsticks over their shoulders because there weren't any rifles—they had all been left behind rusting in the surf at Dunkirk. They shoved World War I bayonets into long pipes and practiced using them as pikes, or they brought along their own shotguns and hunting rifles and air pistols. They dug ditches across their farms and dragged felled trees into their pastures so that German gliders would have no level place to land. They took down every road sign from every highway and village lane, so that the invaders that were daily expected might lose themselves in the twisting convolutions of the English countryside; they took down every place name from every railroad station and from every pub in England, and woe to the stranger who had to ask where he was. More than one person was escorted to the local constabulary at the point of a pitchfork.

The church bells were silenced; they would ring again only to announce the start of the invasion, to call the populace to arms. The beaches were mined and crisscrossed with barbed wire, and a battalion of Royal Engineers prepared to set the Channel on fire by pumping oil into the waters and lighting it when it floated to the surface—one of Churchill's favorite ideas. The "syphilitic old bastard," as Hitler called him, wasn't fooling: they *would* fight on the beaches and in the streets and in the hills and in the woods.

The hell they would, Hitler cursed. Send over the Luftwaffe, he told *der Dicke,* the Fat One. Give them a taste of it.

Across the Channel Fighter Command was waiting, untried, uncertain, and unsure. The performance of the individual Chain Home stations depended to a discouragingly large extent on the nature of their positions. Height determination relied on radar echoes bouncing off the ground in front of each tower and reaching two separate antennae placed at different heights, the relative strength of the two signals indicating angle of elevation of the target. But for this to work properly the towers needed a stretch of flat land in front of them, and such land is not easy to come by along the coast of England. Although the sites had been chosen with this requirement in mind, most of the towers sat behind intermittent obstructions such as bullocks, hills, forests, or scarps. The accuracy of their height determinations would vary wildly as each of them looked in one direction or another. Two towers sat on the edge of cliffs high over the sea, and could not report heights at all.

Their performances depended also on the quality of their personnel, most of whom had been trained only hurriedly and incompletely, and on the quality of installation and maintenance. But it was wartime, and every technically experienced man in the country was wanted by a dozen different services and companies; the radar equipment was "often badly installed and inadequately maintained." With the prototype equipment at Bawdsey, Watson Watt had claimed an accuracy of one or two degrees in direction, but at the operational stations test flights indicated errors ranging up to ten or even fifteen degrees, while occasional mistakes of even thirty degrees "were not unknown."

These problems had been forseen, however. Perhaps Watson Watt's most important contribution to radar was his insistent philosophy of pushing ahead to develop a working set as quickly as possible rather than taking time to produce the perfect set—which might be ready too late. What made his philosophy work was a sharp awareness of the limitations and problems of the equipment they were producing, and of the reporting and analytical organization that would be necessary to turn the raw, error-laden data into useful information.

They developed what might be called the world's first computer, although it was really more of a complicated calculating machine. They called it the "fruit machine," the British name for what we call a slot machine. The radar operator would measure the range and direction of a target from the blip on her radar scope, and enter this data into the fruit machine, which would then automatically apply the corrections that had been painstakingly devised for each individual station from dozens of test flights, obtaining an estimate of the target's true range and direction. The corrected range would be automatically multiplied by the sine and cosine of the corrected direction vector, the map coordinates of the receiving tower would be read in, and within a second or two the map coordinates of the target would come out. The radar operator would then switch connections from direction finding to height finding, and the process would be repeated. In another second the altitude of the blip would be produced.

These data, from each CH station, were sent to the Filter Room at Fighter Command Headquarters, Bentley Priory. At this point the problems and inaccuracies associated with each individual station were known and the resultant data were evaluated, or "filtered," and merged into final plots. Several yards underground, a large room had been excavated, in the center of which was set a table on which was laid out a map of England. A horde of plotters, mostly women, fitted with telephone headsets connecting them each to a particular radar station, surrounded the table, waiting for information to come in. As each sighting was reported, the WAAF would place a marker on the table in the proper position. Since the radar coverage of the stations overlapped to provide Watson Watt's impenetrable radio curtain, many of the sightings were multiply reported—and since each station had its own problems and inaccuracies, these plots were often not situated at precisely the same point. It was the job of a filtering officer to study the plots as they were placed on the table and to decide if two close markers represented the same raid as seen by two stations, or two separate raids. As he did so he would replace the markers with others carrying his own best estimate of friend or foe, altitude, speed, and size of the formation.

On a balcony which ran three-quarters of the way around the room sat another echelon of officers looking down on the plots being moved across the map of England and on a vertical board lining the opposite wall, on which was listed the state of readiness of each available fighter squadron. These officers then decided which raids were to be followed by which radar stations, a decision based on marrying the known qualities (good and bad) of each station with the characteristics of each raid, such as height and size.

At the same time the established plots were telephoned to the operations room, where the radar data were combined with sightings phoned in by the Observer Corps. Basically the radar looked out to sea and was blind inland, while the ground-based observers could see the raids only after they crossed the coast. Where both sightings overlapped, final corrections to positions were made. These combined plots were then sent to similar operations rooms at each group headquarters, which would follow only those plots within and along its borders and which would direct appropriate responses, and from there to sector operations rooms which would translate the group's orders into individual scrambles.

It was a brilliantly devised, tightly organized system—and inevitably it was a shambles. If each piece of information coming from the CH stations had been perfect, the complex process of interweaving all the data—each piece of which changed every second as the incoming bombers changed direction and altitude precisely in order to confuse whoever might be watching them—would have rendered the whole process difficult. With all the experimental errors compounding the difficulties, it was a maddening, bewildering complexity of split-second, life-and-death decisions made irrevocably by officers who were still learning the business. To the technical problems, malevolently grinning gods added equally frustrating nontechnical difficulties: for example, there was not enough room around the table for all the headsetted plotters, who tangled wires as they bumped into each other and who might upset one plot as they leaned across the table to place another. (One wise but unknown genius suggested buying a set of croupier's rakes from the London gambling establishments, an arrangement which obviated this last difficulty.) New problems were discovered daily; many of them were conquered, many were not. In the beginning, the total lack of experience led to many errors; by the end, overwhelming numbers of enemy formations swamped the system so that it had to close down and simply scramble every available fighter with orders to attack anything they saw.

It was a complicated, quietly frantic, muddling-through type of operation. It was a shambles. But God help us, it had to work.

"I pray for radar," Dowding said. "And I trust in God."

Dawn. Monday, 1 July. 0600 hours. Blue Section of 72 Squadron, 12 Group, based at Acklington, just north of the Tyne River on the eastern coast, is scrambled. Bogey on the radar scope, position near a convoy eight miles off the coast at Sunderland, heading south. Reaching the convoy, the fighters circle and spot a black shape moving low over the water; diving down they recognize a Heinkel 59 floatplane. This is not a combat airplane, it is an old two-winger, painted white and showing Red Cross markings.

"Line astern," Flight Lieutenant Graham calls on the intercom. "Fighting Attack number one...go!"

One after the other, they dive and riddle the Heinkel. It crash-lands on the water and sinks. Its crew is rescued by a convoy escort vessel. There is no doctor or other medical personnel on board, and why should there be? There was no possible Red Cross mission it could have been on. It had been sent as a reconnaissance plane to locate the convoy.

Later that morning over the Channel there are numerous bogeys reported, but they turn out to be RAF bombers or flying boats; the IFF has not yet been installed on these. Just after noon another blip is picked up and three Hurricanes are scrambled. As they are in the air and on their way, convoy Jumbo radios that she is under attack. By the time the Hurricanes reach the area, the Stukas have completed their attack and are gone.

The towns of Wick and Hull are bombed. The fighters arrive too late to intercept the attackers.

That night Dowding confers with Park, and the next morning at dawn several flights of fighters are moved to coastal satellite airfields. These are fields which are either too small or incomplete to be suitable as permanent squadron headquarters; but fighters can be landed and scrambled from them, and they are close to the coast and the fighting. Each morning assigned flights set down on them and wait for orders; each evening they return to their home bases.

From the Ultra intercepts of Luftwaffe radio signals and from his own understanding of the situation, Dowding realized what Goering's plan was. He wanted to find out how strong the RAF defenses were; by attacking the convoys which were Britain's lifeline, he aimed to bring up the fighters, to flush them out.

Carefully husbanding his resources, Dowding refused to cooperate.

The Battle 186

A dozen Stukas would come over, and he responded with three Hurricanes; twenty Dorniers might attack, and he sent up four Spitfires.

Across the Channel *Kanalkampführer* Fink sent in his daily reports, and Goering smiled, chuckled. Surely with such weak responses the British were down to their last few fighters. Another week or two of such tactics and the end would be in sight.

The week came, and the week went. And the next week, and another after that. Throughout July the pattern of the Luftwaffe attacks remained the same, and the attacks met with the same RAF response. The pattern remained the same, but the tempo was picked up; the size and frequency of the attacks escalated from day to day.

Each morning at dawn the operations rooms would be quiet. The WAAFs would sip their morning coffee in the pit, waiting beside the map tables. On the balcony the controllers would sit, hands on knees, waiting.

Shortly after first light, the operators in the forward radar huts would see the first blips rising out of the constantly buzzing noise; single plots of reconnaissance aircraft snooping along the Channel, sniffing out the convoys at sea. From the first week of July whole squadrons had been flying to the forward aerodromes on the coast, leaving their own bases in the darkness so as to be ready, refueled and armed, when these early reports came in. Flights of three aircraft would be sent off, but though the German recco flights suffered appalling losses, they couldn't all be shot down before they radioed their reports back across the Channel to France. And a few hours later the raids would begin.

On July 12 Fighter Command lost two Hurricanes and a Spitfire in action, and another eight Hurricanes in accidents; they shot down eight German bombers. On July 18 they lost four Spits and three Blenheims, while destroying three bombers. On the nineteenth they lost four Hurricanes and seven Defiants; they shot down three bombers and one Messerschmitt.

On the sixteenth Hitler had decided to begin plans for invasion. He issued Directive No. 16 to the *Oberkommando der Wehrmacht* (*OKW*, the Army High Command):

> As England, despite her hopeless military situation, still shows no sign of willingness to come to terms, I have decided to prepare, and if necessary to carry out an invasion operation against her.
>
> The aim of this operation is to eliminate the English motherland as a base from which war against Germany can be continued, and, if necessary, to occupy the country completely.

The Luftwaffe was now to destroy all coastal defenses and any naval forces that dared steam out of Scapa Flow. The Channel was to be German, not English.

Goering reported to Hitler that the RAF was already finished; they were surely down to their last few Spitfires. One overwhelming effort would destroy them. Once Fighter Command was eliminated, German bombers could roam the country at will, thundering death and destruction until the populace rose in revolt against their willful masters and insisted on peace. Or, if he would rather, the Fuehrer might send his army unmolested across the Channel to sweep up the remains. Goering's men were straining at the leash, confident of victory, eager to gather the spoils, to deliver England into Hitler's lap. But still Hitler held them back from total warfare, restricting their targets to convoys and coastal defenses, hoping for an easy peace.

During these July days the fighting was concentrated in the areas of the Channel and along the parts of the coast protected by No. 11 Group, among which were several 600-series squadrons of the Royal Auxiliary Air Force. This was a service (similar to our National Guard) that had been started by Trenchard in 1924, based on the concept of the yeomanry that had been created as a defense against possible invasion by Napoleon a century and a quarter earlier:

> Recruiting, organisation and command were upon a county basis, the county gentlemen officering the force, the farmers and yeomen serving in its ranks, and all alike providing their own horses.

As in the original yeomanry, the officers of the Auxiliary were composed solely of gentlemen, the only difference being that they were not required to provide their own horses, that is, Spitfires. The original four squadrons of the 1920s were 600, City of London; 601, County of London; 602, City of Glasgow; and 603, City of Edinburgh. Others followed a decade later when the German threat was recognized, despite the protestations of Winston Churchill, among others, who very much doubted whether gentlemen flying on weekends could ever be more than weak-kneed dilettantes. At the same time a working-class force, the RAF Volunteer Reserve, was also being formed. The Auxiliary (AAF) was made up of the gentry, even the aristocracy; the Reserve (VR) was made up of the lower and middle classes. It was sarcastically said at the time (by regular army officers, most probably) that the RAF was composed of flyers trying to be gentlemen, the AAF were gentlemen trying to be flyers, and the VR were neither trying to be both.

601 Squadron, County of London, had an unusually high percentage

The Battle

of millionaires, even for the Auxiliary. They were flying Blenheims when war broke out and they suddenly realized that somebody might shoot at them; it was a sobering thought since their aircraft had no armor plating. They hadn't worried about that before, but now they raised a howl; in the ruckus of wartime no one listened. So one of them simply called the Bristol Aircraft Company, which made the Blenheims, and put in an order for armor plating to be installed in all the aircraft of the squadron and charged to his personal account. When this was done, they found the armor was too heavy and slowed the planes down too much, so another fellow ordered the job to be totally redone by the Wilkinson Sword Company; this was more satisfactory, and they toasted the fellow's health.

But the real crunch came when they realized that fuel rationing would soon be ordered; how then would they be able to drive their motorcycles, MGs, and Rolls-Royces from Biggin Hill into London for a little recreation? They decided to build up a reserve. Willie Rhodes-Moorhouse was appointed petrol officer and sent out to scrounge around the countryside.

"Well," he was asked the next day. "How much have you got?"

"Enough to last the war."

"What have you done?"

"I've bought a garage."

And so he had. Why buy a quart of milk when you can buy a cow? While asking about the fuel he had suddenly had that great idea, and when the garage owner named a price, he had taken out his checkbook and bought the place right off.

The next day, when the squadron drove out to the garage to look it over, they found that the tanks were nowhere near full. They wouldn't last out the year. But then Loel Guinness said, "I'm not sure, but I think I'm a director of Shell."

"What do you mean, you *think* you are? Telephone your bloody secretary and find out!"

He did and he was, and so the tanks were filled and the County squadron flew off to war. In July they were flying from Tangmere, while 609 (West Riding, Yorkshire) Squadron was operating out of Northolt. Not quite as rich as the London squadron, they too had their share of the aristocracy, among whom was the squadron adjutant, the Earl of Lincoln, who later in the war succeeded to the Dukedom of Newcastle. (On D-Day he was stationed on an LST off the coast of France, commanding an advanced radar unit. The LST was sunk by enemy action, and, as Group Capt. James McComb later told the story: "Fished out of the Channel, dressed in an ordinary seaman's gear and very dirty, he was taken to Naval HQ in Portsmouth, where the C-in-C, observing a figure resting in the sunshine on the grass outside his office, said to me: 'What is that filthy object? Bring it in!' I duly obeyed and made the presentation: 'Wing-Commander His Grace the Duke of Newcastle, sir.'")

On the ninth of July a section of 609 Squadron flew at dawn to their forward station at Warmwell, flying so low because of filthy weather that as they zoomed over one hill two bicyclists in their path were scattered off their cycles into a muddy ditch. This was the kind of flying that had brought complaints and reprimands a year before, but no one was complaining now.

They spent the next eight hours in a leaky tent under heavy drizzle, and then finally at 1830 the telephone rang and Green Section was ordered off: radar showed a large blip out to sea. In the air they received a further radio message from sector control: there were "fifteen plus" bombers coming in from the southwest. Despite the heavy clouds, they found them, but as they dove in to the attack they were in turn jumped by a formation of Messerschmitts which suddenly appeared from behind another cloud. Number 609 shot down one bomber and one fighter and disupted the bombing raid; they lost one of their own. Two days later they lost two more pilots when, bounced by enemy fighters, their radio sets failed to work and they didn't hear the warning call. They shot down no one that day.

The German attacks continued and intensified throughout the weekend. Radar continued to plot them as they came in, but there were still lessons to be learned. Later in the battle, sector operations would plot on their map tables the position of the sun as well as of the airplanes, so they could direct the interceptions upsun. But in July the old refrain "Beware of the Hun in the sun" was still a doggerel of death. "Each combat," Squadron Leader Darley of No. 609 Squadron said, "produced its own lessons." The ones who lived learned fast.

Dowding's strategy was to send an advance flight from the No. 11 Group squadrons to a forward aerodrome near the coast every morning at first light. When radar would pick up a plot indicating a bomber attack, the flight would go off to intercept and disrupt the raid. The rest of the squadron would then move up to the forward aerodrome, or even go right in to the attack if the raid was large enough to warrant it.

The danger of this scheme was twofold. First, the initial interceptions were generally made by a single flight of three aircraft, whereas the bomber raids were usually composed of dozens of aircraft. This overwhelming numerical inferiority was unavoidable; the Luftwaffe simply had many more aircraft, and they had the additional advantage of choosing when and where to attack. The second danger was that the RAF airfields of 11 Group were left unattended while their planes were attacking. This danger was unacceptable, and was counteracted by 12

Group which was detailed to stand in readiness to intercept any further raids which might be aimed at 11 Group's airfields. The system would work since, although the 12 Group fighters would have to fly from their own aerodromes to those which would be under attack further south, the radar system would give sufficient warning—barely sufficient, but sufficient if 12 Group's fighters were scrambled efficiently and followed the sector controller's orders. Finally, it was necessary to hold 12 Group in reserve because of the continuing threat from *Luftflotte 5* in Scandinavia: although the distance from their airfields was too great for the Messerschmitt fighters, the bombers based there could hit the industrial north at any time. Dowding knew, from the Ultra intercepts, that Goering was waiting until he was sure that all of Fighter Command had been pulled down to the southern sector to counter the convoy attacks. Then, when the north was unprotected by fighters, he would send the Scandinavian bombers to sail unopposed over the whole of the north counties, bombing at will. Goering hoped, and not without reason, that such an awful, continuing, unopposed attack would finally break the British spirit.

The Ultra intercepts were too secret to be disclosed, and so this final argument could not be made either to the commander of 12 Group or to his pilots. They were told simply that they must stay where they were in case of enemy attack to the north (which they didn't believe) and to protect the airfields of 11 Group, which they did believe and deeply resented. Every day they read in the newspapers of the desperate, heroic struggle going on in the south; and every day they sat on the grass in the bright sun and waited.

The London Times, Monday, July 15, 1940:

HISTORIC WEEK FOR THE R.A.F.
HEAVY PRICE PAID BY THE
LUFTWAFFE

The past week has written another brilliant page in the history of the R.A.F's successes over the enemy. No fewer than 85 German raiders were certainly shot down over and around this country, an average of rather more than 12 a day, at the cost of only 13 British fighters. The actual total is almost certainly much higher, for at least 50 other German aircraft are known to have been so badly damaged that few of them could have reached home.

Though the actual totals were somewhat more modest, sixty-six German planes destroyed for the loss of thirty-four British, the statistics still represented a furious action from which 12 Group was systematically excluded. The pilots began to fidget. Leigh-Mallory was furious. Douglas Bader was going out of his mind.

At the end of June, 242 Squadron had come back from France totally beaten and demoralized. It would take someone special to get them ready for combat again. The pilots were all Canadians, sick of the war, sick of Europe, sick of England. They thought of themselves as wild colonials, superior to the stuffy British, above any thought of discipline or military nonsense. They were assigned to 12 Group, out of the fighting, to give them a chance to rest up, reorganize, and learn to fly and fight again. They were also given a new squadron leader.

Bader drove to his new command in his MG. Coltishall was a brand new base, having become operational just the day before; 242 Squadron had come back from France just three days earlier. The aerodrome was situated three miles north of the village, about ten miles north of Norwich near the east coast. To this day there are no proper roads leading in, only a confusing series of winding lanes from the B1150; when Bader drove down, the situation was even more confused by the blackout and the complete absence of signs. Just finding his new command proved to be difficult. Reaching Norwich at about midnight, he got directions from a policeman; but a few twisted miles later he was lost, and nobody in that wartime Kafkaesque countryside would answer his questions.

"Can you tell me where Coltishall is?"

"'Oo wants ter know?"

"I'm looking for the RAF base."

"'Ow do I know yer nowt a spy, then?"

"I'm the new squadron leader."

"Then yer should know where 'tis, shouldn't yer?"

"But—"

"Get stuffed, mate."

And they would fade back into the darkness. One woman actually ran away from him, probably looking for a phone to call the police. He drove 'round and 'round for hours, doubling back on his trail again and again, although he never recognized anything because he couldn't see anything in the blackout. Finally he saw a dull red light in the distance and, driving up to it, found a barbed wire gate and a sentry on duty. Thank God.

"Password, sir?"

That was just too bloody much. "I don't know the damned password! I've just got here!"

No password, no entry. He sat out there in the darkness for another half hour before the sentry was able to contact the duty officer, who let him in and showed him to a bed. Next morning, while he was talking with the Base Commander prior to meeting his new squadron, the pilots were naturally curious.

"What's he like?"

Shrug. "English."

"Christ. Still, what can you expect? Can he fly?"

Shrug. Pause. Then, "Don't see how. He's got no legs."

Silence. Glances exchanged. Then, "Shows you what they think of us, don't it? When's the next bus home?"

When Bader stumped out to meet them, they stood around and stared at him gloomily. When he asked them questions, they didn't answer. It wasn't so much that they were rebellious as extremely young, barely into their twenties, and had been knocked about terribly in France; they were sullen, dispirited, beaten and disappointed.

He stared back at them. He glanced down at his legs. He knew what they were thinking. Kicked to tatters in France, then come home to be given a cripple as leader.

Right, then. He turned away from them and stumped across the grass to his Hurricane. With his powerful arms he pulled himself up onto the wing. He leaned against the fuselage and lifted his right leg with both hands and pushed it over the side of the cockpit, braced himself against the canopy and lifted his other leg in his hands and pulled it in, half-falling in with it as he always did. He was a cripple, all right—until he was in the cockpit. He strapped himself into the parachute and then into the seat harness, started the engine, and taxied down the runway. As the wheels left the ground he tucked them in but kept his nose down, inches off the ground, while the air speed built up. As he reached the perimeter fence he pulled her up steeply, rolled off the top, and came screaming down again straight at the group gathered to watch. He knocked their hats off with his slipstream and zoomed back up into the sky.

He did in that Hurricane what they had never seen done before. He looped low over the grass and at the top went straight on into another loop; if the first had not been precisely perfect, if he had lost any altitude in it, he would have inevitably crashed on the second. He didn't crash. And then he went right on into a third, and a fourth. They looked at each other. They knew that not one of them would have dared try that.

He came roaring back down to grass-top height and rolled upside down: his old specialty. He held it there right across the aerodrome in front of their wide-open eyes, then he flipped over upright again, and took her straight up in a tight, looping climb. At the top of the loop he flicked into a roll and dropped off into a spin, coming right out of that into another loop—a Hendon air show specialty, but they were too young, they had never seen it done before.

When he came down a half hour later, dropping the Hurricane softly into a perfect landing as if it were a feathery bird instead of a ton of metal and canvas, they were still standing there at perfect attention. He pulled himself heavily out of the seat and hobbled across the grass back to them; he gave them a quick glance and then went on into his quarters.

He had won; they were his forever.

Nasty weather that July in Norwich country. At 0700 the clouds were down nearly touching the ground, with drizzling rain and blowing mists scudding through the few feet in between. The pilots were sitting around, playing cards, talking, napping.

In dispersal, the phone rang. Bader picked it up.

"Sector Control calling. Radar's picked up a plot on the screen off the coast. Looks to be a single, stooging around looking for something. Angels one. Can you put up a section to have a go?"

No, Bader said, he couldn't. Weather was too filthy. Couldn't fly in this muck. Wasn't going to prang his pilots for the sake of a bogey they couldn't find, which would probably turn out to be a thunderstorm or a British flying boat anyway. Angels one didn't make any sense, either, because that'd put the supposed German right in the clouds and what could he see from there? Still.... "I'll go have a quick look myself," he said.

He took off heading due east and stayed right down on the deck, under the lowering clouds. As he left the coastline behind and roared out over the water, beating it to a froth with his prop wash, he found the clouds lifting. He pulled up to the base and found it just over a thousand feet; so maybe the radar was right, and maybe it was a German—

Suddenly there he was, appearing out of the drizzle right in front of him, less than a thousand yards ahead. He pushed the throttle forward and tried to sneak up close, but the rear gunner fired off a quick burst at him. He cursed; it wasn't the bullets that bothered him, but once they saw him all they had to do was pull the nose up into cloud and they'd be gone.

Incredibly, another few seconds passed and the Dornier stayed straight and level. Just one more second, please God, thank you, and now just one more, one more—

At 200 yards he pressed the red teat on the round handle of his fighter and felt it shudder with the backlash of eight machine guns. He saw the tracers sizzle out from his wings and disappear right into the Dornier's body, and then suddenly the German got his finger out and the nose lifted and the next second it was gone, vanished like a ghost into the murky clouds.

He pulled up into the clouds himself, but it was hopeless. There was no top to them, they continued on up forever, and he couldn't see his own wingtips in them. He circled around and came back to base.

Back at dispersal the telephone rang again.

"Thought you might like to know. Observer Corps at Cromer reports a Dornier came tumbling out of the clouds on fire. Went straight in to the sea. Good show."

Again the slow days settled in, and 242 Squadron sat in the sun outside dispersal, or in the gloom of a rainy day they sat inside, reading the newspapers which told them of the fighting going on day after day down south. Throughout East Anglia and Yorkshire 12 Group squadrons sat around and waited, and groused, while their brothers died in glory down south. In Group Command Headquarters Leigh-Mallory fumed, and wondered what he could do about it.

On the nineteenth of July, Hitler broadcast to England once again; this time it was a "final" offer, one last appeal to reason:

> If this war continues it can only end in the annihilation of one of us. Herr Churchill thinks it will be Germany. I know it will be Britain. I am not the defeated one begging for mercy. I am the *victor!* But I see no reason why this war must continue. I want to avert the sacrifices that will claim millions of lives....

On Monday, the twenty-second of July, Lord Halifax delivered His Majesty's Government's official reply. When, the previous evening, the announcement that Halifax would give the official answer had been made, there was worry in some quarters. At the beginning of hostilities the previous autumn, Halifax had been the man to whom the appeasers had looked for leadership. He was the head of a faction in the Government that, it was popularly thought, wanted peace at any price. And so Hitler waited eagerly for his speech, while in England many waited with apprehension. At RAF stations across the land, all work stopped when the BBC chimes rang. The tension built higher and higher as Halifax began to speak, and then burst with a bang as he said,

> ...Hitler's...only appeal is to the base instinct of fear, and his only arguments are threats....If Hitler were to succeed it would be the end, for many besides ourselves, of all those things which make life worth living. We realize that the struggles may cost us everything, but just because the things we are defending are worth any sacrifice, it is a noble privilege to be the defenders of things so precious. We never wanted the War; certainly no one here wants the war to go on for a day longer than is necessary. But we shall not stop fighting till freedom—*for ourselves and for others*—is secure.

That last sentence was the core of the matter. That was what they had all been waiting for, an echo of Churchill's earlier pledge that "freedom shall be restored to all. We abate nothing of our just demands; not one jot or tittle do we recede. Czechs, Poles, Norwegians, Dutch, Belgians.... All shall be freed, all these shall be restored!"

At Bentley Priory the lights stayed on late that night behind the blackout curtains, as they would every night for the rest of that interminable summer, preparing for the inevitable tomorrow. In No. 12 Group they drank bitter beer, wondering how they could get into it.

Chapter Twelve
Attack of the Eagles: "Adlerangriff!"

> Things fall apart; the centre cannot hold;...
> The blood-dimmed tide is loosed...
> W. B. Yeats,
> "The Second Coming"

As July petered out into the first hot days of August, the Luftwaffe attacks slackened off. The immediate reason for this was simply that the British had stopped sending convoys through the Channel; their losses were too high. But the coastal defenses were still there, targets enough still existed. And yet the tempo decreased markedly.

Were the Germans beaten? Had they been hurt too much for them to bear? According to claims filed by Fighter Command, the Luftwaffe had lost twenty-five bombers on the twenth-ninth of July and another dozen the next day, but Dowding distrusted these numbers. (He was right to do so: the actual loss figures were thirteen bombers on the twenty-ninth, seven on the thirtieth.) Contributing to his distrust was more than his intuition: Ultra intercepts decoded at Bletchley Park indicated a massive reorganization on the other side of the Channel, with extra forces being brought into attack position, more bombers and fighters being crowded into forward airfields, bombs and bullets and gasoline streaming from all corners of Germany to the coastline...

On August first, Hitler met with his commanders to decide on an appropriate response to the Halifax speech. He summoned to the *Berghof* Generals Keitel and Jodl of the *Oberkommando der Wehrmacht* and Admiral Raeder of the *Kriegsmarine*. The result was Directive No.

17 from the Fuehrer of the Third Reich to his Luftwaffe commanders, entitled *"For the Conduct of the Air and Sea War against England:"*

> I intend to intensify air and sea warfare against the English in their homeland. I therefore order the Luftwaffe to overpower the Royal Air Force with all the forces at its command, in the shortest possible time. After achieving air superiority the Luftwaffe is to be ready to participate in full force in *Fall Seelöwe*.

The code name *Fall Seelöwe,* Operation Sea Lion, had already been identified by the code breakers at Bletchley. It was the plan for the invasion of England.

The plan depended on the safe transport of the Wehrmacht across the Channel. Once across, they were not worried about the effect of pikes and shotguns on Tiger tanks. The Wehrmacht proposed to send twelve divisions against the Sussex and Kent coasts. The Kriegsmarine protested that such a wide landing area was too difficult; they wanted the forces concentrated in one spot. The Wehrmacht insisted it needed a broad front.

Goering came to the rescue. The Kriegsmarine's problem was the difficulty of protecting a wide band of transport vessels against the Royal Navy. Goering promised that there would be no Royal Navy. The skies would be swept free of Spitfires, and his unmolested bombers would sink any British ship that ventured down from its northern haven in Scapa Flow.

The other German commanders were getting used to Goering's promises. "If a single bomb ever falls on sacred German soil, you can call me Meyer!" he had once proclaimed, promising to abase himself with a Jewish name if his Luftwaffe's umbrella over the Fatherland were ever pierced. But nightly now the antiquated bombers of RAF Bomber Command, Whitleys and Hampdens and Wellingtons, dropped their irritating loads on the Ruhr and the nearer German cities, and there seemed to be little the Luftwaffe could do to stop them; the invaders were invisible at night. Of course, the targets were invisible too, and the British bombs scattered widely and did little harm. Still, the bombs fell. And now here was "Meyer" promising to destroy the Royal Navy.

Well, maybe.

Reluctantly, egged on by Hitler, the admirals accepted his word and proceeded with their plans. Twelve divisions to be landed on a wide front in Sussex and Kent.... But first, those last few Spitfires had to be destroyed.

* * *

Throughout the first week in August the air fighting was light and sporadic, as the British held their convoys in port and the Germans prepared for *Adler Tag,* Eagle Day, and the unleasing of the Luftwaffe in a final full-scale assault. On the eighth, the English decided to risk another convoy, sending Peewit from the Medway through the Straits of Dover under cover of darkness. But when dawn broke, the Luftwaffe found the convoy. Three major raids took place throughout the day, and although each one was picked up perfectly by the radar station at Ventnor and met head on by the Spits and Hurris, there were too many Messerschmitts; the Stukas got through to their targets. The RAF destroyed nearly a dozen German bombers and fighters, with the loss of only two of their own, but the convoy was decimated. Four ships out of twenty reached their destination at Swanage.

Then once again the air war faded away. No more convoys were sent, and the Luftwaffe went back to getting ready. A few scattered fights took place on the ninth, and on the tenth—the date set for *Adler Tag*—it rained.

One bombing raid set off, aiming to hit an aircraft factory near Norwich. Douglas Bader's 242 Squadron got the call. Bader by now had had plenty of time—too much time—to sit around and read the newspapers and study what was happening down south. He decided that the problem with the RAF was that they sent off such small replies to each raid. He argued that the basis of all military tactics was to "get there fustest with the mostest," as an American cavalryman had once put it. When the call came through on the tenth, he responded by scrambling his entire squadron, on his own authority, against Dowding's prescribed concept of having interceptions organized and directed by sector control.

There were two problems with Bader's theory. Getting there with the "mostest" was playing into Luftwaffe hands, since the German strategy was based on bringing the RAF into pitched battle, while it was the basis of Dowding's strategy to frustrate this aim and continue to conserve his forces in order to remain prepared to attack the planned invasion force. Secondly, it wasn't possible to get there "fustest" with a whole squadron. The airfields weren't large enough to enable an entire squadron to take off together, so flights or sections would take off, with the first ones airborne having to circle while waiting to form up with the flights launching after them. This took time, and radar wasn't yet good enough to provide enough time for such aerial organizing. On this day in particular, because of the abundant thunder squalls throughout East Anglia which showed up as blips on the radar scopes, several false directions were given as the controller sought to bring the fighters onto an interception course. Combined with the slow takeoff and formation time of the squadron, this caused them to miss the invaders altogether. When they finally arrived over the Boulton Paul factory, they found it cratered and burning, the bombers long gone.

Bader reported to Leigh-Mallory that they needed quicker and better radar information. He had, of course, no suggestion as to how it might be obtained—that was the scientists' job, not his. But he was sure that if he could hit the enemy with an entire squadron at one go, he could destroy them.

Leigh-Mallory agreed.

FROM REICHSMARSCHALL
GOERING TO ALL UNITS OF
LUFTFLOTTE 2, 3, AND *5:*

Adlerangriff! You will proceed to smash the British Air Force out of the sky. Heil Hitler!

Adler Tag was now set for August 13. The strategy was simple: destroy the RAF. Destroy it in the air and on the ground. Overwhelm the defending fighters in the air with hordes of Messerschmitts, and destroy those that don't rise to the attack by bombing their airfields and burning them on the ground. And, oh yes, knock out those funny little wooden shacks with high antennae that lined the coast.

The concept was good, but the detailed plans were not. On the eleventh and twelfth, under cover of heavy fighting, Luftwaffe reconnaissance planes sailed all over southern England to bring back pictures of the targets to be hit. They were nearly all misidentified.

They found Tangmere aerodrome, but only after all the Spits and Hurris had been scrambled, so that their photographs showed nothing but the night-fighter squadron of Blenheims on the grass; they misidentified them as bombers and so dismissed Tangmere as a bomber station rather than as one of the crucial Fighter Command fields. They never knew that the Woolston factory, sitting exposed and vulnerable on the south coast, was a primary producer of Spitfires, and so it had no target priority. The aerodrome at Upavon was one of their highest priorities, although it hadn't been used for fighters in over ten years. And when they did finally begin to knock out the radar stations, they didn't know it. Goering himself told them not to bother with those funny little shacks anymore.

It was clear by this time that the high antennae lining the coast were radar installations, since their electromagnetic emissions were clearly discernible on the other side of the Channel, and General Martini's radio technicians were competent enough to identify them for what they were. The Germans, in fact, had developed their own radar to a technically sophisticated level, but had not yet managed to integrate it into their strategic planning. Goering never understood the extent to which Dowding had done this; he never appreciated the true significance of those flimsy, swaying steel towers that rose over the White Cliffs of Dover and saw everything he did, almost everything he thought. Throughout July and the first week of August the Luftwaffe had ignored the radar stations.

But slowly, reluctantly, they began to recognize a pattern in the intelligence reports they gathered from the returning crews: wherever and whenever they attacked, there was always a flight of Spitfires or Hurricanes waiting for them.

Goering shrugged it off. *Der Herr Gott* has a weakness for taking care of drunks and idiots, that was all it was. Luck. Keep hitting them, and their luck will turn.

Still. . . .

General Wolfgang Martini, who had led the *Zeppelin* mission before the war to detect radio emissions from England, had not been idle since. He had been measuring the radar emissions and was impressed with the inescapable curtain with which they covered the coast; putting it together with the intelligence reports, he was sure that the British system worked. And look at the pictures, he pointed out to Goering, laying on the table in front of him some clear shots of radar towers taken by reconnaissance aircraft. The towers were flimsy, the wooden shacks which obviously housed the equipment and operators were unprotected. An easy target for the Stukas.

Goering lifted his eyes to the ceiling. All right. *All right*. Enough already. He would attack the little wooden shacks.

Monday, August 12. As a prelude to *Adler Tag,* scheduled for the next day, the morning began with a feint by Me 109 fighters prior to attacks on the radar stations. At Biggin Hill, No. 610 Squadron was at dawn readiness. The weather was clear; this early in the summer morning it was still cold. The pilots were taken out to the dispersal hut in a small van. The erks warmed up the Spitfire engines while the pilots got out their flying kits, parachutes and helmets and gloves, and laid them on the wings or stashed them in the cockpits. Then they sat around the small stove in the waiting room, and waited.

The telephone rang. Eyes blinked open, legs tensed to run. One or two of the newest kids (few were over the age of twenty-one) actually jumped up out of their chairs. The duty corporal yanked the phone out of its cradle on that first ring. "Yes, sir,.... Right!" He hung up. "Breakfast's on the way over," he announced.

0700 hours. The radar operator at Dover picked up a plot coming out of France, heading over the Channel toward them. The information was transmitted to Bentley Priory, and from there to Group Headquarters and Sector Control, which was Biggin Hill. A marker was placed on the table; minute by minute it was pushed along while the controllers tried to figure out where they were headed. In the dispersal room the fighter pilots knew nothing of this; they began to doze off again.

0720. The plot was moving inland. Time to do something. At 11 Group headquarters, Keith Park quietly spoke to the controller.

0721. The phone rang again in dispersal. "Yes.... Squadron scramble!" the duty corporal shouted. "Angels ten over Dungeness!"

0723. As the first Spitfires rose into the air, the squadron leader called in: "Dog-rose Red Leader calling Sapper. Squadron airborne."

"Hullo, Dog-rose Red Leader. Receiving you loud and clear. Vector 120 degrees. Nine bandits approaching Dungeness. Angels ten. Good hunting."

There turned out to be two formations, one at 10,000 feet as radar had reported, but another 6000 feet higher. As the Spits turned in to the first, they were bounced by the second, but the Me 109s were spotted in time and the squadron turned in to them. In the fight that followed two 109s were shot down, one Spit was lost.

An hour later sixteen twin-engine Messerschmitt 110 fighter-bombers of Kesselring's *Luftflotte 2* took off. Flying parallel to the English coast, just offshore, they cruised past the Chain Home stations at Dover and Rye.

In the wooden shack at Dover, the WAAF operator, as she had done with the earlier raid an hour before, was following them, reporting them heading west, passing by. At Sector Control headquarters their path was followed by a marker pushed along the map table while the controller stood guessing at their target. Then suddenly they swerved, breaking into four sections.

"Sir," the WAAF in the Dover shack beckoned to the officer on duty. He strode over to the radar screen. "Change of course."

He stared at it. "Good Lord. They're coming straight for us."

"It certainly appears so."

As he called in the news to Sector Control, she calmly reached up to the hook over her head and put on her tin hat. Then she sat at the screen and continued to read off the coordinates as the bombers dove in to the attack.

The first section of four 110s hit them moments later. The second section struck at Rye, the third at Pevensey, and the fourth at Dunkirk. Bombs burst throught the small compounds, breaking electronic cables, shattering power supplies, blowing apart wooden huts and people within them. Dover, Pevensey, Rye stations all went off the air and,

The Battle

coordinated perfectly with this blinding of the radar eye, other raids swept in unseen and hit airfields at Hawkinge and Lympne.

But the towers still stood. Fragile, skinny, exposed as they were, it turned out that their open structure was practically impervious to blast: they could be knocked down only by a direct hit, which was unlikely because of their skinniness. By late afternoon the damage had been repaired, the WAAF operators who survived had had their cups of tea, and the stations were back on the air. When further attacks were launched that afternoon, on the assumption that the RAF had been blinded, they were met once again by Spitfires and Hurricanes in precisely the right place at the right time.

One of the afternoon attacks was directed at the Ventnor radar station. Fifteen Ju 88s dived on it and destroyed every single building in the compound; although the towers still stood, the connections between them were broken and the power suppliers were demolished. Repair crews rushed to the site, but they had to start over practically from scratch. The aerials on the towers had to be taken down and fixed and reinstalled, a mobile power generator had to be brought in, all the ground equipment had to be replaced. For the next three days the Ventnor site was totally blind, leaving a pathway miles wide within which the German bombers could have slipped without being seen. But they never knew it. The unexpected rapidity with which the Dover, Pevensey, and Rye sites resumed action that day indicated to the Germans that the radar sites were impervious to bomb damage.

Goering was disgusted. "There doesn't appear to be any point in continuing attacks on radar sites," he wrote to his commanders, "since not one of those attacked has been put out of action." It was not necessary anyway, he continued, since his intelligence confirmed that the RAF was down to its last few Spitfires. One or two days of really intense action would finish them off. The attack of the eagles would end the history of England.

13 August. *Adler Tag*. The day's dawning was lost behind a curtain of heavy clouds and rain. But *Oberst* Johannes Fink, commanding seventy-four Dornier bombers of KG 2, wasn't worried; he could deal with the weather. At 0500 hours, just before dawn, they took off, met their fighter escort over the French coast, and headed for England.

Nobody else took off. At dawn Goering decided to postpone the attack of the eagles. A radio signal was sent off, recalling the KG 2 bombers and their escort. But radio communications were difficult; the fighters received the message but the bombers did not. As Fink led his bombers around a large cloud formation, he found that the fighters, which had gone around the other side of the cloud, had abruptly disappeared. He continued on.

Attack of the Eagles: "Adlerangriff!"

Radar picked up the bomber formation heading towards Eastchurch, but the radar operators misjudged the number of bombers and the Observer Corps gave faulty directions as they passed overland. Sector Control sent five squadrons, but four of them went off late and to the wrong place. One fighter squadron found them, but there were too many bombers to stop, and they got through.

Luckily, Eastchurch had been abandoned as an operational fighter airfield weeks before. Unluckily, however, a squadron of Spitfires was using it temporarily. On their return *Oberst* Fink reported the squadron destroyed. In fact, one Spit was lost; but the airfield was badly damaged.

At 3:45 that afternoon the weather cleared sufficiently for the attack to begin in earnest. At Bentley Priory, Dowding had hoped that the morning's attack on Eastchurch was an isolated event; now he watched in dismay as the plots on the operations map table showed the raids coming in one after the other for the airfields rather than the coastal defenses. This was what he had feared. His argument against sending the fighters to France had not been based on fear of what losses they would suffer in the air so much as that they could be destroyed on the ground. Here in England they had radar warning, but they couldn't stay in the air continuously. If the Luftwaffe were able to overwhelm the defenses and catch them refueling on the ground, the fight would be lost. Even if the bombers found only empty airfields, they would be able to destroy the hangars and runways and supporting equipment; without these necessities, his chicks could not fly, his fighters could not fight.

Other bombing raids roamed up and down the countryside. The vital aerodrome at Middle Wallop was the object of one large raid, but they hit the empty field at Andover instead. They did catch Detling, and practically destroyed it, along with twenty-two parked airplanes—but none of them were fighters.

At the end of the day, the Luftwaffe was triumphant. Eighty-four British fighters had been destroyed in the air, the returning Messerschmitt pilots claimed, while the bomber crews chimed in with more than fifty more destroyed on the ground. The actual figures were thirteen lost in the air, and although nearly fifty planes had been bombed on the ground, only one of them was a fighter. The Luftwaffe had lost sixty-four.

The next day's fighting was again heavy, despite worsening weather as the day progressed. By evening the clouds had closed in and the forecast was for worse the following day, the fifteenth, which had been planned as the culmination: a synchronized force of *Luftflotte 5's* bombers was to strike against the north of England from their bases in Scandinavia, since it was clear to Goering that the previous two days' heavy fighting in the south must have drawn all of Dowding's fighters down there. But the bad weather indicated that no such attack would be possible that day, and so on the evening of the fourteenth, Goering summoned all his front-line commanders to a conference at Karinhall.

German reconnaissance flights slipped in and out of clouds all morning, investigating the weather even more than the British defenses. For the first time that summer most of them came home safely to report, and that in itself was important information—but it was misinterpreted. They came home because six of the Chain Home radar stations—nearly half the defensive network in that southeast corner of England—were off the air, still trying to repair the damage from the last few days' raids. But the Luftwaffe intelligence, still underestimating the importance of radar, thought the safe return of the planes only confirmed their estimate that Fighter Command was sorely depleted in aircraft. If they could be hit again, hard, that day, before they had time to rest and repair their damaged Spitfires, it would all be over.

Across the Channel Fighter Command waited. Nine o'clock came and went, ten o'clock came and went, and still they waited. The Control Rooms were quiet. The WAAFs stood around the plotting tables with folded arms, leaning against the walls, talking quietly, waiting. The colored counters for the enemy stood unmoved on the plotting tables, waiting. On the balcony looking down on the plotting tables the Sector and Group controllers leaned backward or forward, scratched their heads or their bottoms, lit pipes or cigarettes, waiting. Out at dispersal the pilots stood around or sat around or lay around, looking up at the sky, wondering if those clouds were clearing, was that patch of blue sky growing?

Over the Channel and along the coast, German reconnaissance aircraft buzzed around, ducking into the clouds when intercepted, flying out to sea to see what weather was coming, snooping around. One by one they returned and turned in their reports; one by one they told of weather that was clearing.

With the senior commanders at Karinhall, operations had been left in the hands of the deputies. The air crews had not been dismissed for the day, they had been held at readiness. All up and down the coastline of Europe great air fleets were standing by, waiting for the word that would release them to fly—*Fliegen gegen England*—and to saturate the already battered defenses by sheer overwhelming force of numbers, to show England that there was to be no respite, there was no hope, there was nothing in the future but complete and utter destruction. From bases in France, Belgium, Holland, Norway, and Denmark, in a deadly ring around the Island, the Luftwaffe's noses were sniffing and pointing inwards to the center of that circle.

Oberst Paul Diechmann, Chief of Staff to Gen. Bruno Lorzer, commanding *Fliegerkorps 2,* held the responsibility in the absence of his chief to decide whether the day's air strikes should be launched. Although Goering's cancellation order had been clear, Diechmann felt sure that the operation had been canceled only because of the weather, and now the weather was clearing. He stood by the window with his hands clasped behind his back, looking out at the sky, at the clearing clouds. Finally he turned around.

"Go," he said.

1107 hours. Friday, August 15, 1940.

The radar station at Kent reported a buildup of enemy formations over the Pas de Calais.

"Number, please?"

"Massive."

"How many, please?"

"Very massive. Very bloody massive."

The Sector controllers had orders to hold back from scrambling fighters until they knew exactly where the raids were coming, and then to engage as few fighters as possible. But this time the raids were developing so massively, spread out so widely, that it was impossible to count the enemy or to determine where he was going. The entire air seemed to be filled with Stukas, Messerschmitts, Junkers and Dorniers and Heinkels. It was like a massive flock of bees gone mad, they were everywhere. So many blips of such huge size appeared on every radar screen that the controllers lost all control, simply ordering every available squadron into the air with orders to "patrol" and attack anything that wasn't a cloud. It was a last-ditch, desperate sort of defense, and although the British fighters scored heavily, the bombers got through—to Hawkinge, where hangars and repair sites were destroyed, as were underground cables carrying power to the radar stations at Dover, Rye, and Foreness, adding these to the list of nonoperational sites—to Lympne, where practically every building was destroyed and the runways were so cratered that no one could land there for a week—to Manston, where two Spitfires were caught on the ground and burnt to cinders.

And then, while "the last few Spitfires" were desperately defending their bases in the south, Goering's masterful and long-planned coup de grace started off from Scandinavia. *Luftflotte 5* gunned up its engines and prepared to attack the "undefended" north of England.

Escorted by long-range Me 110s instead of the more deadly (but short-range) 109 fighters, they took off on the long flight across the North Sea.

Into the waiting arms of Fighter Command.

The Battle 206

Taking no chances on fighter interception, the Luftwaffe sent a flotilla of Heinkel seaplanes first on a course to Dundee, with the intention of drawing off any British fighters that might possibly have been left up north. When they reached the coast of Scotland, they would turn away and return home. A hundred miles further south, seventy-two Heinkel 111 bombers of KG 26, flying out of Stavanger in Norway and escorted by the long-range but ineffective Messerschmitt 110 fighters, would hit the Tyneside industrial area; carrying 3000 pounds of bombs each, they would be enough to devastate the entire region.

As it happened, however, a navigational error put the bombers right behind the seaplanes. The radar station at Anstruther picked up a blip which they reported as three-plus; a few minutes later they called in a correction: thirty-plus. In Newcastle the 13 Group controller began for the first time to bring his forces to battle. Number 72 Squadron was scrambled from Acklington and No. 79 Squadron was ordered to Readiness to take their place in reserve. Number 605 Squadron was scrambled from Drem. Number 49 Squadron roared into the air from Catterick, and now No. 79 Squadron was sent after them....

At Watnall in Nottingham, Leigh-Mallory couldn't believe his ears as he listened to the staccato conversations between radar stations, Bentley Priory, and Nos. 11 and 13 Groups; but there on the map table in front of him he saw the plots being placed one after the other. To the south of him 11 Group was scrambling every plane it had, and now to the north of him 13 Group was sending their men off—and *his* men stayed on the ground with their fingers up their noses, waiting for orders. His Spitfires stood quiet and immobile, squatting on their haunches, noses pointed to the sky, and the only sound heard on his airfields was the eternal slap of cards on the tables as his pilots played pontoon. He stared at the map table and saw his career sinking under the inexorable weight of inactivity while Keith Park and now even Richard Saul gathered the laurels of glory. *What the bleeding hell was going on*?

At Coltishall, Douglas Bader waited with 242 Squadron again, spending this day as he did every day in the dispersal hut, staring out the window, ignoring the constant clawing in the stumps of his legs, wondering what was happening.

Eleven Spitfires of 72 Squadron caught them first. Chasing after the radar-reported raid of "thirty-plus," they ran head-on into more than 100 bombers and fighters. Circling around to get the sun behind them, they dove in to the attack. Five minutes later five Hurricanes from 605 Squadron found the fight, and then 41 and 79 Squadrons joined in. The

Messerschmitt 110 escorts were twin-engined fighters, powerful and well armed but clumsy; they were no match for the nimbler Spits and Hurris. Together with the bombers they were supposed to protect, they began to fall flaming from the skies.

At the same time, another raid of forty-plus (Ju 88s of KG 30 from Denmark) was plotted by radar headed for 12 Group's territory—but before they reached it they were intercepted by two squadrons of 13 Group. 12 Group was ordered to scramble one squadron—to guard a convoy in case the bombers swerved to attack it. The convoy was not attacked, and that was the closest 12 Group came to action that day.

It was the first day of action in the north of England, and the last. *Luftflotte 5* was destroyed; never again would they attack in force. The "undefended" north had been shown to be alive and bristling with fighters, and the fishes of the North Sea fed well and leisurely.

Not a single British fighter was lost.

In the south, where the 109s flooded the skies, things were not quite so easy. At 1415 hours Kent and Essex radars reported new buildups along a wide front emanating from the Pas de Calais. It was difficult to discern the target, and since every squadron of fighters had seen action that morning, it was important to give them as much time to rearm and refuel as possible. So they waited, waited....

And suddenly a fast, low-flying squadron of 110s burst over Martlesham. Three Hurricanes of 17 Squadron managed to get off the ground; the rest were destroyed, as was the airfield. The 110s escaped without loss.

Meanwhile, more than 200 bombers and fighters came in, headed for Deal, and a fighter sweep of more than sixty 109s flashed in over Kent. Forty fighters were scrambled from Biggin Hill, but they were swamped by sheer numbers, and the bombers got through. Eastchurch and Rochester were wiped out. Another raid of 100 plus hit Middle Wallop, Worthy Down, and Andover. Portland was bombed by another hundred-plus raid, and two final raids headed for the heart of the defense: the airfields at Kenley and Biggin Hill. The Kenley attack got through, but hit the wrong place; they smacked Croydon, which, though important, was not as vital as the Sector Control aerodrome at Kenley. The Biggin Hill attack did the same, hitting West Malling—a new, as yet unused airfield.

And then finally the day was over. More than 2000 German airplanes had attacked England (counting those that came more than once), and Goering was ecstatic; though his forces had been hard hit, they had done their job, they had finished off the RAF. The twelve most important airfields of Fighter Command had been destroyed, together with 99 fighters in the air and uncounted numbers on the ground. The London newspapers, in their turn, proclaimed a gigantic victory for England: 182 German planes knocked down. At Bentley Priory Dowding was more cautious; he waited for the wrecks to be counted.

They found 75 wrecks, not 182. Fighter Command had lost 34 fighters, not 99, and though their airfields had been hit hard they were a long way from being wiped out.

Goering thought the fight was over. Dowding knew it was just beginning.

Leigh-Mallory was furious. Why hadn't his group been called in to action?

In a sense, he was right; they should have been used. True, 13 Group had savaged the Scandinavian attacks and had turned the raids back, but there was "plenty of meat on the table" and more than half the enemy had made it back home. Why hadn't the 13 Group Controllers called on 12 Group for help in dispatching the Dorniers? Quite simply because they were so inexperienced in regard to enemy attacks that the radar operators had been unable to recognize the full extent of the numbers of bombers that were coming their way. The initial raid had been estimated at thirty-plus, which meant there was a distinct possibility of other raids coming in from Scandinavia; in accord with the plan, 12 Group had been held in reserve. By the time they realized how large the raid was and that all the airplanes of *Luftflotte 5* were committed, it was too late to call for reinforcements.

Down south the reason was, in a sense, the opposite: they were not inexperienced, but they had been swamped to such an extent that the Controllers had lost control of the action. With hindsight one can say that they should have scrambled some of the 12 Group squadrons to back up the defense of 11 Group's aerodromes, but there were so many planes in the air that they simply couldn't keep track of them all, and they failed to do so.

Dowding should have stepped in at this point. If he had his finger on the pulse of the action, he must have realized the frustration of an eager, ambitious commander surrounded on both sides by combat while he was isolated and alone in the middle. He should have rectified the situation and informed Leigh-Mallory that he had done so. That night he should have talked to Leigh-Mallory. But he did nothing. By this time he was more concerned with comforting his dead fliers, with soothing them and sending them on their way to the better land he knew.

The next day was more of the same; the battle continued with intolerable fury. No. 266 Squadron, which had been flying Spitfires out of

Tangmere throughout the battle, lost five airplanes and pilots in less than five minutes. Less than half an hour later, while the survivors were refueling, radar reported a large formation headed straight for them. Together with the other Tangmere squadrons, they scrambled just as a squadron of Stukas came tumbling out of the sun. The fighters wheeled and caught the Stukas as they pulled out of their dive, destroying seven and badly damaging another half dozen, but the aerodrome was hit hard; every building was destroyed or set afire and those airplanes which hadn't got off in time were ruined.

While the bombs were still plummeting down, one of the Hurricanes came circling back to land, its pilot injured in the combat above. He was Billy Fiske, one of London's prewar colony of high-living Americans, and the first American to join the RAF. As he came in to land he saw through his oil-splattered windscreen a bomb erupt right in front of him; he slipped his plane at the last minute and touched down between craters, steering carefully as he rolled to a stop. He pulled back the canopy and tried to raise himself out of the cockpit, but he was hurt, he couldn't quite manage it—

The bursting of the bombs was nearly continuous, the roaring of airplane engines filled the background; it was a cacophony of noise in which the ground crews watching from their slit trenches saw events as in a silent movie, the sounds disconnected from the objects they saw. Thus they saw a *Gruppe* of four black-crossed planes come swooping in over the stopped Hurricane, saw the splatter of their bullets slipping over the grass, saw the bullets pass through the shuddering Hurricane, saw the pilot slump and disappear into a sudden roaring cascade of flames that enveloped him.

They burst from their slit trenches then, ran through the continuing rain of bullets and bombs straight into that inferno, climbed onto the burning wings and reached with their bare hands into the flaming pool of oil that filled the cockpit and pulled Billy Fiske out of hell and down onto the still bullet-splattering grass and rolled over and over on him to put out the flames. They lifted him onto their shoulders and half-ran, half-stumbled back across the dangerous open grass, through the shrapnel and bombs and the splattering of bullets and fell back with him into their slit trench.

When the raid was over an ambulance rushed him to hospital, where he died the next day.

Although Goering had decided that the radar stations were not worth the effort of destroying them, his *Luftflotte* commanders were not so sure; though placed low on the priority list of targets, they were not left off altogether. Five Stukas hit the crucial station at Ventnor again, and again put it out of action. The damage this time was so great that it wouldn't be operational for a full week. It left a gap so wide in the defenses that it had to be plugged. Luckily they had planned for such eventualities, and a mobile van was brought into place, powered by a gas generator, and plugged into the system. Unluckily, there had never been time to test this concept of mobile stations operating with lesser equipment, and it didn't work; in fact, it turned out to be a distinctly negative contribution, as the information it fed into the system was erroneous and simply tangled up the plots.

And still the Germans came. Again and again they came. On Saturday and Sunday, the next day and the day after that, they came back again and again, and the radar towers shook and fell, and the wooden shacks shuddered and splattered apart; the aerodromes were pockmarked with craters and the hangars burnt, the wrecks of Spitfires and Hurricanes littered the runways and the countryside.

And still the fighters rose into the half-blinded air to defend their land, still they came slipping out of every cloud and diving down out of the sun; whenever the Luftwaffe came, wherever they intruded, they found those "last few Spitfires" somehow there, waiting for them, clawing at them and dropping them flaming and tumbling out of the British sky.

On Monday, 19 August, it rained.

Thank God. All across the devastated south of England weary fighter pilots rose at first light and looked out through their bomb-shattered windows with uncomprehending puzzlement at the darkness outside, looked again at their wristwatches, sat up in bed half-asleep trying to understand why it was still dark out at six o'clock in the morning. And then the orderly came in with a cup of tea and the news: "Bloody awful out today, sir. Nothing but rain."

At last. Thank God. The cups of tea went untasted as the pilots slid back down under the covers and back to sleep.

Keith Park took advantage of the lull to review the situation. Nearly 200 Luftwaffe planes and crews had been shot down in the last four days, losses which no air force could long endure. But Fighter Command was even worse off. Their airfields were being systematically erased, their fighters were being lost, and worse than that, their pilots were being killed. A new Spitfire fresh from the factory was as good as the one lost that day, but a new pilot fresh from operational training was *not* as good as the one burnt to death in the lost Spitfire. A new pilot was barely better than none at all; it took weeks of careful nurturing *after* training before one could throw a Spit around the way it had to be done while all the time searching the skies for ambush, lining up a target, and finally actually hitting it.

Perhaps worst of all, the radar stations were going blind. One by one they were being knocked out of the fight. No concerted attack had been made against them, and thank God for that, but they were so extremely vulnerable in their unprotected locations that the few attempts that were made were succeeding all too well.

Taken all in all, the situation was desperate.

Park was busy all that day supervising the repairs, the reorganization, the reshuffling of squadrons, and reviewing the work of the radar system and his controllers. He recognized finally, now that he had time to think, that 12 Group had been wasted in the preceding fights. Late in the afternoon he issued a new directive, ordering that "If all our squadrons around London are off the ground engaging enemy mass attacks, ask No. 12 Group or Command Controller to provide squadrons to patrol [our] aerodromes."

This message was passed on to Leigh-Mallory and from him to Bader, who replied that when he was called on by 11 Group he wasn't going to respond with the snippety penny-packets of two or three fighters that 11 Group was fighting with so parsimoniously; he was convinced that his own idea of how to fight a war was the right one. Though it hadn't worked the first time, he still insisted that if only he could get sufficient warning when called upon by 11 Group, he'd have time to form up all the planes in his squadron and bring them into action as a powerful unit, smashing into the enemy with enough force to knock the bloody Hun right smack on his arse.

Leigh-Mallory nodded. That sounded good to him.

The rain continued throughout the week. A few small raids were mounted, mostly tip-and-run stuff just to remind the English that there was a war on. The Luftwaffe pilots were as grateful for the rest as were the British, but to their commander it was a different story. Goering was impatient to finish off the British; ever since the beginning of July he had been telling Hitler all he needed was a few days' fighting, and now that victory was within his grasp he chafed at the delay. Dowding, on the other hand, saw the rains as coming not from a cyclonic weather pattern over the North Atlantic but as having been sent straight from God: "I pay my homage to those dear boys, those gallant boys, who gave their all that our nation might live; I pay my tribute to their leaders and commanders; but I say with absolute conviction that, but for God's intervention, the Battle of Britian would have been lost."

On Saturday, August 24, the sun came out again.

The raids started at 0830. By lunchtime a series of raids was hitting the airfields at Hornchurch, Manston, Ramsgate, and North Weald. By

1600 hours every squadron of 11 Group was engaged, and radar picked up yet another force heading for North Weald. The message went out to Leigh-Mallory, asking for help. In minutes Bader's 242 Squadron was scrambling its first sections; but as the German bombers bore in toward their target, as the precious minutes ticked away, instead of racing off to interception, these first airborne planes turned and circled around to pick up the next section, and again to pick up the next....

When they arrived over North Weald there wasn't a German in sight; there was only a sickening black cloud of smoke hanging low, completely hiding from sight what they knew to be the aerodrome. As they circled helplessly around it they saw in the distance similar clouds rising from Hornchurch and Hawkinge, from Manston and Croydon.

When darkness settled that night the drone of aircraft engines over London did not cease. For some weeks now the Luftwaffe had been coming back at night to bomb and harass, and though the damage they did was minimal, it portended evil tidings for the future as they slowly learned to find their way around and about England in the night. But on this night, 24 August, they changed the tide of battle...and nobody realized what had happened.

What seemed to happen was a simple violation of Hitler's orders. From the beginning of the air war, he had specified that London was not to be bombed; he wanted to leave a path free for the Government to surrender without the egg on their faces that a devastated London would bring. Though civilians had been murdered from the sky before this in Barcelona and Madrid, in Rotterdam and Brussels, in Portsmouth and Dover, Hitler had ordered that the citizens of London were to be spared simply because of the symbolic status of their city. But on this night nearly a hundred Luftwaffe bombers dumped their load of high explosives and incendiaries on the city, without Hitler's permission. A navigational error was blamed.

The effect was unforseen. Though the raids were not repeated, lending credence to the navigational error supposition, Churchill was furious. He would not allow a group of psychopathic murderers to commit their foul offenses on the hallowed city of London. By God, he would not! He ordered a retaliatory raid on Berlin the very next night. Bomber Command argued against it: they did not want to waste their still slender resources by scheduling a political or terror raid against the civilian population of a city rather than hitting a military target. (They were under the impression, quite mistaken, that their navigation and bombing accuracies were good enough to destroy specific military targets at night; in fact, the bombs they aimed at particular factories or depots or railway installations often did not even hit the cities in which the targets were located.) They argued further that it would be suicide to challenge the Germans to a tit-for-tat bombing policy, since the German bomber force was so much greater than their own, and since the distance from England's bomber bases to the German cities was so much greater than that to the English cities from the Luftwaffe bases in France. If they were to initiate a war of retaliatory terror raids against civilian populations, they would only bring even worse retaliations against their own women and children. London was, in Churchill's own words of twenty years before, "a fat white cow" tethered to the stake. But the only time that a military commander had won an argument against Churchill was when Dowding had forced him to cancel his order sending more Hurricanes to France. And so on August 25, eighty-one Wellington and Hampden bombers set off for Berlin. Eastern Europe was covered by cloud that night, but Churchill would not brook a moment's delay: the Hun must be taught that he cannot bomb London with impunity, and the lesson must be taught immediately and firmly.

The night bombers flew singly rather than in formation, each finding its own way across the blacked-out countryside; it is a tribute to them that of the eighty-one who took off as many as twenty-nine found the city. For three hours they flew over it and dropped their bombs. The immediate results were rather more comical than military in nature. Not one Berliner was killed, not even by any of the toilets that were dropped in addition to the bombs. But Goering flushed red in embarrassment, Goebbels screamed like a madman, and as the British bombers came back again and again the following nights, Hitler's temperature began to rise.

It would soon burst the always tenuous boundaries of his reason, and in so doing change the result of the war. But not quite yet. After each night's raid by Bomber Command he would rant and rage to his subordinates and fuss to himself, but he did nothing else, he issued no new orders. And meanwhile the Luftwaffe bombers continued to chip away effectively at the RAF defenses. Fighter Command began to reel, teetering over the edge of final defeat. In the ten days after the weather once again turned good on August 24 they were battered and pummeled day after day from dawn till sunset, and though they continued to give better than they got, they were now falling back on their last resources.

The radar operators stayed in their wooden shacks while the bombs dropped outside—and sometimes inside, with disastrous results. When that happened, the maintenance personnel came rushing to the scene before the smoke had drifted away. They worked under unrelenting pressure; as soon as a station was put out of action they would be resplicing wires and replacing shattered tubes and driving mobile generators down winding roads and climbing swaying steel towers 300 feet high to restring the antennae. One by one the stations were knocked out, and one by one they were brought on line again. And still day after day when the Luftwaffe came slanting down out of the sun or banking around a huge cloud they would find another penny-packet of Spitfires waiting for them. Day after day those "last few Spitfires" clawed their way into the air and were directed straight to where they were needed most.

But when they staggered home to their bases again, more and more often they would find there black smoke rising from a devastated ruin. Again and again they would have to search through the pockmarked grass for a safe lane to put their tired fighters down, and no sooner would they climb wearily from the cockpits than they would hear the Tannoys shouting again: "All personnel take cover! All squadrons scramble! Base is about to be attacked!"

The Luftwaffe tactics had put Germany on the edge of victory, and just barely in time. The purpose of this whole air war, remember, was to clear the RAF fighters out of the sky in order that the Channel might be prohibited to the Royal Navy and thus made safe for the Kriegsmarine to transport the German Army across to England, which lay there virtually unarmed and ripe as a plum for the plucking. Through all the waterways of France and Germany a constant stream of barges and small ships was wending its way to the French coast, where they were being prepared and stockpiled as transport for the army. Materials, supplies, men, and horses (for all their vaunted panzer divisions, the Wehrmacht still relied heavily on horse transportation and motive power) were piling up along the coast, pointed at England.

But the RAF was not yet defeated, and the invasion preparations couldn't be postponed much longer. Already the date had been pushed back from mid-August to the beginning of September, and was nearing the limit of possibility. By the end of September, October at the latest, the weather would break; the North Atlantic would bring in the cold arctic winds which would mix with the latent humidity and bring down fog and rains and soon snow onto the ground. With the loss of clear air the Luftwaffe would be grounded, releasing the Royal Navy; simultaneously the Channel waters would begin their winter chops and swells, which could swamp and overturn the small boats on which the Kriegsmarine had to rely.

The German admirals complained that "preparations for Sea Lion are being affected by the inactivity of the Luftwaffe, which....for reasons not known to the Naval Staff, has missed opportunities afforded by the recent very favourable weather...." Goering replied that, on the contrary, the job was all but done; another few days would see the end of the last few Spitfires. Hitler announced to his generals and admirals that the final date for the invasion would be decided by September 14.

The aerial attack continued with increased frenzy. Keith Park's Controllers had every squadron in the air every day, and now they called on 12 Group for help. On August 24, Bader was scrambled, but while he was assembling his squadron, the raiders destroyed Hornchurch and North Weald. On August 26, 11 Group itself broke up a new attack on these two nearly destroyed airfields, but when more bombers came in towards Debden they had nothing left to reply with. They called on 12 Group. Again they came too late; Debden was devastated.

Bader argued angrily that if he were given more time, if he were warned earlier, he could destroy the Luftwaffe. Park answered that there *wasn't* more time to give: the radar system and communications network was being strained to its maximum. Incredibly, Leigh-Mallory told Bader he was right.

At this point Dowding should have stepped in and settled the bickering between Park and Leigh-Mallory, but he did not. He said after the battle that he was not even aware of the controversy at the time. It was a very sad statement; a commander *must* know these things. But Dowding was not concerned at this stage with the day-to-day war; he was more concerned with divining the mechanisms by which his God was intervening to save them, and with comforting his poor dead fliers and shepherding them on to their reward. We can only be grateful that his mind lasted as long as it did, that he prepared the defense system as well as he did, that in the end and by the narrowest of margins it worked.

On August 30, 12 Group was called in to defend Biggin Hill. It was not Bader this time but another squadron using his tactics; they circled and circled while forming up into squadron strength before departing, and like Bader they were too late. Luckily, most of the bombs missed the airfield. But in the afternoon the raiders were back; by this time the radar system was overloaded and they snuck in without warning, appearing suddenly overhead, and this time they did not miss. One bomb made a direct hit on a slit trench crowded with ground personnel, other bombs hit the gas and water mains, the electric cables, hangars and depots and workshops. Once again, and at the must crucial stage of the battle, Biggin Hill was out of action.

At the same time, 12 Group was called upon to defend North Weald. This time it was Bader's turn, with 242 Squadron. As they circled, wasting time, the radar instructions came in through Bader's headset. "Vector 190 at angels fifteen. Patrol North Weald. Bandits seventy-plus."

It took them fifteen minutes. And all the while they climbed to altitude Bader was thinking. It was late afternoon: if the Jerries had any sense they'd come in from the west with the sun at their backs.

He wheeled the squadron around.

"Vector 190," the Controller said, heading them due south.

Bader took them out on vector 260, bringing them around to the west.

"Angels fifteen," the Controller said.

But Bader knew that height estimation by radar was rough, and he wanted to be above the enemy. He took his lads up to 20,000 feet.

And there he circled, twenty miles westerly and 5000 feet above where the Controller had sent him....

"Tallyho! Bandits at ten o'clock low!"

"Right, I see them." It was a group of thirty Heinkels with Me 110 escorts, maybe fifty or sixty in all. Bader's squadron of twelve was in perfect position for a textbook bounce, and he wheeled over and took them down.

It was beautiful. The Hurricanes tore through them and although they weren't able to scatter the bombers or deter them from their bombing run, they claimed twelve aircraft shot down without loss. Bader was triumphant. As his biographer, Laddie Lucas, claimed, "In a few lethal moments all Bader's long-held beliefs had been confirmed." Reporting afterward to Leigh-Mallory, he made no secret of what he could have done had he had two or even three squadrons flying in formation rather than only his one. He could have decimated the enemy formation.

Leigh-Mallory agreed. From now on when called into action by 11 Group, Bader would fly his squadron to Duxford first, there to form up with 19 and 66 squadrons, and fly them into combat as a Wing. This, of course, only made the interception situation worse, as it would take even more time to travel to Duxford and form everyone up into battle formation.

They didn't know till after the war that on this first battle 242 Squadron, despite its perfect position and classic bounce, had shot down only two of the Germans, not twelve. This was to remain throughout the war a feature of air actions involving the big wings: with so many people attacking at once, several planes would be aiming at the same target, and when it went down they would each claim it individually.

They should have known, or at least Leigh-Mallory *must* have, that their interception was pure luck: Bader had guessed right that time as to the enemy's intentions, but if the Jerries had come in low and straight and hard as they often did, Bader's squadron would have been sitting all alone up at 20,000 feet fifteen miles west of where they should have been. The entire system of defense that Dowding had set up depended on the fighter pilots following the instructions of the radar-directed controllers; not only did that put them where the enemy was, it also meant that the controllers knew where their own aircraft were so they could direct reinforcements where needed. Bader's intuitive interceptions were romantic, dashing, and only very occasionally effective. They were the very antithesis of the system being used by the rest of Fighter Command, and if followed by more squadrons would have lost them the war.

It's easy to understand and sympathize with Bader; it's always hard for the man doing the fighting to realize that he doesn't have the whole picture in front of him and to obey orders when he thinks he knows better. It's harder to forgive Leigh-Mallory. He was so taken by Bader's success on this day, so impressed with the overblown claim of twelve enemy aircraft destroyed and so hypnotized by the vision of hundreds of Jerries falling to the concentrated guns of No. 12 Group—*his* group—that on September 9 he went even further, giving Bader another two squadrons to add to his Wing.

The results were predictable. When the Wing got into combat it did well. Of course it did; it had overwhelming numbers and Bader saw to it they always had the advantage of position. But because of the time it took to form up and the effect of following his own guesses as to where the enemy might be, Bader often was late or missed the attack entirely. And day after day the all-important aerodromes were being destroyed, with airplanes and support and repair facilities and communications cables and radar stations going up in smoke with them.

Biggin Hill, the pivot of 11 Group's work, was virtually deserted; one squadron still operated from there, but the others had been moved and all the supporting operations were nonfunctioning. Another five of the most used airfields were either totally out of action or virtually so, and the six sector stations were all heavily damaged and barely managing to limp along. The coastal stations Manston and Lympne had been totally shut down. By the beginning of September the intricate system of interlocking fighter support controlled by radar and fed in from the spiderweb of stations was falling apart, crumbling. Gaps were appearing in the defenses wide enough to drive whole *Staffeln* of enemy bombers through.

The whole future of western civilization hung by the most slender of threads as the first few days of September slipped over the edge of time.

Chapter Thirteen
London Calling...

> Oh it's a long long while from
> May to December,
> But the days grow short when
> you reach September.
> —*Maxwell Anderson*

In the war of nearly infinite devastation which began in September 1939, laying waste to the continent that was the cradle of our civilization and reducing it to a buffer area between two snarling giants, the Bomber Command air raid on Berlin the night of August 25, 1940, was little more than a minor action. Not a single German was killed, no factories were destroyed, nothing of any military significance was even damaged. And yet the action that night was a fulcrum on which the fate of civilization creakily turned.

From the beginning, Goering had wanted to hit London. As the summer waxed and began to wane, as he saw the awful hint of autumn weather beginning to slip into the meteorological reports, as his pilots came back day after day to report that those last few damned Spitfires were still there, he began to feel an irresistible swell of angry desperation creeping up under his bulging belly. Somehow he was continually frustrated in his efforts to bring those last British fighters up into the battle, somehow Dowding always seemed to have a few in reserve. But if he could attack London—?

They'd have to defend it. They'd have to send up everything they had. And then in a few days, a week at most, his Messerschmitts would destroy them, would sweep the skies clear, and the British and their Empire would crumble at his feet.

But Hitler said no. He had not bombed Paris and he would not bomb London. Somehow in his twisted mind the splayed and splattered bodies that lay crumpled in the streets of the cities and villages of France and Poland and Belgium had nothing to do with European civilization; they could be forgotten. The rich history of Europe lay rather in the majestic architecture of Notre Dame, in the wide, stately boulevard of the Champs Elysees, in the Kurfürstendamm and Unter den Linden, in Buckingham Palace and St. Paul's and the West End: in the streets and monuments and offices and palaces of Paris, Berlin, and London. These would sail untouched and untouchable through the chaos of war, a monument to the spirituality of the Aryan race.

And then the bombs fell out of the night on Berlin. It is difficult for sane minds to grasp that in this cataclysmic eruption of a war which snuffed the lives of millions of people and crushed their cities into rubble, a turning point whose effects will be with us till the end of our times occurred when one petulant politician stood up to a psychopath. Like two angry little children they faced each other across the Channel as if it were a line drawn in the dirt, daring each other to step over it.

"I'll bomb your city!"

"Oh, yeah? Then I'll bomb yours!"

And so they did. Because the Luftwaffe hit London—by mistake or otherwise—on August 24, Churchill sent the RAF to Berlin. And the next night he sent them again, and the night after that, night after night until finally on the fourth of September Hitler's anger broke its bounds and gasping with rage and frustration at Churchill's outrageous effrontery, at his inability to comprehend the sanctity of Berlin, he personally promised in a furiously raging speech to "wipe London from the face of the earth."

Into the radio microphones he screamed: "If they intend to attack our cities, then we will raze *their* cities to the ground! We will stop the murderous activities of these air pirates, so help us God! If the British air force drops 3000 or 4000 kilos of bombs, then we will drop 300,000 or 400,000 kilos!"

To Goering that same night he gave the final orders: Destroy London. Smash it into rubble. Erase it from the map of the world and scatter salt over the ruins.

Go!

September 7, 1940.

The radar screens were quiet. Since early dawn when the sun had first come up to reveal a relentlessly clear day, the operators had been waiting. The personnel in the operations rooms at Bentley Priory, Uxbridge, Biggin Hill, Kenley, at every sector aerodrome throughout the south of England, had taken their first and last look at the sun this day and had gone into the windowless plotting rooms. At Uxbridge, 11 Group headquarters, they were deep below ground, bombproof and safe, although when you were buried in there and the bombs were falling and the earth around you was shuddering, you weren't quite sure. At Kenley the room was below a few feet of earth, reasonably safe from anything except a direct hit. At Biggin Hill they had already been bombed out and were operating now from a butcher's shop in the neighboring village. At other airfields they were in wooden huts, exposed to whatever attack might come.

Early in the morning the usual single reconnaissance flights had been sighted by radar and a few sections had been scrambled. A Dornier was plotted and intercepted and shot down coming home from Liverpool. And then nothing.

On Thursday, September 5, Ultra had decoded a message from Goering to Kesselring ordering a massive raid on London, but neither on that day nor the next had the attack materialized. Instead the Luftwaffe had continued to batter Fighter Command's aerodromes. Losses on these two days had been exceptionally high, too high to be maintained for much longer. Dowding's chicks were on the verge of exhaustion.

At Bentley Priory he strode silently back and forth. The previous day the King and Queen had come for a visit and had had to remain underground in the Ops Room for an hour because a raid was in progress. There had been raids in progress all day long, striking again and again at his dying airfields.

Today there was nothing.

Three weeks before, Fighter Command had been on the brink of exhaustion and defeat. Then too they couldn't have lasted more than another few days without a break, a respite, a rest of some kind. In that hour of need God had intervened, Dowding remembered, sending four days of rain and fog, keeping the Luftwaffe on the ground and enabling him to reorganize his tired squadrons, repair his battered aerodromes, and get his bombed radar stations back on line.

That rest had enabled them to fight another two weeks. But now they were again exhausted, and the aerodromes and radar stations were in even worse shape than before. They couldn't possibly last more than another few days, and his reconnaissance and intelligence reports left no doubt that Hitler's invasion was cranking up and getting ready to move. In another few days, if the Spits were driven out of the air, the armada would cross the Channel and all would be over. Hitler had broadcast his warning: "In England they are filled with fright and wonder. They keep asking, 'Why doesn't he come?' Be calm. Be calm. *Er kommt! Er kommt!*"

In England they believed him. That very morning the Air Ministry had issued to all commands its Invasion Alert No. 1: *Invasion is imminent and may be expected within twenty-four hours.*

Dowding opened the tall French door in his office at Bentley Priory and stepped outside onto the terrace. He walked a few steps up and down, looking out over his formal gardens and drumming his fingers nervously. There was nothing else he could do. Two or three more days of this constant hammering and he wouldn't have an aerodrome or radar station left operative in southern England; he'd be helpless to prevent the invasion.

Three weeks ago it had rained. Today he stood on the terrace and looked up at the sky and saw that it was blue, pale blue, flecked with white cumulus clouds, perfect. A lovely September day.

Where was God today?

Dowding stood on the open veranda and listened to the quiet. The only sound was the muted buzzing of the bumblebees as they hovered over the budding flowers which covered the gentle incline below him. Not the slightest hint of the rumble of aircraft engines broke the silence. Come to think of it, where was the bloody Luftwaffe?

They were on the Pas de Calais, at a dozen airfields strewn throughout that land. The fuel tanks were being topped off, the pilots were being briefed, the bombs were being loaded. *Kampfgeschwader 1 und 2 und 3 und 26 und 76* were being bombed up, the fighters of *ZG 2 und LG 2 und JG 2 und 3 und 51 und 52 und 54 und 77* were being gunned up, a thousand aircraft were being readied to *fliegen gegen England*.

On the cliffs of Calais, on the coast looking out across the Channel to England, Hermann Goering arrived with a full entourage of photographers and reporters. He had taken personal command, he informed them, and from this vantage point he would stand and watch his chargers sail overhead as they flew to destroy the British Empire.

At six minutes before four o'clock the controllers seated in the balcony in the operations room saw the first purposeful movement on the floor below. As the headphoned WAAF moved, it was immediately obvious that she was not just going for another cup of tea. She fetched a counter and placed it on the map table. It was labeled "20+."

In the next few minutes other plots were quickly telephoned through, and soon the map table was covered with its normal assortment of colored arrows and labeled counters moving across the Channel. Dowding was fetched from his office, and by the time he reached Operations, the table was figuratively shuddering with the weight of the counters: There were more than ever before, and minute by minute as he looked down on Armageddon, the calm, efficient WAAFs responded to the voices in their headphones by placing new counters beside the old. From every airfield along the French coast the Luftwaffe bombers and fighters were swarming into the air now, like the bumblebees around the flowers outside in the garden, swarming toward England. On the squadron readiness boards lining the wall, lights began to come on as every squadron in 11 Group was ordered to varying states of readiness.

"Better get them into the air."

Radar was indicating altitudes of 15,000 to 20,000 feet, unusually high; it would take extra time to climb up to them. But where should the fighters be sent? The German hordes were coming straight as arrows across the Channel, headed it seemed for London, but their usual tactic was to break direction at the last moment and scatter for the targeted airfields. The British fighters were sent to guard the precious airfields, patrolling the skies to the south of London.

Though it was late in the day, it was still summer and it was a Saturday; throughout the south of England people were relaxing and trying to forget the war. White-trousered players in every cricket patch along the Thames Estuary were making full use of the last of the hot, dry weather. Bowlers and batsmen stopped as a growing roar reached their ears. They stood still and looked up, hands shielding their eyes against the bright and terrible sun which a few moments before had been a bright, warm, loving relic of summer, as they watched the stream of bombers file slowly, majestically overhead.

The German formation was twenty miles wide and forty miles long. It took a full ten minutes to fly by. For ten minutes the sky overhead was black with strange airplanes, the air shook and trembled for fully ten minutes with the power of their passing, the sun was blotted out by a horde of locusts with black crosses on their wings, covering eight hundred square miles of English soil with their shadow. And finally, finally the end of the procession appeared and swept inexorably overhead and disappeared up the Thames Estuary towards London.

The cricketeers watched them go, then turned to look at each other, then shrugged, and the bowler hopped two steps forward and flung the ball and the batsman stepped forward smartly and swung at it, but their concentration was no longer on the game.

And up the valley the Dorniers and the Heinkels and the Messerschmitts roared. Motorists in Sittingbourne heard their angry sound and pulled off the road and put their heads out the window and stared up to watch them pass. All those foreign wings in a British sky!

"Where are our boys, then?"

Families picnicking in the fields outside Bearsted saw them appear over the horizon and march steadily up the estuary towards the city that lay at its mouth. Parents lay back on the grass, where a moment before they had been savoring the last days of summer, and held their children tight and watched without words the slow 200-mile-an-hour progress of the heavy, lumbering birds, obscene with the vile load in their bellies, pulling themselves like terrible maggots across the English countryside.

"They'll never get to London. Never will they get through."

"But where are our boys, then?"

They were all committed. They were in the air patrolling their vital airfields, waiting for the bombers to turn. And still the bombers came on.

"Notify 12 Group," the command went out.

And they waited, watching the counters being pushed inexorably straight toward London, watching more and more counters being placed on the board. Finally there were more than 300 bombers and, worse, more than 600 fighters streaming up the Thames: the largest aerial bombing attack in history. Within half an hour of the first sighting, every one of Park's 11 Group fighters was in the air and now, as the immensity of the attacking force was apparent on the map table, its objective was equally clear. They were coming straight as an arrow for London; they were not going to swerve, to break into multiple attacks aimed at the aerodromes; they were going to go straight in.

Quickly the orders went out from the ops room, bringing the defending fighters in from their wide orbits. At first only four squadrons were close enough to engage the enemy....

"Hello, Hornet Leader," came the voice of the controller, sounding in the Spitfire cockpit three miles above the face of the threatened earth. "This is Big John. Vector zero nine zero. Forty plus bandits angels eighteen just a few miles in front of you. Do you see them?"

"Hello, Big John, this is Hornet Leader. Sighting negative. Listening out."

They opened up to search formation and then, just passing Sheppey with the estuary right in front of them, they wheeled around a towering cloud and slipped past a sharp bank of haze and broke into the clear and there, stretched out in front of them for miles and miles and forever were a million German planes. One moment the Spits were sailing through a white haze, checking their gun sights and buttons and the fuel-tank selectors and the engine temperatures, making sure that everything was running smoothly, and the next second the whole sky opened up to them as if a curtain had been pulled and there was a feast laid out for them with row after row of thick black knives on a blue tablecloth. Row upon row, stacked up one on top of the other, two miles thick, Dorniers and Heinkels in the center and around them and pulled up high above them swarms and swarms of Messerschmitts, hundreds of them, stretching without end as far as the eye could see, filling the sky with their thin wings and black crosses, nearly blotting out the sun as they stretched in from the coast.

The Spitfires wheeled and without hesitation dove into that black mass, but they couldn't stop them. What followed was a catastrophe unprecedented in the history of human warfare, though it would be eclipsed in the coming years as the combined British and American air forces struck back at Cologne, Hamburg, and Berlin; it was the beginning of a catastrophe that would be climaxed five years later at Hiroshima and Nagasaki. The German bombers got through, as Baldwin had warned they must; they came through the fighter defenses and others came streaming in behind them, and London began to explode and to burn.

There was no stopping them. There were too many of them and the British defenders were too few and too exhausted and had been kept too long patrolling over their scattered airfields. They came tearing into the German bomber formations as quickly and ferociously as they could, but the 650 Messerschmitts drove them off again and again. They were too few and too late. Up at Coltishall, Douglas Bader's 242 Squadron was scrambled, but they did not head immediately into the fight. Leigh-Mallory was going all the way with the Wing concept. He had told Bader that the next time they were called on for help he was to take the three Coltishall squadrons and rendezvous with two more from Duxford and lead them all in one massive attack. He was to ignore any specific requests or orders from 11 Group controllers; he was instead to take the radar information they gave him about the Jerry attack and decide himself how best to use it.

As they climbed away from Duxford toward London, Bader tried desperately to guide them all into a proper formation, but such large numbers were simply too cumbersome for the rapid maneuvering that was necessary. One of the squadrons, No. 303, was Polish; they were survivors of the Warsaw massacre who one by one had slipped across the map of German-controlled Europe and found their way to the last country still free. Despite their atrocious English, which made them difficult to control in the air, they had recently been made operational. Now, as had been feared, they conveniently forgot what English they had and ignored Bader's orders, leaving the convoluting, half-formed Wing behind, climbing full-throttle and alone to find the Germans.

While the German bombers were setting London on fire, Bader kept trying to form the Wing properly, until finally he sighted a large formation of bombers—too late. They had already unloaded their bombs and were on their way home. He led his half-formed Wing into a diving attack, shooting down one bomber and losing two Hurricanes. The other two squadrons of the Wing never arrived in time to do any fighting at all.

11 Group hit the bombers as hard and as quickly as they could. They hurt them, but they couldn't stop them. Trailing individual plumes of smoke and flame as one bomber after another dropped out of formation and fell to the ground, the black phalanx turned up the estuary and

crossed over the city itself. The ugly bellies opened and bombs dropped through 20,000 feet of empty air onto London. They exploded in the streets of Westminster and Kensington; they cracked the pavements and tore down the docks at Rotherhithe, Limehouse, and Millwall; they split open Woolwich Arsenal.

Fire spurted into the air and along the streets and into the Thames. A sugar warehouse spilled blazing melted sugar in a fiery stream, as from a volcano, that spread out and covered the surface of the Thames itself with fire.

Bombs fell and ignited the oil tanks of Cliffe and Thameshaven, the gasworks at Becton. The towns of Croydon and Barking, West Ham and Silvertown blazed from end to end. Despite all of Fighter Command's tightly hoarded fury, the Luftwaffe broke through and dropped their bombs. Beneath their thunder London shuddered and trembled and burnt...but south of London the ground crews were repairing the damage at Kenley and Biggin Hill and Tangmere. At the aerodromes in Manston and Ramsgate and Croydon and North Weald they filled in the craters and cleared away the rubble, repaired the telephone lines and the maintenance shops. At the Chain Home radar stations in Ventnor and Dover and Rye and Pevesey, new generators were installed and new wires strung between the aerials, and they came back on the air. And when the fighters returned home after shooting down nearly fifty German aircraft, they found no towering black fumes rising from the cratered grass; they found instead their airfields in better condition than when they had left them; they found a quietly busy workplace waiting to greet them and refuel them and rearm them.

London had been hurt. But England had been saved.

The invasion alert sent out by the Air Ministry that morning had been premature, based on their estimate of aerial intelligence which showed the buildup of channel-crossing barges and the loading of equipment on the other side of the Channel. They were wrong, but in fact they were not very far off the mark; when *Adler Tag* had been planned in mid-August it had been correlated by Hitler with a precise timetable for invasion. The day had been set for September 21.

Now, after his bombers had battered London and had found only scattered and badly coordinated defensive measures by the RAF, Goering reported that the invasion could take place as scheduled. The RAF had been defeated and the way lay open.

But the Kriegsmarine yet demurred. After all, despite Goering's happy smile, nearly fifty Luftwaffe planes had been lost; obviously there were at least a few British fighters still around.

And so it was settled: the Luftwaffe would hit London again.

The Battle

* * *

They followed up the September 7 daylight assault with continued attacks throughout the night, and they came back again on the next night, but it was the daylight assault that mattered most. At night the city was defenseless, the bombers invisible; but precisely for this reason the night raids could not destroy Fighter Command. This was the beginning of what the English called, with their gorgeous, condescending ignorance of other languages, the blitz: the beginning of fifty-seven consecutive nights of air raids on London. It was the beginning of what Dowding called another miracle: a chance for Fighter Command to breathe.

On the eighth of September there were only sporadic attacks, but on the ninth the Luftwaffe made once again a concerted effort to destroy those "last few Spitfires." The invasion timetable was set to begin within forty-eight hours: For invasion to take place on the twenty-first, the initial dispositions such as the laying of a mine field across the Channel were scheduled for September 11. Hitler was waiting only for the results of this day's fighting to issue the order to begin.

But this day was to be a shocker. Aware now of the target, the controllers had no hesitation in assigning their fighters properly as soon as the first radar information came in. The two-day respite from bombing had enabled nearly all the important Chain Home stations to be brought back on line, and the information they fed into the system was perfect. The bombers were intercepted on their way in, and although the attack was not as heavy as the initial one two days previously, the RAF response was much better and nearly as many Luftwaffe planes were destroyed.

Goering was stunned. He had been sure the Luftwaffe would find the skies empty that day over London. Hitler was quietly furious. He postponed Operation Sea Lion until September 24, nearly the last day the invasion could possibly take place. The Kriegsmarine insisted that the landing be made on an ebb tide in order that the landing craft be securely beached, while the Wehrmacht insisted that the landing be made at dawn—which meant that a night with a full moon must be matched with an ebb tide coincident with the following dawn. These conditions would be met September 19 to 26 for the last time before the winter rains and snow would come. Postponement beyond that date would be irrevocable.

At the beginning of July Goering had promised to destroy the RAF "within a few days." That was precisely what he now had left.

On the tenth the Luftwaffe took stock, and on the eleventh they attacked again. The fighting was fierce and heavy, marked with victories on both sides. London was at the limit of the Me 109's fuel endurance, and one squadron of bombers was left unprotected when their escorts had to return home; six squadrons of Hurricanes and Spitfires then raked the bombers and left them a bedraggled mess. But when a radar plot was given with the wrong height, another two squadrons came in below an enemy formation, were bounced by the Me's, and lost a dozen fighters. At the end of the day the losses were upwards of forty aircraft for each side, indicating to Goering that the RAF had at the beginning of the day more fight left in them than he had thought, but at the end of such a day *surely* they were nearly finished. The problem was that he was running out of aircraft of his own, and he was running out of time.

He now had a choice of sending over whatever bombers and fighters he had as quickly as he could, or holding them back to reorganize and refit before attacking in strength. He chose the latter course, and the twelfth and the thirteenth went by quietly enough. On the fourteenth the weather deteriorated, and for one more day Goering held his hand. He sent out a few raids to hit the radar stations in preparation for a massive assault set for the following day; just in case Dowding had any fighters left, he wanted them blinded. On Sunday, 15 September, he would hit them with everything he had. His fighters were rested, his bombers loaded and ready to go. He informed Hitler that the invasion could proceed as planned.

Hitler nodded, and said nothing. Goering had promised as much before. When he left, Hitler called in *Generalfeldmarschall* Wilhelm Keitel, chief of the OKW, and informed him that he had decided to postpone once again, and for the last time, his decision on Sea Lion. He would wait until Goering's supreme effort was over, and then he would evaluate the situation. He would make a final decision on September 17.

Sunday, 15 September.

The dawn sun came up into a cloudless sky. The white mists left over from the damp night were quickly burnt away, huddling about the wheels of the Spitfires as they were trundled out to their dispersal points, blown back over the grass as the engines coughed and caught and the propellers whirled and spun around, heated and dissipated by the sun as it climbed steadily higher into a blue September sky.

The pilots stretched out in their deck chairs, faces lifted with closed eyes to the heat of the sun. They had had a few days off, they were young, they were ready again. The radar antennae sent off their invisible rays, the receivers waited patiently for some of them to come scattered back. But the morning hours passed, the sun climbed higher, the pilots began to doze, and the radar beams disappeared over the horizon.

The Battle

1050 hours. The Chain Home station at Rye reported an echo. The first marker was put on the plotting table. Twenty-plus forming up ten miles southeast of Boulogne.

Keith Park ordered two squadrons to readiness.

More plots began to come in. Ten-plus, twenty-plus, thirty-plus. It was going to be another big one. Clearly, even then, it was going to be the climactic big one. The coming autumnal weather was no secret on either side of the Channel; everyone in England knew the invasion had to come soon or not at all. And it was no secret that the Royal Navy could prevent the invasion if the Luftwaffe could be held at bay. The only secret was whether or not Fighter Command still existed.

The secret would be out today.

Park ordered every one of his squadrons to readiness, and then waited. There was something strange about the radar reports: The clouds of aircraft were not immediately moving across the Channel, but were rather circling as they gained height and built up their numbers. Then they moved off westwards, staying safely over the soil of France while they picked up their fighter escort.

Unbelievable. Park had to smile. Up till now they had moved over the water as rapidly as possible, picking up their fighters on the way. But today, evidently to make sure there was no hitch in the rendezvous, they were forming up before they started across the water—never realizing that they were being watched all the while by the radar eyes of the British. They were giving Fighter Command what it needed most: time.

"Notify 12 Group," Park ordered. The luxury of having time to think was pleasant; even more pleasant was the luxury it gave to 12 Group to form its Wing and climb to altitude in time to intercept.

That first raid was met by two squadrons from 11 Group as it crossed the coast over Dungeness, and at first it was the same old story: twenty Spits diving into the attack against hundreds of Germans. But soon another squadron joined them, and within minutes three more dove in. At a few minutes before noon the German bomber pilots had fought their way to within sight of London, when another sight suddenly burst upon their eyes. Four squadrons of Hurricanes came diving out of the noonday sun in a wild, head-on attack. As the bombers tried to turn away, Bader's Duxford Wing came upon them; five whole squadrons of fresh fighters blasting their way through the escorting Messerschmitts before they could react and sweeping like a scythe through the bomber formation.

The Luftwaffe broke and fled. They dropped their bombs where they were and turned and ran. The bombs scattered all over London, all over Kent. They fell on houses and pubs in Battersea and Beckenham, they fell on Camberwell and Clapham and Chelsea, on Wandsworth and Westminster and Victoria Station and Buckingham Palace, and it seemed that for every bomb that fell a bomber fell with it.

When the remnants returned to the airfields with their reports of the morning's action, Goering flew into a rage. He was on his microphone without delay, screaming at all units to get gassed up and bombed up; they were going right back again to catch the British on the ground and unprepared. He was a madman.

He was doomed. The British radio interception unit picked up his message and the decoders at Bletchley passed it on to Dowding within the hour. As soon as the Spits and Hurris returned to base they were rearmed and refueled, the pilots were stoked up on tea and biscuits, and they were ready again; and for the first time the scent of victory was in their nostrils. As they finished draining the thick mugs of strong tea laced with sugar and milk, the first telephone calls were already coming through from the Chain Home stations, reporting yet again plots of twenty-plus and thirty-plus building up all over northeastern France. By two o'clock hundreds more bombers and escorting fighters were sweeping in once again over Kent, heading one more time for London.

The first Hurricanes met the armada over the coast, but were unable to break through the clouds of fighters around them. Then a squadron of Spitfires waded in, taking on the Me's in an effort to free the Hurrithings for the bombers. But there were too many escorts; they maintained an unpassable barrier. The German formation fought its way through the Kentish skies toward London.

And then another squadron of Spits came sailing over the clouds, and another of Hurris came under them. Squadron after squadron came flying into the fray as Park followed Dowding's orders and sent up everything they had; and little by little the Messerschmitt screen was penetrated. One and two Spitfires here, a flight of Hurricanes there, pressed through and twisted among the bomber formations, often in a determined head-on attack at the lead *vic* of a formation that broke up the entire group. Once the tight formations were broken up, it became nearly impossible for the escorting Messerschmitts to shepherd them all. As they approached London, the bombers began to scatter, individuals falling away from the edges of their formations, but still there was a central mass that edged forward into the sky over London itself.

And there the final force burst upon them.

Bader's Duxford Wing was as unwieldy as his critics all said; it had not been possible to keep it in formation and climb quickly enough to reach the enemy throughout the vicious fighting of the summer; it had been unable to respond in time to save Park's aerodromes. But on this climactic day, the radar stations had seen the enemy early enough to give sufficient warning, and the target the enemy was heading for was clear and obvious, and there was no tomorrow, there was no thought of holding anything in reserve, there were no airfields to protect. And finally, for the first time, it worked. As the bombers edged over East London they saw the five-squadron Wing come wheeling around the edge of a cloud and disappear into the blazing glare of the sun, and then moments later the entire Wing came crashing into them, pouring through the suddenly overwhelmed escorts like a flood through Holland, eight machine guns in the wings of every one of them filling the skies with tracer bullets and with burning Heinkel engines and falling Dorniers and flowerlike parachutes.

The bombers faltered as the Messerschmitts turned and twisted, trying now more to save themselves than the bombers. The British assault was simply too furious, too overwhelming, too much. To the German bomber crews who had fought and bled their way across England all that summer, who had now reached London once again and were opening their bomb bays, this overwhelming onslaught by squadron after squadron of Spitfires spilling out of the skies where no more Spitfires should have existed was quite simply too much.

They flinched, they broke, they scattered. Their bombs began to fall at random, and so did the bombers themselves, with wings raked off by machine-gun fire and cockpits shattered by bullets.

And then two more squadrons reached them from 10 Group, plowing through the now disorganized and bewildered escorts that began to circle helplessly, not believing what they saw. Where were all these British fighters coming from?

On the ground below, the radar stations were showing the rest of the skies clear: There were no other raids coming in, there were no more building up over France. Park gave his orders to send in everything, hold nothing back.

The Luftwaffe couldn't believe it. They had been told there were no more fighters left in Britain. And now still more came! Six full squadrons of 11 Group found the fight and sailed into it, and it became a rout. The bombers turned and fled. They jettisoned their bombs without any hope of aiming or of concentrating their damage. Bombs dropped over East Ham, over Eric and Lewisham and Hackney, bombs dropped without purpose or reason as the bombers turned and fled headlong for their lives.

The Messerschmitts rallied as best they could and tried to protect them, but London was too far from the bases they had taken off from and the fight had gone on too long. They had been twisting and turning at top speed for minute after precious minute and now in each Messerschmitt cockpit a small red light began to wink. Their fuel was low, their fuel was nearly gone, and they still had fifty miles of England and another twenty miles of cold, wet Channel waters to cross. They began to dive from the fight and streak for home, with Spitfires spitting tracers at their tails and with Hurricanes turning their attention to the bombers they were forced to leave behind.

After all these weeks of brilliant summer, in the last few minutes of the Battle of Britain the terrible struggle became a slaughter.

On the next day, Monday the sixteenth of September, every squadron in Fighter Command was at dawn readiness as the sun broke over the horizon and climbed into a clear blue sky, but nothing happened. The radar operators sat tensely at their screens, waiting for the first echoes to come beeping back at them, but nothing returned. Their antennae sent out continuous streams of electromagnetic waves across the Channel, and none of the waves bounced back. At the aerodromes, the erks kept the Spitfire engines warmed up, the pilots dozed in the sun or chewed their nails or vomited whenever the telephone rang, and nothing happened. They waited.

Across the Channel the Luftwaffe commanders were meeting with Goering. It was a somber meeting. They had gone into yesterday's battle thinking they had it all won, but Fighter Command had risen up and bashed them and broken them. They were bewildered and beaten: Wherever they attacked, there was always a squadron of Spitfires waiting for them. Surprise was impossible to achieve. The radar stations seemed indestructible, and they saw everything. It was hopeless.

Nonsense, Goering answered. He tried to cheer them up. They would destroy Fighter Command yet. He smiled around the table. No one met his little pig's eyes.

When the conference broke up he sent a message to Hitler that he was confident that the Royal Air Force would be destroyed in "another four or five days."

That was exactly what he had said at the beginning of July, at the beginning of August, at the beginning of September.

He received no reply from Hitler.

What he did receive were messages from one *Fliegerkorps* after another reporting missing aircraft, damaged aircraft that would never fly again, wounded aircrew who could not fly again.

A few reconnaissance Junkers came over England later that day to see what damage had been done. Wherever they came, they found the Spits waiting for them. Nine were shot down.

On the next day, Tuesday the seventeenth of September, the day of Hitler's final invasion decision, a signal went out from the German general staff to a particular officer in Holland. The British wireless interception unit was listening, and they picked it up and sent it to the Ultra decoding experts at Bletchley Park. When they finished reading it, they leaned back in their chairs, rubbed their eyes, and smiled; and then they sent it directly to Churchill.

The Battle

Back in July, when the invasion of England was first being organized, they had intercepted messages which gave authority to this particular officer to set up on airfields in Holland a unit whose job was to be the organization of quick turnarounds for supply and troop-carrying aircraft. The significance of this unit was apparent. During the invasion, paratroopers and supplies would be ferried across the Channel; upon completing the round trip, the transport planes would be quickly reloaded and sent back.

Today the signal went out informing this particular officer that the rapid air-loading equipment on the Dutch airfields was to be dismantled, and he himself was to report back to headquarters for reassignment.

The Battle of Britain was over.

Chapter Fourteen
Pride and Petulance

> The Captains and the Kings depart...
> Lord God of Hosts, be with us yet...
> —Rudyard Kipling,
> "Recessional"

The August 25 bombing of Berlin was a terrible tactical mistake; presumably that is why Churchill doesn't mention it in his history of the war.

The Air Staff foresaw the German reaction when Churchill ordered out the bombers; they warned that the Luftwaffe would retaliate on London, which had so far been spared the horrors of aerial attack. Further, they warned Churchill that the Luftwaffe could bomb London more easily and more destructively by far than they could hit Berlin, and, given the barbarous savagery of the Nazi government, they warned him that he was inviting the very assault they had all feared throughout the thirties: If he insisted, he would bring destruction and misery down upon London on a scale unthought of by previous generations.

But he did insist, he would not give way, he would not listen to their arguments, and the bombers flew to Berlin. He insisted out of pride and petulance, and the calamitous consequences he had been warned of followed: Terror and fire rained down by day and by night onto London. The daytime attacks were followed by two solid months of bombing every night, and because of the relief afforded by that diversion, Dowding's reeling aerodromes and radar stations were saved—and so was England. That petulant outburst of Churchill's—which the Air Staff argued against as useless—was the most important single action he ever took in his life. But it was too much to take credit for, too much to take blame for, and so in his history that decision is simply ignored.

It was the final "miracle" of the Battle of Britain.

*   *   *

In September 1940, Dowding was knighted; two months later, in the second week of November, he answered a phone call. It was from the Secretary of State for Air, who told him that he was to relinquish his command immediately. Having forgotten in the excitement of battle and in the flush of victory the aggravating instances of periodic dismissal that had been inflicted on him in the months and years previously, he was somewhat taken aback by the suddenness of this new order. He asked what was meant by "immediately." The answer was that it meant "within twenty-four hours." He protested: "It was perfectly absurd that I should be relieved of my Command in this way unless it was thought that I had committed some major crime," but no explanation was given, no protest accepted; the decision was final.

Looking back on it now, it seems as unavoidable as it was uncharitable. Although he was not to write his books about the spirit world and his visits with dead fliers and with his dead wife for another couple of years, he was beginning to talk about such things to his associates, and it must have been embarrassing for the Air Ministry. People were beginning to smile.

And then there were the intrigues that had been festering all summer. Trafford Leigh-Mallory, the No. 12 Group commander, had nearly lost the Battle of Britain with his refusal to follow Dowding's strategy. By allowing Douglas Bader to form his fighters into Wings, he had effectively ruined any chance of getting them to 11 Group's airfields in time to prevent the disastrous bomber attacks on them. In his makeup Leigh-Mallory had more of Macbeth than Hamlet: His ambition overrode any waverings that might have been born of loyalty to his commander. He was well connected politically and socially and, as he threatened to Park at the time, he did not hesitate to pull what strings he could—to "move heaven and earth"—in order to get Dowding fired. He made shameful use of Bader's ingenuous and naive belief that his own tactics were superior to Dowding's, even though he had no recourse to, or even knowledge of, the information that the Ultra intercepts brought to the AOC, nor had he Dowding's years of experience and intricate knowledge of both the extent and the limits of the radar system. It is easy to forgive Bader's youth and impetuosity, but Leigh-Mallory's maturity and sophistication made his actions monstrous. Using Bader's leglessness, bravery, and glamour, he brought him along to high-level conferences as an "example of the men who were actually doing the fighting," to protest against Dowding's incompetence. In actual fact, the men who were doing the fighting were overwhelmingly behind Dowding—but they never got into the conferences and libraries and clubrooms where things were decided.

Douglas Bader himself had never kept his thoughts private; all that summer he vociferously damned Dowding's policy of responding quickly with "penny-packets" of fighters instead of sending up whole Wings to attack the raiders. Although he was only a squadron leader, one of the very lowest echelons of operational officers, he was one of the Battle's most glamorous aces—and his squadron adjutant happened to be a member of Parliament. Peter Macdonald was as subject to Bader's charismatic leadership as was anyone else who served with or under him, and at the height of the Battle he was so overwhelmed by Bader's frustration at being kept out of the fight while others were dying down south that he took it upon himself to discuss the matter with Harold Balfour, the Under-Secretary of State for Air. Balfour recognized the nastiness of a squadron adjutant making use of his political connections to attempt to bring pressure on his military superiors, and refused to discuss the matter. Macdonald then utilized his parliamentary prerogative and asked to see the prime minister.

It is impossible to follow the thread that led from this meeting out through the corridors of Whitehall and the Air Ministry, although it is recorded that Churchill took a keen interest in Macdonald's arguments and followed up by discussing the matter with "various group commanders." It is unlikely that we shall ever know the truth of the spider's web that was woven with that thread; the official records have long been open to public scrutiny and all the principals are several years dead, so what is going to be known is already known, and still it remains an incomplete story. Churchill himself has claimed ignorance of the affair, saying that he was "surprised and disappointed" when he heard—after the fact—that Dowding had been dismissed. But Churchill was *never* ignorant of appointments and dismissals at senior levels under his administration: One of his strongest points—or weakest, depending upon one's point of view—was a compulsive need to keep his finger on the pulse of the body he was directing, whether it was the navy as First Lord or the nation as prime minister. There are no memoranda in existence which suggest that he asked for or demanded that Dowding be fired, but neither are there any which suggest that he protested the decision. The claim that he did not know about it until after it took effect is difficult to accept.

When Dowding took the floor at the War Cabinet meeting on May 16 and demanded that no more fighters be sent to France despite Churchill's personal promise to the French prime minister, and when he followed up by protesting Churchill's later decision to send them anyway—writing the famous letter that today hangs framed on the wall at the Air Staff College at Cranwell—he was warned that Churchill did not take kindly to subordinates arguing with him once his mind was made up, and that he never forgave what he perceived as a public insult. Dowding was told that, though he might continue to run around as a chicken can do when its head is cut off, from that moment he was just as dead.

Well, we shall never know. Whether it was because he talked to his dead wife and comforted his dead pilots or because he believed in mysterious emanations and "etheric doubles"; whether it was due to Leigh-Mallory's social/political maneuverings or to Douglas Bader's scathing arguments or to Peter Macdonald's parliamentary suggestions; or whether it was due simply to Churchill's long memory—in the afterglow of the greatest military victory since Waterloo, the victor was summarily fired.

And subtly denigrated. He was elevated to the peerage as Lord Dowding of Stanmore, but was denied promotion to Marshal of the Royal Air Force. He was asked to write an official report on the Battle of Britain, which was then never published, little circulated, totally ignored. The Marquess of Londonderry, a former air minister, wrote a book published in 1943, *Wings of Destiny*, which purported to give the history of the Battle of Britain as viewed by those officially in the know; unbelievably, in that entire 300-page book Dowding is mentioned en passant in just two sentences. He himself wrote a book in which his views of world harmony and military thought were entwined; Churchill refused to allow him to publish it, on the basis that "the Air Ministry are anxious about the technical aspects"—which was utter nonsense. The book, *Twelve Legions of Angels*, was published after the war, when Churchill was no longer prime minister. In it there are no "technical aspects" at all; it deals only with the most basic points of aircraft design and purpose, and this in a philosophical rather than technical or even strategic sense. He writes also of the need for one world religion and language, of the "brotherhood of man." It is all totally harmless poppycock.

At any rate and for whatever reason, Dowding was out. And Leigh-Mallory was in. As soon as Sholto Douglas took over Fighter Command, he turned Keith Park out to pasture with command of a training field, and brought Leigh-Mallory in to command 11 Group which, under Park's leadership, had become the most prestigious in the RAF. Leigh-Mallory rose in time to take over Dowding's position as Commander-in-Chief of Fighter Command, and in 1944 he was appointed chief of the newly created Allied Expeditionary Air Force, assuming supreme command of all Allied air activity during the invasion of Europe. Sent to take command of all air forces in southeast Asia the following year, he and his wife died when their airplane crashed.

Keith Park was eventually sent to Malta, where he repeated his brilliant use of the same radar-based tactics that he used in the Battle of Britain and kept that island free from invasion by fighting off a Luftwaffe which was again overwhelmingly superior in numbers. Aerial and naval strikes sent out by the British from that island redoubt severely hampered the German flow of arms and reinforcements to Africa and played an important part in the Allied victory there and the subsequent invasion of Europe from the south. In the bitter infighting between him and Leigh-Mallory, Park was granted the last move when, in a quietly ironic turn of fate, he was appointed Allied Air Commander-in-Chief, Southeast Asia Command—the post that Leigh-Mallory was flying to accept when he was killed. He fought the rest of the war in Burma as effectively as he had done in Europe. He died quietly in 1975 in his native New Zealand.

Douglas Bader collided with a Messerschmitt while leading an offensive fighter sweep over France. His Spitfire was torn in half, and one of his artificial legs was jammed and caught beneath the rudder pedal. As he struggled to get out of the cockpit it held him back, while the splintered Spitfire flipped over and over, falling through the air. Finally, with a great wrench, his leg came off like that of a chameleon escaping from a trap, and he fell out into the open sky.

He was taken prisoner. The Germans, astonished at this legless airman, were so impressed with his bravery that they searched the wreckage of his plane and dug out the missing leg. When they presented it to him, he pointed out that it had been badly smashed, and so they took it away again. The next day they brought it back not only repaired but cleaned and polished. That night he strapped it on and crawled through a window, climbed down a sheet-rope, and escaped. He was recaptured and, though he tried again, he stayed a prisoner until he was liberated by the advancing Allied armies. He died peacefully a few years ago.

And so the Battle of Britain ended and the invasion was called off. By the time the warm weather came back the following spring, the RAF was stronger than ever and the cross-Channel invasion was no longer possible. Instead, Hitler attacked Russia. The island thorn in his back remained; British planes nightly bombed his cities as a continuing reminder of their stubborn presence.

In 1942 a large American force sailed across the Atlantic and invaded Africa. They could not have done so if there had not been an English fortress at Malta splitting the Mediterranean and an English army in Egypt fighting Rommel from the east.

In 1943 the climactic struggle at Stalingrad was fought down to the last inch; if the Germans had been able to use their Afrika Korps and the Luftwaffe units that were tied up in Africa or those that were back home defending Germany from the RAF bombers, the scales would certainly have tipped in the opposite direction.

And during that year and the next, in the ancient scepter'd isle, in that seat of Mars, that fortress built by nature against infection and the hand of war, in that precious stone set in the silver sea, there was building up a repository of American and Commonwealth forces that in 1944 would cross the Channel to invade the enslaved continent and end the war.

But in September of 1940 all these events were hidden in the impenetrable mists of the future; in the present tense, nothing is inevitable. Though badly burnt with the loss of the Battle of Britain, the German military staffs still had the greatest bomber fleets, the biggest submarine forces, the largest modern armies in the world. They still stood astride the continent of Europe like a black colossus. Resolutely they turned their attention to other ways of winning the war.

Part Four
The Blitz and the "Boot"

From things that go bump in the night,
Good Lord deliver us....
Child's prayer

Chapter Fifteen
The Blitz

> Tyger! Tyger! sliding light
> Through the forests of the night,
> What immortal hand or eye
> Could find thy fearful symmetry?
> —*After William Blake, "The Tyger"*

The German strategy after the Battle of Britain was a continuing two-pronged attack, the less deadly of which became much the better known. The blitz went on night after night that fall and winter and spring. London was bombed for fifty-seven consecutive nights, Coventry was destroyed, Liverpool and Manchester and Birmingham and every other major city were hit hard and repeatedly. The English learned to sleep under their dining-room tables or in their cellars or in the tube stations of London, but never for a moment were they in danger of losing the war because of this nighttime bombardment.

Though it was frightening and deadly, the night bombing threat was not conclusive for several reasons. It was too early in the war, for one thing, and the techniques had not yet been perfected. Neither the Luftwaffe nor the RAF could yet find each other's cities consistently by night, and so a high percentage of both German and British bombs exploded harmlessly in fields and meadows on both sides of the Channel. But, even more than that, the Luftwaffe was simply the wrong instrument for such a purpose. Its planes had been designed as tactical weapons, as an extension of the army's artillery, and in that role they were awesome. They carried small bomb loads that could be placed accurately by dive-bombing on armies in the field, rather than the tremendous loads of giant bombs that had to saturate a city in order to destroy it. In the years to come, the RAF and the American Eighth Air Force would hang a curtain of doom over the Third Reich with thousand-plane fleets of four-engined bombers carrying up to 14,000 pounds of bombs, but in the winter of 1939–1940, the Luftwaffe had only twin-engined bombers carrying a maximum of 4000 pounds, with the largest bomb weighing, at most, 550 pounds. The attacks were simply insufficient to cause the massive destructions that were later laid on Hamburg and Cologne—and even those catastrophic bombings were not quite enough to break the spirit of a people and end the war.

Though the night bombing raids of the Luftwaffe were frightening and deadly enough to the people in the houses they hit, the British leaders knew that the more terrible, long-range threat was the strangling of supplies brought in to the island fortress as the U-boats sank her merchant shipping. But to the people crouching every night in their cellars or hiding under thick, oaken dining tables as their houses shook and crumbled and burst into flames, those invisible night bombers sailing untouched and untouchable above their heads seemed to be calling to them with their rumbling engines in the voice of implacable, unavoidable, unrelenting doom.

It was to be radar that would show the way to defeat both the night bombers and the U-boats. The technical problems that needed to be licked were the same for both threats. The Chain Home stations guarded the English coastline and gave warning of enemy airplanes as they approached, but it was capable of guiding interceptors only to within a few miles of them; in daylight this was good enough, but at night it was hopeless. A night fighter pilot could not be expected to see the enemy as a black shadow against the background of stars above or burning cities below until he was within a few hundred feet of him. Throughout the summer and fall of 1940 keen-eyed youngsters sat each evening with dark goggles protecting their night vision, then were scrambled and directed by radar as well as they could be toward the path of the invading night bombers and turned loose to find them on their own; invariably they failed. While they circled and peered helplessly into the darkness, the lumbering bombers sailed by a mile or two away, a few thousand feet above or below, and the keen-eyed youngsters saw nothing except the sudden splash of light as the German bombs suddenly exploded upon the darkened streets of London, Manchester, and Liverpool.

At sea the U-boats sailed out across the Bay of Biscay, dove below the sea, and gathered unseen into terrible wolf packs that lay in wait for the convoy streams. Shadowing the merchant ships beneath the waves by day, they would surface at night and slip in among them (their speed submerged was not sufficient for such maneuvering), and one by one the tankers would explode and burn. Lying low in the water, the U-boats skimmed among the waves as invisible at night on the surface as they were by day submerged.

*   *   *

The answer to both these problems was a radar set that would be small enough to be carried on a ship or an airplane and that could be tuned as finely as a searchlight. The difficulty in producing one seemed to lie in basic and insurmountable laws of physics. It is difficult to understand these concepts except by working laboriously through the mathematical equations which describe the generation and transmission of electromagnetic waves, and these differential equations are not accessible without a good deal of previous study of higher mathematics. But basically we can think of the problem as intrinsically related to wavelength: We are interested in generating a beam that remains thin and concentrated rather than spreading out into a diffuse glow, and we want equipment small enough to be carried in an airplane or on a destroyer, and mobile enough to be swiveled around rapidly. But as a wave promulgates outward, it spreads and diffuses in a manner proportional to its wavelength, so that long wavelength radiation cannot be confined in a narrow beam. And the generation of a beam depends on equipment with dimensions roughly the size of the wavelength, so that to create a radar beam of twenty-five meters (the dimensions of the radar used by the first Chain Home stations) takes aerials of about that length—much too big to be carried in an airplane or swiveled around like a searchlight.

To understand this without mathematics, consider the analogy of waves of water. Think of a piece of wood floating on a calm surface. We can generate waves in the water by gently pushing down on the wood: As it is pushed down, it pushes the water away from beneath it, and this water spreads out radially in a wave. If we repeat the motion periodically, a series of waves will be generated, separated from each other by a distance (the wavelength) which is determined by the time between pushes (the frequency with which we push). A string attached to the wood and floated out over the water surface would show a sinusoidal wave characteristic of both this type of experiment and of electromagnetic wave generation.

In the first radar sets, the outgoing beam of radiation was generated by bouncing electrons back and forth in a wire of fixed length. It is a necessary consequence of the mathematics of Maxwell's theory of electromagnetism that a bouncing electric charge must emit such radiation through the aether; the bouncing electric charge is analogous to the wood in the above example, and the aether is a hypothetical medium analogous to the water. (At the turn of the present century much vigorous effort was devoted to determining the true physical existence of the aether. The failure of these efforts helped lead to the theory of relativity, but a discussion of this work is beyond the bounds of this book.)

If a wire is connected to a source of electrical power such as a battery or a dynamo, and if it forms a closed loop, the electrons can be pushed by the source and will flow around the wire, forming a *direct* current of electricity. But if the wire is closed off at one end, the electrons will come up against this barrier and will bounce back; if the power is applied cyclically so that each time they bounce back they are pushed up against the barrier again, we have an *alternating* current. Each time they hit the barrier and bounce, they generate a wave of electromagnetic radiation which spreads out from the wire through the air surrounding it, just as the bobbing wood spreads waves out over the surface of the water.

Since the electrons in such a circuit travel at a constant speed, the frequency and the resulting wavelength are both determined by the length of the wire—by the time it takes the electrons to bounce back and forth from one end to the other. In the illustration shown here, electrons in the horizontal wires W are given a jolt by a voltage from the dynamo D through the lead-in wire. This pushes the electrons from one end of the wire, A, to the other, B, where they bounce off and return. When they come back to A, another jolt sends them back again to B, and each time they bounce back another burst of radiation is emitted. The frequency with which they bounce back and forth is linked to the wavelength of the waves they generate through Maxwell's equations, such that the frequency ν times the wavelength λ is equal to the speed of light, or

$$\nu = c/\lambda$$

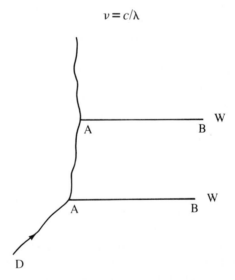

The frequency can also be defined in terms of cycles, one cycle being the time it takes an imaginary typical electron in the wire to make one complete trip; in the diagram above, it would be the time it takes an electron to go from A to B and back again. Since the imaginary electron moves at the speed of light, it will complete a cycle every $2L/c$ seconds, where L is the length of the wire ($2L$ being the distance covered in one complete cycle). Since the frequency is the number of cycles per second, it follows that

$$\nu = c/2L = c/\lambda$$

or

$$\lambda = 2L$$

The wavelength of the generated waves is always just twice the length of the wire. To generate the twenty-five meter wavelengths desired, wires of 12.5 meters are needed. To gain sufficient power, a series of such wires is stacked up in an array; this is the transmitting antenna. The first Chain Home stations along the coast of England were made this way, but obviously antennae of this size could neither be carried on airplanes nor swiveled around like a searchlight.

Luckily, it was soon discovered that the requirement that the wavelength of the radar beam be roughly the same size as the objects being searched for was not necessary: Smaller wavelengths could do the job without much loss of sensitivity. This was a great step forward, since a shorter wavelength meant both that the waves could be concentrated into a narrower beam and that the antenna could be made smaller. Unluckily, however, they found that the power output was inexorably linked to the wavelength: The shorter the wavelength, the less the power output. It was simply impossible to generate enough power at the short wavelengths to make a useful set.

The problem was understood nearly at the outset: While the planned Chain Home system would give adequate warning for day bombers, it was useless for raiders flying at night or for the detection of submarines. And so, simultaneous with early developmental work, a separate group of scientists headed by Dr. E. G. "Taffy" Bowen began work on the design of an airborne set working on short wavelengths. (The nickname is one of the characteristics of the RAF: The river Taff, which flows through Cardiff, is erroneously supposed by many Englishmen to mark the boundary with Wales. Anyone born on the "wrong" side is automatically "Taffy," just as someone born on the wrong side of the Atlantic would be "Yank.")

It's interesting to recall the character of Watson Watt, and then to read the words Bowen wrote in 1985: "I first heard of the idea of detecting one aircraft from another from Watson Watt, and for some time I thought the concept originated with him." That was much the same impression Watson Watt gave when he talked about the origin of the idea of radar itself, neglecting to mention Skip Wilkins. "However," Bowen goes on, "in later years I saw a great deal of Sir Henry Tizard and it became quite clear that he was the one who originated the idea.

He realized that the fledgling warning system...would be instrumental in beating back a daylight attack and that the enemy would then turn to night bombing. What could be done about this? What was needed was a miniature radar in a night fighter...."

Work began in the late summer of 1936 at Bawdsey Manor. It was impossible to produce enough power at meter wavelengths, so Bowen hit upon a compromise. They built a large transmitter which was installed in a tower at the manor house, and a small receiver that could be carried aloft in an aircraft. Known half-jokingly as RDF 1.5, it enabled the first detection of one aircraft by another to be made. Since it could detect airplanes at distances of up to twelve miles (so long as they were close enough to the transmitter), Bowen urged that it be put into service with the RAF, "but Watson Watt would have none of it. With hindsight it is now clear that this was a grave mistake." It might have provided an interim device, and certainly would have been useful as a training device; without it, the first operational sets were shoved onto the RAF pilots and crew without any real training at all. The results were predictable.

That was, however, still in the future. In 1936 the work to produce a completely airborne setup continued, and by 1937 they had managed to construct a set that would give 100 watts of power at a wavelength of one-and-a-half meters. In order to tell where the reflected echoes were coming from, a receiving aerial was put out near the tip of each wing so that the returns would generate blips alternately on each aerial. If the two blips were of equal height, the target was directly in front; if the target was off to one side, the aerial on that side would show a bigger blip. The pilot would then turn in that direction until his radar operator told him the blips were of equal size, when once again they would be heading straight for it. Using the same principle, another two antennae were mounted above and below the wing, so that they could tell if the target was higher or lower in the sky. Known at first as RDF 2, the system was clever, simple, and not quite effective enough.

There were three major problems. Although a few determinations were made at distances of several miles on large objects such as naval warships, something as small as an airplane or a submarine simply couldn't be seen: the 100 watts of power wasn't enough. It was like searching for bombers in the night sky by shining a flashlight at them—the "light" just wasn't bright enough.

The second problem was that the radar beam spread out and hit the ground, and the returning ground echoes swamped the receiver. This was because the one-meter wavelength was not yet small enough to be focused like a searchlight; instead, the waves that were generated spread out spherically ahead of the night fighter, bouncing off anything that might be flying in front of it but also bouncing off the ground below. This limited the range in front of the bomber to a distance equal to the vertical distance from the ground, so that if they were flying at 5000 feet they couldn't see anything further away than one mile. Since the Chain Home stations couldn't bring them to within one mile of a night bomber, the RDF 2 sets weren't much use.

Finally, even if by some chance they did manage to find anything up there at night, the rather long duration of the pulses emitted meant that the receiving aerials were swamped by the outgoing pulse during the first thousand feet of emission; the result was that they couldn't see anything closer than this. Since they couldn't expect to get a visual sighting much further away than a few hundred feet, it was clear that even if by chance they managed to find a night bomber on radar they would lose it before they ever saw it visually.

So they couldn't see further than one mile, and ground control wasn't good enough to bring them within better than about three miles. And if by chance they did find anything, they'd lose it at about 1000 feet—and they couldn't see it visually until they were less than half that distance from it. All in all, not too good. By the beginning of the war they realized that they needed some totally new insight into the problem. The solution—if there was to be one—would have to be based on a theoretical principle not yet thought of.

Well, not exactly. As is usually the case in science, the pertinent principle was thought of years before anyone ever thought of it. What that curious statement means is that most great discoveries have been presaged by the work of lesser men who didn't quite put it all together. The concept of gravity was envisaged by both Robert Hook and René Descartes before Newton worked it out, and the basic equations of special relativity had been derived by Lorentz before Einstein. We have already seen how the mechanism of radar had been visualized and even how working equipment had been constructed by several different groups in several different countries years before Watson Watt began his work. And now here too we find that the answer to the problem had been found many years before the problem itself was thought of.

In 1916 a Connecticut farmboy, who had graduated from Yale with a major in Greek literature and had then gone back to school to learn physics, invented the magnetron. Albert W. Hull was working at the General Electric Research Laboratory in Schenectady, New York, and had been assigned the task of finding a practical alternative to the ordinary vacuum tube, for which GE's patent control was being challenged in the courts by AT&T. In the vacuum tube, the flow of electrons is controlled by an electrostatic charge applied to a grid between the cathode (which generates the electrons) and the anode (which accepts them). If the electrostatic charge is adjusted, the electron flow can be allowed, inhibited, or shut down at will. Hull was an extremely inventive engineer/physicist and now, following up earlier suggestions for a device that no one had been able to make work, he began studying the possibility of a tube in which the electron flow was controlled magnetically instead of electrostatically.

The GE group had a curious fondness for Greek names—a fondness that others sometimes ridiculed as "Greco-Schenectady"—and so the Greek scholar felt right at home there. The lab had already invented and named the dynatron, the kenotron, and the pliotron, so it was natural for Hull to name his invention the magnetron. The basis was that a variable magnetic field was induced inside the tube, so that electrons given off by the cathode were directed in a spiral motion instead of being allowed to flow straight to the anode. If the magnetic field was varied, the electrons could be made to spiral around inside in such a manner as to oscillate between the two diodes, which were made of concentric cylinders, and in this manner the electric current could be controlled as effectively as in a normal vacuum tube.

But when the vacuum-tube litigation was settled satisfactorily, neither GE nor anyone else showed much interest in the new invention. For another decade Hull continued research on the characteristics of the magnetron, working out the physics in detail; it was the sort of physical research that is interesting to a small group, is published in the technical journals, but is never thought of as terribly important by anyone outside of the active researchers. It is analogous to the poetry written by the starving Bohemian in a puccinic garret, read aloud to his friends and greeted by them with rapture, but quietly ignored by the rest of the world as it whirls by uncomprehendingly.

In 1939 war broke out and Watson Watt, along with nearly everyone else who was working on radar, realized that the system they were trying to get to work could never, because of the basic laws of physics, be carried usefully in an airplane to look for bombers in the night or submarines in the sea. It was decided that an entirely separate line of research should be started in an effort to find some new solution to the problem. In the poetry of bureaucracy a typically innocuous name was coined, and the Inter-Services Committee for the Co-ordination of Valve Development was formed.

Frederick Brundrett, chairman of the committee, had to decide who should work on the problem. Lindemann claimed the work for his people at the Clarendon Lab in Oxford but Brundrett, aware of the fuss that had accompanied Lindemann during the past few years of radar development, assigned the work instead to a new group at the University of Birmingham. Lindemann flew into a rage, but Brundrett stuck to his decision, and for the first time the scientific stewardship of Oxford and Cambridge was diluted by outsiders. And not only outsiders, but colonials! Professor Mark Oliphant had come from Australia (by way of Cambridge, it is true) to take over the physics department at Birmingham just two years previously, and had impressed Tizard and Brundrett with the quality of the young men he was producing and hiring there.

Among these were a professor named John Randall and a graduate student named Henry Boot. Together with other members of Oliphant's department, they spent the summer of 1939 touring the radar establishment to gain firsthand experience of what was actually working and what was needed. In September the department returned to Birmingham and set to work to invent out of whole cloth an electronic valve capable of generating high-frequency radio waves with sufficient power to find objects as small as airplanes or submarines.

The first stage was the usual one for a brand-new research project: No one had any idea how to proceed. So they sat around and talked for a while, and quickly reached the next stage: where everyone knows how to proceed, but what they each "know" is quite different and divergent. A valve with the Greco-Schenectady name of "klystron" seemed to most of them the best bet, and they decided to devote their energies to working on it. But Randall and Boot decided to try another tack, and they began to study a type of tube in which the electron flow was modified by charges on the positive grid. After a couple of months they came to the conclusion that their idea wasn't going to work; they thought they had better pitch in with the rest of the group on the klystron—but then they looked at each other and each read in the other's eyes a reluctance to do so. Somehow they felt the klystron just wasn't the answer. They decided to take one last quick look around to see if there wasn't anything else that might do the trick, and they found a paper on rotating-field theory and the magnetron. They hadn't time to read all the scientific literature on the magnetron—if they had, they later said, they wouldn't have decided to work on it: there were too many negative aspects—but they didn't have time, and they didn't like the klystron, and there wasn't much of anything else, so they thought they'd give it a shot.

Almost immediately, before they had time to discover for themselves all the things about the magnetron that made it unworkable, they came up with a new idea that was to turn their work around. The basic concept of the magnetron is to use a magnetic field to herd the electrons along, to guide them where you want them to go. Randall and Boot combined this with a totally different concept—the concept of a penny-whistle.

A whistle consists of a loose, hard object in a cavity into which you blow. The force of your blowing makes the object inside rattle around in the cavity, generating sound waves which reverberate inside and then escape. The frequency with which they bounce around inside governs the frequency of the sound emitted, and that frequency is in turn decided by the dimensions of the cavity, so that a large whistle emits a long-wave, low-pitched hoot while a tiny whistle gives out a short-wave, high-pitched scream.

Ergo, the cavity magnetron. Boot and Randall constructed a small solid block of copper, which would conduct their variable magnetic field. In this they carefully scooped out a precisely measured cavity. When an electric current was made to flow through the copper and a magnetic field was applied, electrons were caught in the cavity—which served as an anode—and were made to bounce around in there by the magnetic field. As they bounced back and forth, resonating, they emitted electromagnetic waves just as the bouncing ball in a whistle emits sound waves. The size of the cavity was constructed so that the electrons moved only a few centimeters between bounces and therefore produced electromagnetic waves just a few centimeters long.

It was brilliant, and it worked. On February 21, 1940, after borrowing transformers from the Admiralty and building rectifiers themselves, sealing off the internal cavity with whatever glutinous crud they could find in the lab and then pumping it down to a reasonable vacuum, they turned it on for the first time. They knew from the design that the wavelengths would be short; the dimensions of the resonating space settled that. But they had no idea what the power output might be. In order to get a quick estimate, they hooked the circuit up to two automobile headlights, hoping enough current would be generated to make the lights glow.

They threw the switch, and the headlights glowed and lit up brightly and blinded them all, and then burnt out. Excitedly they ran out to the parking lot and brought in a bigger set from a truck. Again the current was too strong and they burnt these out too. Time after time they ran out to get bigger lights that could take a greater load of current, and time after time they burnt them out. Finally they hooked up a set of neon floodlights; the circuit held and they were able to measure the current they were producing. They found that their cavity magnetron was putting out radio waves at the expected small wavelength of 9.8 centimeters and at the incredibly high power level of 400 watts. The baby was born squalling and kicking and ready to go into action. Within two and a half months, that first prototype had successfully detected airplanes in the sky and the periscope of a submerged submarine at distances of more than seven miles.

From that first moment when the automobile headlights burnt out it was clear that the cavity magnetron was going to be the key to the development of airborne radar, but it took nearly two more years to get a workable production set into blueprints and out of the factories. In the meanwhile the Battle of Britain began, and while the focus of action was on the daylight fighting, it became increasingly obvious night by night that the invisible after-dark raiders were developing from a nuisance into a real threat. This was precisely what Tizard had foreseen, and what Taffy Bowen and his group had been working to combat. But they were not quite ready.

The previous summer Bowen had demonstrated his RDF 2 airborne equipment to Churchill and Dowding, and it had been approved for possible use after further testing. Then, on August 1, 1939, "with no preliminary warning," Bowen was told to equip thirty Blenheims by the end of the month; these would go into service immediately. He had a staff of twenty at his disposal, and they worked eighty- to ninety-hour weeks attempting to design and produce fittings as well as getting the sets themselves ready and installed. When the Blenheims came to them, straight from the factory, they found that they were a different model than they had been told to expect, and so the fittings they had designed didn't fit. By the end of the month, worn to a frazzle, they had installed six complete sets rather than the thirty ordered. When war broke out three days later, No. 25 Squadron, with no previous training, began night patrols over London in the Blenheims.

"There were no instruction manuals, no maintenance staff, no test equipment, no nothing," Bowen wrote. New meanings were found every day to the expression "muddling through." A Fighter Interception Unit was organized at Tangmere, primarily for the purpose of studying the new techniques of night fighting and to pass on what they learned to operational squadrons. One of the first of these was No. 600 City of London Squadron, one of the Auxiliary Air Force's "millionaire" units. Their Bristol Blenheims with the pack of four machine guns strapped under the fuselage were equipped with the rudimentary RDF 2 airborne radar sets. The squadron was just learning to use them when Russia attacked Finland; the radar sets were ripped out of the planes and the squadron was suddenly transferred to Scandinavia to help the Finns fight in daylight. Led by Lord Carlow, they flew off in their Blenheims, which were totally unequipped for a Finnish winter; not one of them ever returned.

In April 1940 the squadron was reformed with new personnel and again began to train for night fighting, but the next month Germany invaded France, Holland, and Belgium, and the squadron was sent to defend Rotterdam (again after removing their newly installed radar equipment so that it wouldn't fall into enemy hands in the unlikely event that they were shot down and that Rotterdam would fall). As it happened, Rotterdam did fall—and they were shot down: Only one aircraft returned from the slaughter.

* * *

The first operational radar was officially designated AI (Airborne Interception) Mark II, the Mark I version having been the first lash-up Bowen had put together on his own. It was not a very useful set. Aside from the ground return, which meant that there wasn't much use even trying to catch a bomber lower than about 10,000 feet, there was a symmetry aspect: The set did not give direction unambiguously unless the target was nearly dead ahead. A bomber far off to the left and above could just as easily be interpreted by the radar operator as being to the right and below. The radar's 600-pound weight and jutting antenna didn't help the flying characteristics of the Blenheims, either. The radar-equipped model could manage about 250 miles an hour in level flight—enough to catch a Heinkel 111 with full bomb load, but not enough to catch it after it had dropped its bombs, lowered its nose, and was speeding home. Nor was it fast enough to catch the Junkers 88 or Dornier 17, even with their bombs aboard.

By October of 1939 the improved Mark III radar had appeared, and the symmetry problem was largely licked, but all the other problems remained. The Fighter Interception Unit which had been formed at Tangmere became the official unit responsible for night-fighting instruction and organization, traveling to other squadrons to show them how to use the equipment. It was a good idea, except that neither they nor anyone else could possibly have used radar effectively: It simply wasn't good enough.

The pilots who had been trying to find bombers in the night skies since the war had begun tended to distrust the technical people who now tried to show them how to use the "magic mirrors," as they were at first called. As the months passed and the magic mirror performed no magic, it became known simply as "the thing." Sergeant C. F. Rawnsley was one of the first to try to learn to handle it:

> The Thing itself appeared to be subject to the most amazing and infuriating number of faults. The operators knew only that these fell into two categories: those which one could do something about by administrating a gentle kick in the thing's vital parts; and those which produced the Awful Smell. The latter was always followed, unless one switched off at once, by volumes of smoke and anxious inquiries from the sharp end of the aircraft [the pilot].

There was one brief, shining moment of success. On the night of July 22–23 a blip was picked up by the Chain Home station at Poling. A patrolling FIU Blenheim was vectored toward it while ground control was passed to the operations room at the FIU's home base, Tangmere. The FIU commanding officer, Wing Commander Chamberlain, took over as controller; in the Blenheim, Sergeant R. H. Leyland was operating the AI for the pilot, Flying Officer G. "Jumbo" Ashfield; while stretched full-length in the nose and straining to see something out there in the night through the plexiglass was observer Pilot Officer G. E. Morris.

The task facing Chamberlain was so difficult as to be virtually impossible; it was much more than merely bringing together the two blips. If he simply headed the Blenheim toward the oncoming bomber and they met successfully, they'd come together at a relative speed of more than 500 miles an hour and zip past each other before any possible radar contact could be recognized, let alone homed in on. By the time the Blenheim would be able to turn around, the two planes would be many miles apart, and the bomber would be lost again. So first he had to bring his fighter around in a wide circle, approaching the bomber from behind and coming up on him slowly.

Next there was the question of height. On an overcast, pitch-black night it really was impossible to get a visual sighting no matter how you approached the bandit; the situation was hopeless. On a moonless but clear night, everyone agreed, the best technique was to come in low and try to silhouette the bomber against the background of stars above. Or if you were tracking him over a burning city like London, again there was general agreement: It would be best to come in high and spot him against the glow from below. But tonight there was no burning city below and no stars above; instead there was a full moon. In this case the consensus was divided: Some said to come in low and spot him against the moon, others said to stay high and look for the reflected moonglow on his wings below. If you knew precisely where he was and could maneuver to put him between you and the moon, the former strategy would be best; if you were searching nearly blindly in the wide sky, the latter was probably the best you could do.

The problem, then, was to guide the interceptor in a wide, curving turn so as to bring him in behind the bomber at whatever was thought to be the proper height, bringing him in slowly so that he just barely overtook the prey and had time to pick him up on his own radar scope. On the plotting table in the Ops Room there was one counter to mark the German and another to mark the Blenheim, but everyone in the room knew that the true positions of each of these could be off by factors of a mile or more, with possibly thousands of feet uncertainty in the bomber's height. With such intrinsic errors it was certainly possible—it had indeed happened before this and would happen many times again—that a superb job of ground control would bring the two markers on the table together, but in the air the interceptor would find no trace whatever of the enemy; the two airplanes would actually be miles apart, each for all intents and purposes quite alone in the dark night sky.

Chamberlain decided to guide them high, bringing them in above the bomber and hoping they could spot the dull moonglow on its wings. On the plotting table in front of him, the WAAFs moved the two counters: one for the bomber which sailed on straight across the English countryside, unaware that he was being observed and tracked, the other for the British fighter swinging around in a great arc to home in on his tail. Inch by inch they moved across the table, gradually coming closer and closer—and all the while Chamberlain was wondering if the two airplanes themselves were also coming closer together, if their plotted positions on the map table matched at all their real positions in the air.

Somehow, unbelievably, the ground control got them close enough, and Leyland excitedly called out on the intercom that he had a blip on his scope. With a mixture of satisfaction and anxiety, Chamberlain signed off from the ground, passing control for the first time over to the airborne operator. Patiently the Blenheim tracked the blip—they had no choice but to be patient, since they hadn't the speed to zip around and try to catch up to it quickly—and slowly Leyland saw the blip moving up the cathode screen as their distance closed.

Then suddenly Morris called out that he saw it: a dull black shape crossing in front of them, slightly above—and then it was gone. Leyland shouted that he had lost it from the radar scope because they were too close. Morris had caught just a glimpse of it, heading off at ten o'clock he thought, maybe twenty degrees high. But they had thought they were *above* it. Had Morris seen a shadow in his own eye, a reflection on the plexiglass, a ghost? Ashfield had just a fraction of a second to decide.

He turned sharply and pulled the nose up—and there it was, silhouetted darkly against the glowing moon, nearly directly in front of him. He held the climbing turn a moment longer, the gun sight tracked around as the nose turned, and now he saw clearly that it was a Dornier 17. The next moment it filled the gun sight. He pressed the tit on his control column and the belly-slung packet of machine guns began to rattle. The Blenheim began to shake and they smelt the smoke from the guns swirling through the cockpit. The Dornier turned sharply and dove away. Ashfield followed it around with his guns still blazing and then there was a sudden burst of light and flame. He yanked the stick back and banked sharply as pieces flew off from the German and clanked against them. Then time seemed to be suspended and the world was topsy-turvy as they watched the blazing bomber fall away from them—but straight upward instead of down! They hung there for a long moment and watched in amazement and without understanding as it turned into a ball of fire, falling *up*. Then in a quick burst of realization, Ashfield knew that they themselves were upside down, that they had turned over in their anxious pursuit of the Dornier and were about to spin into the ground. Luckily he realized it in time (they wouldn't have been the first night-fighter crew to dive into the ground under the impression they were climbing), turned right side up again, and circled as the Dornier fell properly now. When it had splashed against the ground in a blossoming shower of sparks and fire, they set course for home.

That was the first radar night kill of the Battle of Britain.

It was also the last.

The story of Ashfield's interception circulated through the night-fighter squadrons as evidence that "the thing" was alive, that the magic mirror actually worked, that it was possible to shoot down bombers at night. But then night after night passed, and all night long the bombers droned overhead and the interceptors climbed after them, and nobody else shot down anything at all. Though the Luftwaffe flew more than 5000 night sorties a month during the first few months of the blitz, it was not until November 7—well after the Battle of Britain had been won, though the Blitz continued—that another radar-guided interception claimed a kill. Interestingly enough, and though by this time three complete squadrons of radar-equipped Blenheims existed, it was again Flying Officer Ashfield who scored.

Perhaps "the thing" really was magic, and it took a magician to make use of it. Certainly it took a crew with uncommon skill and devotion to follow the blinking blobs of green light on the cathode screen through the convulsive turns and gravity-wrenching maneuvers of a nighttime chase in the pitch-black sky after an unseen opponent; and then, at the moment of kill, it took uncommonly sharp eyes and quick reflexes to catch the target in that momentary transition from a cathode-ray blip to a fleeting dark shadow across the plexiglass canopy. Just as certainly it took uncommonly clever guidance by the ground controller, plus a good deal of luck, to bring the interceptor into a position where it had any chance at all of meeting a bomber in those vast expanses of empty air. More often, nothing would be seen but black, empty skies.

Several of the night-fighter teams have written about those nights, and their descriptions of the frustrations are all much of a pattern. A typical mission would begin with the scramble call, telling them to climb toward the coast where there were indications of enemy activity.

"Indications." They have been told that before, have flown off into the dark night time after time chasing after "indications," have flown around and around in silly circles, finally returning home no more bloodied than the most innocent of virgins. "Indications" means nothing more than fairies dancing in the impenetrable night.

As they climb up into the cold upper air, the warmth of the summer night is sucked out through the many holes and slots, replaced by icy breezes which bring tears to the eyes and dripping mucus to the nostrils. The cold creeps up through their bodies and brings with it an almost undiscerned sleepiness; eyes stare out into the baffling darkness and lose focus, nearly lose touch with consciousness....

"Hello Starbright two-five, this is Homestead," Sector Control breaks in. "What angels?"

"Hello, Homestead. Two-five at angels three and climbing."

"Right-ho. Climb to angels twelve at zero three seven."

That sounds definite enough. They must have something. But is it a raider, or merely one of our own bombers returning home? Or a thundercloud or a night-flying gaggle of geese, or—?

"Hello, two-five. Sighting definitely hostile."

Well now! The eyes unglaze, the blood begins to stir. They look out into that black night in every direction, straining for a glimpse of searchlight beams or bomb flashes—

"Two-five! Flash your weapon! (The instruction to turn on the airborne radar set.) Bandit right in front of you!"

And the moments tick by....

Finally the pilot can stand it no longer. "Do you see him?" he calls on the intercom.

No answer. Perhaps a desperate grunt or two.

"Two-five, this is Homestead. He's right in front of you, less than a mile away. You *must* see him by now!"

The pilot stares through the plexiglass. The radar operator stares into his green screen.

"I can't see anything out there," the pilot says. "Can't you find him?"

"It's—I don't know. There was something there for a moment, but now it's gone. Wait! Hold it.... Oh, fuck."

"Two-five, this is Homestead. He's turning to port now. I can see him clearly. Can't you get anything?"

Stupid bloody question! Does he think we're flying around up here for bloody fun? Christ almighty, there's nothing out there—

And then suddenly a series of bright splashes pricks the darkness below, one after the other in a thin white line. The bastard is out there, all right, and he's already dropped his bombs. He'll be turning for home now, but which way? Almost immediately the vectors come in from ground control, but what is the use if he can't be found up here in the black sky?

They follow the vectors for another fifteen minutes, within a mile or two of him all the time, but never do they catch a glimpse, never does the screen give a discernible blip. Finally the radio tells them to return and pancake, and back they go to base. As the engines slow and they return to the warmer air, the pilot and gunner turn to stare morosely at the frustrated, defeated radar operator: "The magician was still kneeling on his prayer mat of blankets muttering incantations to himself. The green glow from the cathode ray tube flickered on his face. A witch doctor, I thought. A witch doctor, and black magic. And just about as useful!"

* * *

Clearly centimetric radar was the answer to the problem, but just as clearly it wasn't going to get the chance. The cavity magnetron that Boot and Randall designed in February of 1940 provided sufficient power and enough resolution to do everything they wanted: It was powerful enough to pick up enemy bombers several miles away; it could be collimated into a beam tight enough to eliminate ground return from all but the lowest altitudes and to allow the interceptor to home directly in on the bomber without the guessing and interpretations necessitated by the current system; and it could be pulsed at short enough duration so that the target could be tracked in to just a few hundred feet, where visual sighting was possible, without losing it in the outgoing pulse.

But it simply wasn't ready. The people working on the cavity magnetron knew without doubt that it would work, but it would take time to design the special receiving antenna and associated electronics system that were necessary to make it work, and they were running out of time. And the RAF pilots—who were, of course, largely ignorant of the scientific technicalities and thus unappreciative of the tremendous improvement centimetric radar would be—were not at all convinced that a workable airborne system could ever be made, and they were running out of patience. The Mk III system that was in operation simply wasn't good enough—nobody except Ashfield seemed to be able to learn how to use it—and the reputation of airborne radar was snowballing into that of a useless bit of gadgetry like the death ray or Lindemann's airborne mines.

During that summer and fall, while Ashfield was shooting down his two bombers and none of the other Blenheims were doing anything at all but wandering through the night skies and returning to complain about the failure or inadequacy of their equipment, several squadrons of Hurricanes and Defiants equipped with nothing but the keen eyes of their young fighter pilots were shooting down six bombers without the use of airborne radar. Six bombers out of 10,000 isn't much, but it was better than the Blenheims were doing. A movement rapidly grew to throw out the Blenheims with their magic mirrors and technician warriors, and turn the job completely over to the Hurricane and Defiant pilots. It would have been a tragic mistake, and it was narrowly averted with the advent of the Beaufighter, Mark IV radar, and the PPI.

The Bristol Beaufighter was the world's first effective night fighter. Designed originally to fill the role of a long-range bomber escort, the Beau ended up doing nearly everything: laying mines and dropping torpedos, attacking submarines with rockets and cannon, probing the night skies of France in intruder missions, tank-busting in the African desert, and even a very little bit of long-range bomber escorting. But it was as a night fighter that it reached its prominence.

It was a big monster of a fighter, the biggest ever made up till that time. It had two giant Hercules air-cooled radial engines sited in the wings on either side of the cockpit, so close that the whirling propellers seemed to skim the pilot's cheeks and so bulky that they wiped out vision to the side and below. This would have been a critical failing in a normal fighter, but at night the vision would be provided by radar; the only place the pilot would have to look would be straight ahead, and there he had a lovely, thick, bullet-proof, flat, single piece of plexiglass to look through, one which did not distort vision nor reflect interior instrument lights as did the Blenheim's curved windshield.

The pilot climbed up into the monster through a belly hatch; the Beau sat much too high off the ground for the usual fighter pilot's entrance through a slide-back canopy. Behind him in the cavernous fuselage, as big as a bomber's, sat the radar operator, who climbed in through his own entrance at the other end. This position was originally designed for a gunner, but in all except a very few models the gun was never installed; the man's job was to operate the airborne radar.

Actually he did have one other job in the early versions of the Beau, one that he hated. The plane was the most heavily armed fighter that had ever flown, fitted with four 20-millimeter cannon in its belly, firing forward from just under the nose. These cannon, together with a speed of well over 300 miles an hour and a good rate of climb, were the prime ingredients of the Beau's success as a night fighter, but they were hell to operate. Instead of firing a continuous chain of shells, as the airborne machine guns did, the cannons had been fit by their designers with drums of sixty shells which had to be hand-loaded. On the ground before take-off this wasn't such a terrible job, but after a four-second burst in the air the cannons had to be reloaded, and this was an awesome task. They weighed sixty-six pounds each, and nearly always they'd have to be reloaded in the middle of a fight, when it would be pitch black and the Beau would be swerving and pitching, climbing and diving in an effort to stay on the tail of the frantic target, and the pilot would be calling out impatiently for his guns, afraid that he'd lose the target momentarily. The weight of the heavy casings would be doubled and tripled by G forces as the plane swerved and swooped, and they'd be banging against the fittings, crushing fingers and drawing blood, and there were four of them to replace. It was hell.

But it worked. Those four cannon (and the six machine guns that were mounted in the wings a little later) could blast any bomber out of the skies, and the Beau was fast enough and nimble enough to catch anything the Luftwaffe had.

She entered service in September of 1940, desperately hurried into the front lines. The need for her was so drastic that instead of waiting till enough were ready to equip a whole squadron, then withdrawing that squadron from action and refitting them and training them in their new aircraft—the normal way of introducing a totally new aircraft into service—the Beaus were sent out singly as soon as they began to roll from the factory. The first aircraft went to the Fighter Interception Unit, of course, and the next few went in turn to Nos. 25, 29, 219, and 604 squadrons.

And everyone hated them.

Morale is a funny and delicate thing. At the end of September the night-fighter pilots had seen their day-fighter comrades defeat the Luftwaffe in the most glorious aerial battle of all time, while they themselves had been hunting the night bombers fruitlessly for more than two months. Just one radar kill had been recorded for all those hundreds of night sorties flown. In the past few weeks they had seen the night bombing of London begin. They were seeing what looked to be the destruction of their capital, while they stood by helplessly—cruising around in the night air at 10,000 feet, blind and helpless and raging—while the invisible bombers sailed above and below and around them and dropped bomb after bomb into the growing inferno of London. They were in no mood to screw around with something new.

One might think that, on the contrary, they'd be screaming for something new, something that might work, that might make a difference. But they had tried too long with gadgets that didn't work, with Lindemann's aerial mines and propeller-snagging wires, with the recalcitrant "thing" that glowed green and led them around in circles, with Blenheims that were too slow and too lightly armed; they had seen the best the RAF wizards could give them, and they had seen nothing that worked. They had been frustrated too long while their country burned beneath them; they felt too keenly the responsibility to do something about it. They had run out of patience. They didn't want the Beaufighter, they didn't like her ponderous, bulky looks, and they were afraid of her.

She was too big, too fat, too clumsy. She looked more like a bomber than a fighter. And she was dangerous to fly, unforgiving of any mistakes in handling. One of her nicknames was "Whistling Death," officially ascribed to the Japanese ground troops who feared the whistling sound the Beau made as she suddenly appeared low over the treetops with Hercules engines throttled back, strafing with bombs, rockets, cannon, and machine guns; but some claimed that the name was originally given her by the RAF fighter pilots first assigned to fly her, who associated her with spin-ins on final approach, ground loops on landing, and a general nasty tendency to stall and fall out of the sky like the 25,000 pounds of iron and steel she was. This fear and loathing was more true of those who came from flying single-engine fighters than of those who came from Blenheims: It wasn't easy for someone who had learned to fight in the small, fleet, nimble Hurricanes and Spitfires to adjust to this hulking monster with two engines so powerful that if they were not perfectly synchronized on takeoff or landing they could swing you swerving off the flight path and into the ground.

The story is told of the Hurricane squadron who had received their first Beaufighter: Their best hotshot pilot was given the opportunity to put her through her paces, and as he rumbled down the runway for takeoff, passing the point where the Hurricane would have lightly popped loose from the ground, the squadron, lined up watching him, praying for him, began to wonder if he could pull the plane off the ground before he ran out of runway. He did, but only for a fraction of a second. As the wheels came unglued from the ground, the Beau swerved, skidded, dipped its wing, caught the tip on the grass, and cartwheeled into a burning, flaming, exploding wreck.

The second Beau arrived a week later. After familiarization flights with the company test pilot, one of the squadron was called to take the plane up. This time she took off smoothly enough, but when she came in to land she ground-looped, turned up on her nose, and rolled over onto her back. The pilot in his glass canopy was crushed into the intrument panel.

Whistling death, indeed. But London was burning and the plan was inexorable, and a few days later the squadron's third Beaufighter arrived. The sound her Hercules engines made as she approached the aerodrome was distinctive and ugly to the squadron which flew only Merlin-powered Hurricanes, and they all came outside the dispersal hut to watch. Bets were placed as to whether she'd make it to the ground in one piece, as the Beau came roaring in low in a full-speed beat-up over the grass, then chandelled up and slipped into the landing pattern gracefully. She circled the field and then floated in like a feather, hanging just over the runway for a moment, feeling with her long, strutted landing gear for the tarmac and then settling gently onto it and rolling along without a bump or a blunder.

As the plane came rolling to a stop in front of the squadron, the bets were paid off, with mumbles allowing as to how it might be possible to fly her if you were a company test pilot who had been brought up since childhood learning her quirks and fancies, but that she was too dangerous for mere human beings to fly. And then the belly hatch opened and the pilot dropped down, stood for a moment taking a deep breath of fresh air, took off her helmet and shook her flowing red hair loose, smiled, and waved at the fighter jocks staring at her.

That broke the spell. If the girls from the Women's Air Transport Auxiliary could pick up and deliver the Beau without any problems, the glamorous he-men of Fighter Command thought they had better rethink their objections.

The Beau became one of the best-loved aircraft ever flown in combat. She wasn't easy, she had her fancies, but she was big and strong and as tough as they came. She'd go through any maneuver they could put her into, and the body of the pilot would give out before her wings would crack or her mainspar would break. She'd bring them back riddled with bullets and cannon holes, flying on one engine or, as they said, on a wing and a prayer. And before she came home she'd have blasted out of the air with her overwhelmingly devastating cannon and machine-gun arsenal anything she found.

And she began to find plenty.

Radar at centimetric wavelengths was the answer to the problem of finding the hidden night bombers, and the cavity magnetron was the answer to the problem of producing these microwaves. But though the cavity magnetron had been invented and demonstrated to work in February 1940, by the end of that year a complete system was not yet ready. There were those who, once it was clear that the magnetron was the ultimate answer, wanted to stop work on everything else; but again the force of Watson Watt's personality was felt. He argued convincingly for the production of a second-best or even third-best system today rather than a perfect system tomorrow, for if they didn't give the RAF *something* to work with today there might be no tomorrow.

And so Taffy Bowen and his group continued to work with the original bulky 1.5-meter wavelength apparatus, trying by sheer force to improve it a little at a time. It was definitely third-rate compared to microwaves, but because the system was built around existing types of apparatus, it could be built and installed in aircraft without delay, and so it was. At the same time, a series of small triumphs improved some of the more serious deficiencies in the Mk III sets. A new modulator was designed which allowed the duration of the outgoing pulse to be shortened; this in turn decreased the time during which the set was "dead" as the receiver was swamped by the outgoing pulse, and so allowed the detection of echoes much closer in time to the generated pulse and therefore the detection of aircraft much closer to the pursuer. Mk III had a minimum range of about 300 yards, which wasn't quite good enough for a sharp-eyed pilot to get a visual sighting before the target was lost from the scope; the new system, installed as Mk IV, brought this minimum range down to 100 yards—good enough, on all but the blackest nights, for a visual takeover. Another group produced the Micropup, a transmitter no bigger than an ordinary bulb that increased the power output to ten kilowatts and thus increased the maximum range of detection from one mile to three or four (if the fighter was high enough so that the ground return didn't swamp it).

At the same time, the ground controller was given a new toy: the plan position indicator (PPI). This is the form of radar we usually see today on television weather reports. The ground radar antenna, instead of floodlighting the whole area in front of it with radio waves, sends out a fairly narrow beam of short wavelength, rather like a radio searchlight revolving, as it were, on a lighthouse tower. The cathode screen observed by the controller consists of a circle with the outgoing pulse starting from the center, with distance measured radially outward to the perimeter, so that at the instant of pulse generation a bright spot appears at the center of the screen and begins to move outward along a radial line. But the time it takes to reach the perimeter is so short as to appear instantaneous to the eye, and since the radial line is defined by the outgoing radar pulse, it revolves in synchronization with the continuously revolving transmitting antenna; the effect that is seen on the scope is a bright radial line continually revolving. Any target within range returns an echo pulse, which causes a point along the line to brighten at a distance from the center corresponding exactly to the distance of the target from the antenna. Finally, the screen is coated with a chemical phosphor so that the bright point continues to glow while the revolving line travels completely around the circle; by the time it returns to the target the glow, which is just beginning to fade, is relit by the next pulse. The result is that the night bomber is "painted" on the screen in semipermanent form, and can be followed as it moves through the night sky and across the cathode scope.

The night fighter, tracking the bomber, is identified directly on the screen by a characteristic blip given by its internal IFF device. The controller's trick now must be to quickly evaluate the relative positions of the target and the fighter and to direct the Beau around in back of the bomber and onto its tail. Finally, the height of the target must be carefully estimated by a separate radar plot, there being no point in bringing the two blips together on the PPI if the actual airplanes are 5000 feet apart in height.

In September of 1940 the Luftwaffe attacked London and the blitz began to take shape. Throughout the previous summer they had come over at night only sporadically, and just one bomber was intercepted and shot down by radar. The night blitz had been anticipated, however, and when it began, the RAF had six squadrons of Blenheims, about half of them equipped with Mk III radar, plus another three squadrons of Defiants and several isolated flights of Hurricanes flying without radar, to look for bombers caught in searchlight beams or illuminated against widespread fires burning on the ground. As we have seen, this night defense force was helpless.

By November the Mk IV-equipped Beaufighters were reaching the squadrons in increasing numbers: 100 by the end of the year, 200 by the following May. On November 7 came the first Mk IV kill: Ashfield's second victory. Then GCI (ground-controlled interception, using the PPI) began to take over, the airborne radar operators began to get used to their Mk IV apparatus, and gradually the night sky began to lose its impenetrability.

C. F. Rawnsley was now the radar operator for "Cat's Eyes" Cunningham, and things were a bit different than they had been before. In the spring of 1941 they alone would shoot down a dozen night bombers.

A typical evening would begin not with a scramble, but with an assigned flight. The crew would be given directions to orbit a given map coordinate at a given height, there to wait until GCI picked up some "trade" and directed them to it. All along the coast these lonely Beaufighters would be waiting, circling unseen and unseeing, until the nightly German visitors might intrude upon their particular airspace. Then the headphones would suddenly crackle: "Hello, Blazer Two Four, Boffin here. We have a bandit coming in. Angels eleven."

"O.K. Boffin," the pilot would acknowledge. "Orbiting and climbing."

The sound of the engines would change as the Beau climbed to the indicated altitude. As she did, the directions would come in continuously from Boffin. "One five zero...turn starboard on to two one zero...turn port on to zero six zero...."

As the pilot followed the GCI instructions, the radar operator would keep his eyes glued on the tubes of the AI set, all the while building up in his mind a picture of their changing directions painted on the black of the sky.

"Flash your weapon," Boffin would order, and they knew that they must be drawing close. But still they saw nothing. The operator would adjust every control he could think of, and still he'd see nothing. "No joy yet," he'd call to his pilot.

The controller would continue to direct them in a slowly changing series of vectors until suddenly and yet gently a small disturbance would bulge at the very tip of the scope, then grow and separate itself from the cancerous melange of ground returns. Almost before he was sure he was seeing it, the operator would recognize it as a blip, most definitely a blip. "Contact at 15,000 feet," he would call. "Slightly port and well below."

The pilot would swing the Beau to the left and straighten it out; the blip would swell and waver, then settle down again.

"Dead ahead...range 10,000...better lose some height before we get any closer."

You couldn't just push the nose down and dive after him because there were no brakes on a Beaufighter; if you built up speed in a dive and came hurtling up to the target you'd keep right on going and flash past him before you saw him, let alone have time to aim and shoot. So the pilot would pull back on the throttles and hold the nose high, and as the Beau began to lose speed, it would lose lift and begin to sink. But the operator would call out that they were still coming in too fast, so the pilot would drop the wheels. They added a sudden and clumsy air resistance; it felt as if they had run into a giant marshmallow up there in the dark sky, but it worked: The blip was now moving slowly and properly down the beam toward them.

Now the blip might begin drifting to the left. The operator would call out a correction and gently the Beau would turn port. They were in to 5000 feet, and the blip was sharp and clear. But now it was growing to the right. They had overcorrected in the last turn.

"Turn starboard five degrees."

"Starboard five," the pilot would acknowledge.

"Range 3000. Throttle back a bit. It's dead ahead and slightly above."

Slowly, slowly, the trace would move precisely down the scope.

"Range 2000...still ahead...fifteen degrees above. Throttle right back. Range fifteen hundred."

The blip was coming in toward the end of the scope now, in to where it would be lost in the outgoing pulse, in below minimum range. Damn it, he *must* see it by now surely—? "Twelve hundred...one thousand... still ahead and thirty degrees above. Nine hundred...."

And then suddenly the blessed words. "Right, I've got him. Want to take a look?"

The radar operator sat facing rearward in the Beau, but the seat swiveled forward and he could look out through a perspex blister. And there it was, black but distinct, a Heinkel. They were slightly below it, unseen against the black background. Above them, great, billowing cumulonimbus clouds picked up the slight moonlight and seemed to glow in great skyward-straining fingers. The Heinkel was winding along between their ghostly crests, with the Beau sneaking along behind and below in the dark valleys.

Now the great engines would begin their muted roar again as the pilot pushed the throttles in; closer they would creep, closer still, right up under the Heinkel's tail, nudging gently right and then left and then right again as he would try to center the bomber in his gun sight. And then, with the slightest of motions, he would press the red button on the top of the control stick, and with that nearly invisible motion all hell would break loose as each of the four cannon fired twice a second. The sound would be like thunder inside the aircraft, which shook and shuddered and filled with acrid fumes of smoke and cordite. And in the next second the universe would light up and blind them with a searing wall of flame.

Then it was dark again, and quiet again, with no sound but the solid rumbling of the Hercules engines and no sight but the green glow from the cathode ray tube. And then, as the Beau would raise one wing in a banking turn, there was again a red glow out there in the dark night sky, sinking down between the ghostly moonlit clouds, sinking down toward the cold wet waters of the Channel. It was all that was left of a Heinkel bomber and its entire crew.

* * *

In July, August, September, and October of 1940, equipped with Mk III radar, the RAF Blenheims destroyed one bomber. In November, equipped with Mk IV, they got another. During the winter the Beaufighters began to arrive in increasing numbers, the controllers began to learn how to use GCI effectively, and slowly the number of kills began to climb. There were three bombers destroyed in January, 1941, and four in February, losses which the Luftwaffe could laugh at. But then in March the Beaus knocked down twenty-two and in April they doubled that score and they more than doubled it again in May with over a hundred kills. The darkness of the night sky over England had been transformed from a sanctuary into a killing field.

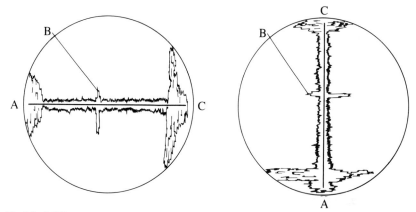

The Mark IV system (as used in the Beaufighter in the spring of 1941) had two displays, one showing height and range, the other direction and range. The display on the left shows the outgoing pulse A, the target B and the ground return C. The distance A–B gives the range, and the relative sizes of the blips above and below the median line show that the target is lower than the fighter. The display on the right again shows the outgoing pulse A, the target B, and the ground return C. Again the distance A–B gives the range. But in this case, the asymmetry of the blips is an indication of direction: The target is to the right of the fighter. As the Beau caught up to the bomber, the blips would move closer to the outgoing pulse; as shown in the right-hand diagram, it would slide down the baseline and finally disappear into the outgoing pulse noise. This set the minimum range at which the target would be visible.

It became clear by now that no victory lay in those night skies, and in the final weeks of May, the great Luftwaffe bomber fleets began to be withdrawn from their French bases, sent east to prepare for the great onslaught against Russia. Although some *Kampfgeschwader* remained in the west and continued their night attacks for another couple of years, mostly in feeble retaliation for the growing Bomber Command strikes against Germany, the blitz was finally over.

Chapter Sixteen
Microwaves

> We are the dark deniers, let us summon Death...
> *Dylan Thomas*

The prime motivation for the original development of centimetric radar was one for which it was never used: coastal defense, or CD. The concept was that if radar of short enough wavelength could be developed, it could be used as an electronic gun sight for the aiming of large coastal defense cannon against an invading armada. By the time it was developed, however, no such invasion armada existed. It fought instead a very different battle.

On the other side of the Atlantic, in that year 1941, America was looking apprehensively and a bit greedily toward the Far East. It appeared as if Britain's far-flung empire might soon be up for grabs—but it also looked as if the Japanese would be doing the grabbing.

Our attitude eastwards was ambivalent at the time, even schizophrenic. We did not want to go to war, but we wanted to conquer the Pacific. Or rather, we were more noble than that—we didn't want to conquer anyone, we simply wanted trade routes and democracy. We wanted, in effect, economic, moral, and political dominance, and we wanted it to come from God and our own unquestioned civilized supremacy rather than having to be vulgar enough to fight for it. We were naive, egocentric, pompous, and arrogant—although at heart, of course, we were very, very good.

When the Japanese, stymied from without and goaded from within, launched their version of the war across the empty pacific wastes of that inaptly named ocean, attacks by night played an increasingly important role, although the fate of the world never hung in the balance—largely because of the single most important gift one nation has ever brought to another.

* * *

In 1940, at the height of the Battle for Britain, it was clear to most people in the British government and to a few such as Franklin Roosevelt on the other side of the Atlantic that the future of our civilization depended on total amalgamation of resources on the part of the democracies. Such a concept, however, goes against the instinct of national self-survival; even those who acknowledged the need found it desperately hard to turn over whatever treasures they held to "foreigners," friendly and helpful though those foreigners might be. Some of His Majesty's Government hoped to weather the storm on their own or with foreign aid, but to remain a viable and independent empire on their own terms. Others realized more fully the threat of total destruction emanating from the other side of the Channel, and were ready to cooperate to the fullest extent with America in the hope that such total and unselfish cooperation would be reciprocated and that together the western democracies might survive.

Henry Tizard saw this right away. In November of 1939 he had urged that Professor A. V. Hill—one of the secretaries of the Royal Society and the man whom Wimperis had first consulted about a death ray—be sent to the United States as scientific liaison with the Americans. Tizard hoped, as he put it, "to bring the American scientists into the war before their government." But when authorization was sought from the Cabinet, the question of exchange of military secrets naturally came up, and when Tizard suggested that Hill be empowered to tell them "everything," the military were appalled: The Admiralty, in particular, insisted that nothing be disclosed except on a strict quid pro quo basis—"an ASDIC for a bomber sight," as they put it—and when Hill was finally sent off in February it was with a tight lid on military secrecy, which he reluctantly accepted. (He discussed with American physicists, for example, a recent American discovery that radio waves could be bounced off stones thrown into the air, producing radio echoes; he nodded quietly and did not mention the British radar.)

In the meantime Tizard continued to advocate the sending of an unrestricted scientific mission, and by the summer of 1940 Britain's plight was desperate enough for his arguments to be accepted by Churchill if not by the entire Cabinet, and Tizard was instructed by Winston to prepare and to lead the mission himself. His Churchillian brief was brief indeed: "I was empowered to tell them what they want to know, to give all assistance I could on behalf of the British Government, to enable the armed forces of the United States to reach the highest level of technical efficiency."

Churchill's aim was not for Tizard to bring back scientific secrets developed by the Americans, but rather to take the English secrets across the ocean so that if Britain fell, America would be able to carry on the struggle for the freedom of the world, the struggle that Britain carried on alone throughout that bitter summer. It was probably the single most unselfish act of any national leader throughout the long ordeals of world history.

Tizard's mission left England aboard the *Duchess of Richmond* in August 1940, taking a rich assortment of military gadgets. The group was composed of an intelligently devised mixture of the scientists and engineers who had helped invent the gadgets, together with the pilots and soldiers who had used them in combat. They wanted to convince not only the American scientists but, even more important, the American military leaders. (The British were aware at the time that though German radar was every bit as technically sophisticated as their own, it was not appreciated by the German military, was not integrated into their war scheme, and so was virtually useless. They hoped to convince the Americans not to make the same mistake. In great measure they failed in this aim: When the Japanese attacked Pearl Harbor, that great naval base was well protected by radar developed to a large extent through the apparatus brought by the Tizard mission; but it was so little appreciated by the military authorities in Hawaii that its warning of the approaching Japanese forces was ignored.)

The mission brought a huge cornucopia. Even before it arrived, the process had begun. The design of the Rolls-Royce Merlin engine—the power under the hood of the Spitfires and Hurricanes—had been sent over; installed in the American P-51 Mustang, it would produce from a marginally effective airplane the best long-distance fighter of the war, the key to victory over the Luftwaffe and the survival of the American Eighth Air Force. In addition, the British previously had sent one of the basic secrets of the atomic bomb to Washington: In February of 1940 (the same month that Randall and Boot had invented the magnetron at Birmingham University, and in that same university) Otto Frisch—who with his aunt Lise Meitner had discovered the phenomenon of nuclear fission—calculated that the critical mass for a nuclear explosion of uranium-235 was on the order of pounds rather than tons, which for the first time made the manufacture of an atomic bomb a realistic possibility rather than a vague threat for future generations. The presentation of that calculation to American scientists had inaugurated the first serious beginnings of atomic bomb research.

Now the mission landed in the New World with plans and a model of a power-driven turret for airplanes. (The American Flying Fortress at that time was anything but a fortress; its handheld machine guns were absolutely impotent for the purpose of tracking and hitting anything flying as fast as a Messerschmitt.) They brought with them news of a new type of aircraft power which they called "jet propulsion"; they brought antisubmarine and antiaircraft devices such as the proximity-fuse and the Bofors repeating cannon with predictor gun sight. And they brought the cavity magnetron.

There is, as we all know, no substitute for good old American know-how. Unless it is good old British know-how, or good old German or Japanese or even, God help us, Russian know-how. The story of the development of the atomic bomb by a coalition of European and American scientists is well known; much more important for the outcome of the Second World War—and equally important to the maintenance of our military security today—is the development of centimetric radar that now proceeded as a joint effort by British and American scientists.

The cavity magnetron was the key, "the most valuable treasure ever brought to these shores," as the official history of the American Office of Scientific Research and Development put it. Our navy had been working on radar but had been able to generate less than ten watts at short wavelengths. Now at one stroke they had a thousand times more power than they had had before, and a whole new field of possibilities opened up before them, strewn with military flowers available for picking.

In 1940 Winston Churchill and Franklin Roosevelt were beginning their unique transatlantic friendship which would have so many consequences for the world. The climate of opinion in America was distinctly isolationist: It had been only twenty years since we had bailed out the squabbling Europeans, and now they were at it again. The 1918 war had been "the war to end wars"; with that title it had been sold to us and we had bought it, and now we felt as if we had bought a bottle of snake oil from a slick con man and the warts it was supposed to cure hadn't improved at all. Those damned Europeans were fighting again, and what else was new?

The Jews wanted us to fight Hitler, but if we fought every country that was nasty to the Jews we'd be fighting nearly everybody except the Arabs, including at times ourselves. (The horror of the Holocaust was not then apparent to anyone, and even if it had been, we had never yet gone to war because of a country's internal policies, no matter how horrendous.)

The British wanted us to fight totalitarianism, to "make the world safe for democracy"; but that was a slogan whose day had not yet arrived. Most Americans saw the British as trying to dupe us into saving the British Empire, and we weren't about to do that, especially when the smart money said it was lost already, and good riddance. Most American politicians—or at least those who looked beyond the east and west coasts of this country, and these were not by any means "most"—

were more concerned about how we might fill the vacuum left by the collapse of Britain, how we might best take over that far-flung empire which the defeated mother country would no longer be able to police or control. Surreptitiously, in darkened rooms and closets or in dark night thoughts, we began to try on the "white man's burden" to see how it fit, and how heavy it was.

When Churchill proposed to Roosevelt, in the summer of 1940, that a British military mission be sent across the waters to discuss matters of mutual interest and to exchange scientific information of a military character, Roosevelt's advisors demurred. They were suspicious: It was neither by chance nor by natural superiority that that small, insignificant island off the coast of Europe had dominated the world for the past century and a half; it was by subtleties, schemes, and secret treaties. They were not a true democracy, they had a King: they were not to be trusted. If we had not stood up to them once, by force of arms, they would rule us still. We must be careful or they would do so again. And never forget: Just twenty years before they had inveigled us into sending our boys to die, and now it seemed that they had died for no other reason than to preserve the British Empire for another twenty years; we must be careful or they would snare us again.

What military advantage of the United States would be served by their proposed mission? Very little, as far as Roosevelt's military advisors foresaw. The American war machine, while in dishabille because of the years of peace following the war which everyone believed had ended war, was basically sound: all it needed was an infusion of money from Congress. Ideas and inventions from Europe it did *not* need, thank you very much.

Across the ocean, Churchill's advisors looked at things from an angle different by a hundred and eighty degrees, as different in point of view as they were in geography and position, and came to much the same conclusions. America (by which was meant the United States) was an untrustworthy ally; she had set the nations of Europe at each other's throats by proclaiming after the war in unhallowed mockery of majesty that she forgave Germany the reparations voted at Versailles—while at the same time insisting that the Allies pay their war debts to her, which she knew perfectly well was impossible unless they in turn collected reparations from Germany. Her president, Woodrow Wilson, had neglected every aspect of the Versailles arguments in order to concentrate on nothing but the setting up of a League of Nations—and then America had refused to join. She sucked in all the wealth of a battered European community, picked up overseas possessions, and then retreated behind her two oceans to live the good life while the rest of the world struggled and slipped in the pockmarked mud of old battlefields.

She was both greedy and impotent, the British advisors argued: She would take the fruits of British military genius, and what could she give in return? They had already visited those purple-majestied shores looking for gifts of war, and what had they found? The Flying Fortress was a joke, indefensible in the air and with a bomb load less than that carried by much smaller British bombers, nowhere near approaching the tremendous loads planned for the upcoming Stirling, Halifax, and Lancaster bombers. The Mustang fighter was lovely but underpowered, the P-39 Airacobra was nearly a hundred miles slower than advertised, and neither it nor the P-40 Tomahawk belonged in the same skies as the Messerschmitt or the Spitfire. The Yanks had obviously never heard of RDF, had no capabilities worth mentioning in antisubmarine warfare, had an army debilitated by budget cuts, an air force in the stone age, and a navy that looked only to the Pacific. An exchange of military information with *them?* What was the bloody point?

While the expert advisors on both sides of the ocean wrangled, Roosevelt privately notified Churchill that he would accept such a military mission and Churchill nodded decisively and sent them off. Official approvals came only much later.

The American attitude changed as soon as the British unlocked their treasure chest. The air force generals expressed polite but reserved interest in the concept of jet propulsion. They appreciated the power-driven turrets. (Although they never admitted for a moment that the Flying Fortress was anything but invincible in the air, still they saw how replacing the handheld machine guns with these turrets would add to the plane's strength.) They resented the Merlin engine. (Curtiss-Wright's Allison engine powered their P-38, P-39, and P-40, and they had been confident—until they saw the Merlin—that the Allison was superior. They eventually installed American-built Merlins into the P-40, where it did not make enough of a difference to transform an obsolete old workhorse into anything much better, and then finally into the P-51 Mustang, which was instantaneously transformed from a low-level ground support fighter into the most successful high-level long-range escort fighter of the war.) And they were astonished by the cavity magnetron.

To the surprise of the insular British, the Americans had independently discovered radar, and aside from giving it the name that stuck, they had made some significant progress. They too realized that it was the production of microwaves that was necessary in order to bring the weapon up to the status of a full-scale, operational, war-winning weapon. But they were stymied by an inability to produce sufficient power at the necessary small (centimeter) wavelengths. Though the magnetron in its original manifestation was an American invention, no one in this country had thought of Boot and Randall's idea of combining it with a resonant cavity. On September 28, the already-established Microwave Committee of the United States National Defence Research Committee met with Taffy Bowen and his compatriots of the Tizard mission at Tuxedo Park, New York, where the millionaire lawyer and amateur (but quite excellent) physicist Alfred Loomis had set up his own private laboratory for microwave research. The committee was absolutely astonished and totally delighted and captivated by the initial demonstration of British results; clearly this was the way to go, and while the Brits were still smiling contentedly at their one-up score, the Americans rolled up their sleeves and got to work. They suggested setting up a new national facility to develop radar, if the British would cooperate. The British beamed: that was exactly what they were there for.

By October 18 the microwave committee had reported to its parent NDRC, which in turn reported to its parent OSRD, and the decision to go ahead was made. By November tenth, bids for the new laboratory from the Department of Terrestrial Magnetism in Washington, the air corps, and Bell Laboratories had been turned down, and work began on the Radiation Laboratory at MIT in Cambridge. That was to be the research center, but different aspects of the work were farmed out all over the country to firms and researchers who had the particular skills needed for the different components. When the first American microwave radar was flashed on less than two months later, returning a picture to the MIT lab of the Boston skyline on the other side of the Charles River, it featured the original British magnetron linked to a Westinghouse pulse generator, a revolving antenna designed and built by Sperry, a receiver built by Bell with identification unit by RCA, and a cathode-ray display from GE. Within a month it was tracking aircraft, and within another month it was itself mounted in an air force bomber and making measurements from the sky.

Cross-fertilization ("mongrelization," as Hitler called it) can be a powerful tool, and indeed it was in this case. Once the basic idea and instrumentation had been brought to the United States, it was decided that further development would proceed separately and independently on both sides of the Atlantic, although with a mixed crew of British and American workers. Taffy Bowen was the strongest influence on the American design, but it evolved into something quite different from what was being built in England.

In that country, in the summer of 1940, while it was clear that the future of centimetric radar belonged to the magnetron, it was also clear that it was going to take time to devise a system based on this wholly new concept. In keeping with Watson Watt's dictum of providing something immediately rather than something good later (when it might be too late), they went ahead with a system based on the klystron. It didn't work. The microwave generator had to be kept under vacuum, and the vacuum pump was so bulky and heavy that it destroyed the aerodynamic characteristics of the airplane carrying it. The set wasn't powerful enough or sensitive enough; its maximum range wasn't long enough and its minimum range wasn't short enough. It simply didn't work.

It wasn't until March of 1941 that the first magnetron-powered airborne set was installed in a Beaufighter. The system had a tiny aerial, less than two inches in length, sitting in front of a paraboloid reflector which focused the microwaves—which were emitted in all directions—into a highly collimated beam, just as the reflector in a searchlight does for optical radiation. This unit, called a scanner, could be rotated to search the skies ahead of the fighter, and the reflected echoes were picked up by the same antenna. The unit was mounted in the nose of the fighter, which projected into a rounded-off shape called a "thimble nose." There was no need for antennae on the wings or above and below the fuselage, for direction in both elevation and azimuth were given by instruments which showed where the scanner was pointing.

The main problem with this experimental unit was caused by the fact that the revolving—scanning—antenna was also the receiving unit; it was resolved in the production model, called AI MK VII, by keeping the antenna fixed in the nose and having only the mirror behind it rotating. This provided excellent radar characteristics, and was followed within just a few months by AI MK VIII, in which the design was perfected, and which became the standard RAF unit for the next few years. Its main flaw was that the scanner couldn't be made to revolve very far away from dead ahead, but inasmuch as the ground control system was intended to bring the fighter onto the tail of the bomber, this wasn't too great a problem. The airborne set had a maximum range of about five miles even at low altitudes (since ground return was now negligible), and a minimum range of less than a few hundred feet. It reached service in April of 1942.

That same month a new British night fighter went on operations. The de Havilland Mosquito was to become one of the outstanding aircraft of the war, but only with difficulty had the RAF been persuaded to accept it.

In 1933 a 15,000-pound prize had been set up for the winner of an air race from London to Melbourne, Australia, and de Havilland won it with a wooden twin-engine racer called the Comet. In the next few years the company developed it into a four-engine transport plane, and in 1938 they suggested to the RAF that a similar wooden design fitted with twin Merlin engines could fly faster than any fighter and still carry a

The Blitz and the "Boot"

significantly heavy bomb load from London to Berlin. The idea was that it would be a radically new type of plane, a bomber that carried no guns and wouldn't need them since it would escape destruction at the hands of enemy fighters by flying so fast they wouldn't be able to catch it. (In Germany, Ernst Heinkel was suggesting the same thing. Luckily for our side de Havilland won and Heinkel lost the arguments with higher authorities; the Mosquito emerged and the Heinkel 119 didn't.)

At first the RAF politely said no. They were interested in bombers made of wood, because metals were in short supply, but they were decidedly not interested in bombers without guns. They were not in business to set speed records and win prizes, they were in the business of war. Wars were fought with guns, not without them.

De Havilland persevered. The RAF said no again, perhaps a bit less politely than before.

On September 3, 1939, war was declared, and on September 6 de Havilland tried again. This time, when the RAF said no, they had a happy thought: They inquired if the Comet/Albatross type might not perform as a reconnaissance plane. Of course it could, de Havilland said; it could fly anywhere and nothing in the Luftwaffe would be able to catch it. All right then, they said, they'd give it a try as a photo-reconnaissance type, which was traditionally unarmed. And, oh yes, they mentioned, as de Havilland left the room, if he could fit in a few machine guns might it not make a good long-distance fighter?

In the end it was the best photo-reconnaissance plane of the war. It was also the best bomber of the war, carrying a load bigger than the Flying Fortress could, further and faster and with a much lower rate of loss. It was also the best night fighter of the war. It was one hell of a machine.

But in the beginning no one believed de Havilland's claims that it could do all these things, and the first MK VIII sets went to the Beaufighters; it took another year for them to reach the "Mossies," but once they did, it quickly became obvious that the combination was the best they could have hoped for, and the Mossie became the standard British night fighter, the best any nation had.

In America, meanwhile, the first set flew in March 1941. They licked the problem of a moving antenna receiving the echoes, and had the antenna attached to the reflector dish which could revolve in a series of modes: The search mode searched the skies ahead from nearly full left to full right while nodding up and down to look above and below. Once the target was found, the scanner could be set for a narrower range, looking nearly straight ahead with great sensitivity, locking onto targets as far away as ten miles. The cathode screens on which they saw the targets were a great improvement on the British model: They had two such screens, one a PPI painting a picture of the skies ahead and giving an overall view of where the target was, the other giving precise information on direction or altitude and range.

By the beginning of 1941 the air corps ordered these AI-10 sets for incorporation into their own night fighter, the P-61 Black Widow, the first aircraft in the world to be designed from the start as a night fighter. Unfortunately, the design ran into a series of problems and the P-61 didn't see action until 1944; in the meantime, the United States used borrowed Beaufighters in Europe and a series of improvisations in the Pacific.

Comparisons of the two production radars, the one in England and the other in the States, showed that while the British had achieved the better receiver, the Americans had produced the better transmitter. These two designs were therefore merged into what became the SCR-720 set, which was then produced for both air forces.

In England by this time there was no longer any great threat from German night bombing, but when the Allied armies invaded North Africa in November 1942, Hitler transferred several Luftwaffe night bombers to bases in Sicily, Sardinia, and Italy, from which they nightly attacked the invading armies with devastating effect until No. 600 Squadron—once composed of the City of London lords and millionaires who had gone to their deaths in Finland and then again over Rotterdam (not the same deaths, of course), and who had once again been re-formed and taught to fight at night—arrived on Malta with their new AI MK VIII radar-equipped Beaufighters. From that island base they intercepted and decimated the night bombers and, later, the transports carrying supplies and reinforcements to Rommel's Afrika Korps. As he ran out of water, out of food, and out of gasoline for his tanks, Rommel finally was defeated.

In the Pacific the Japanese night-bombing effort never reached the calamitous peak it should have. At first they ignored the advantages of the night because they didn't need them: Their Zeros swept the American P-40s and Wildcats out of the skies. When the P-38 Lightnings and Hellcats began to take the daytime offensive out of their hands, the Japanese attempted to shift toward night bombing, but by then the American lead in radar was insurmountable.

At first the air corps tried to fit the twin-engine P-38 Lightnings with radar, but this never worked too well. The navy had ordered a twin-engine, two-man night fighter for carrier operations, but the Tigercat came too late, arriving on Okinawa the day before the war ended. Afraid of the havoc that could be wrought by night bombers against the carrier fleet—which at night stood out for attack, with fluorescent wakes pointing to the ships like luminous arrows on a black velvet curtain, bright arrows which could not be turned off—the navy improvised by putting radar sets into their single-seat Hellcats and Corsairs. The RAF had earlier said this couldn't be done: Dowding had decided that a pilot had too much to do besides looking at a cathode screen, and anyway the luminous screen would destroy his night vision. But the navy tried it, and it worked. The night fighter Hellcats saved the fleet, which was never seriously injured at night.

The navy's radar set, originally named AIA, had become AN/APS-4 (Aircraft Navy/Airborne Pulse Search) by the time it was first put into service in late 1943, when it was installed in the Corsairs. Flying out of Munda, New Georgia, VF(N)-75 Squadron immediately stopped the nightly air raids that had been plaguing the forces since their occupation of the island.

But the navy had never really trusted the Corsair, even in daytime. Someone had early on made the decision that it was too big and powerful to be operated safely from carrier decks, and so the navy contracted with Grumman to develop the Hellcat as its standard carrier-based fighter to replace the slow and unmaneuverable F4F Wildcats and to take over daytime air supremacy from the Zeros. Because the Hellcat did this job so remarkably well, the navy naturally turned to it for its night-fighter requirements; but the pilots were reluctant to have this bulky radar set fitted into their lovely machines. It hung out below the right wing in a streamlined but bulky pod, looking like an unsightly bunion or cancer, and they worried about taking off and landing on a heaving carrier deck with that lump disturbing their aerodynamic flow. As it turned out, they were right to worry for, in addition to slightly lowering their top speed, which wasn't too serious, the pod caused the right wing to drop in a stall; if they flared out a bit too high when landing, instead of dropping heavily onto the carrier deck they'd drop right wing first and literally cartwheel off the carrier and into the sea.

But it worked. Learning from the British, the navy replaced the Hellcat's normally curved front windshield with a flat pane of glass to reduce internal reflections. Directly below this the radar display jutted out, looking like a tin can with a rectangular hole cut in the top, through which the pilot could peer into a cathode-ray screen. The maximum range was nearly five miles and the minimum was down to 100 yards, which was excellent, and the Hellcat, which had been designed to beat the Zero, was easily fast enough to catch any night bomber in the sky and powerful enough to knock it down. The problem, the navy thought, was that the pilot might have too much to do; they were still worried about Dowding's judgment that searching with radar, flying a plane, and shooting the guns all at the same time was too much for one man to handle.

So they came up with a unique solution. They formed "Bat" teams consisting of two Hellcats and one TBF Avenger, which was a three-crew torpedo bomber. On night patrol the team would be controlled by a carrier-based radar controller, similar to the sector controller the British used. The radar on the carrier would search the night skies and pick up any intruding bombers, and the controller would then direct the Bat team to their vicinity. One of the TBF crew would be a full-time radar operator, and, once the team got near enough to the bombers, he would lead them to the enemy and bring them in to the vicinity. Then the Hellcats would pick up the enemy on their own radar scopes and the Avenger would slide out of the way as they attacked. It was a lovely idea and in the first combat a year later it almost worked, except that the Avenger, which wasn't even supposed to fire its guns, shot down three airplanes—but unfortunately one of them was a Hellcat.

This happened on November 26, 1944, off the Gilbert Islands, where the carrier *Enterprise* was leading Task Force 50.2, which had been attacked two nights in a row by Japanese bombers. Lieutenant Commander Edward H. "Butch" O'Hare—who, while flying a Wildcat, had earlier won the Congressional Medal of Honor for shooting down five Zeros in one fight, had played a leading part in designing the navy's air combat tactics against the Zero, and for whom O'Hare Field in Chicago is named—was commanding Air Group 6, responsible for the defense of the carrier group. He took off in a Hellcat as part of a Bat team, and when the carrier radar controller reported a contact of about thirty bombers, he followed his Avenger lead ship onto the vectors provided.

The TBF Avenger found the Japanese, but the Hellcats lost the Avenger. While they wallowed around in the sky, desperately fiddling with those damned sets and pointing their noses all over creation in the hope of spotting something, the Avenger had come so close to the Japanese that the pilot looked up and found he had a visual sighting. The Avenger was not designed for aerial gunnery, but somehow the designers had provided the pilot with one 0.50-caliber machine gun in the nose—really for morale purposes only, so that the pilots might not feel quite so much like helpless sitting ducks as they skimmed over the water in a torpedo attack and found the whole Japanese fleet shooting at them. Now he used his gun, attacking a bomber, but without effect. Frustrated, he pulled out from behind the bomber and moved alongside: In addition to his single machine gun he had a gunner sitting behind him in a power turret with two guns. The gunner poured a long blast into the Mitsubishi, and it exploded into flames.

Somewhere out there in the black night O'Hare saw the bright flash of flame and heard the Avenger's exultant cries over the radio, but even as he turned toward the fight he lost it again: The flame dropped away into the ocean and the night ahead was blacker than ever. He called out for the Avenger pilot to turn on his lights so that he could find him; the reply was that Avenger pilots might be stupid but they weren't crazy.

The Blitz and the "Boot"

He wasn't going to turn his lights on while he was flying among a bunch of Mitsubishis. Instead he got on the tail of another one and actually shot him down with his one machine gun. He then relented momentarily, enough to give his lights a quick flash, and then moved into the attack again.

That one quick flash was enough. O'Hare picked it up visually and sped in, coming up onto him from the rear and peering into his radar scope, trying to pick up all those Mitsubishis. The Avenger gunner saw this dark shape suddenly looming up behind him and fired—and O'Hare disappeared into the darkness below, forever. It was the only time that a single, slow, lumbering, lightly armed Grumman Avenger ever shot down in one combat three aircraft, of whatever nationality.

The Hellcat pilots learned; what had at first seemed impossible to the RAF became possible and even, with practice and practice and ever more practice, "a piece of cake." They learned to land the radar-equipped Hellcats on a heaving carrier deck at night, they learned to find their way around in the night skies, and they found the Mitsubishis. They became so proficient at it that they dispensed with the Avengers, and one squadron actually was called upon to replace an army P-61 Black Widow squadron at Leyte; the big Army interceptor, which had been designed specifically for night fighting, was too overloaded with sophisticated equipment to climb fast enough to catch the night bombers—an early indication of the liabilities of our fascination with high-tech gadgetry, which continues to plague us to this day.

As the Pacific war progressed in a series of island-hopping operations up the western waters towards Japan, the momentum of the night operations swung quickly now toward the Allied side. The Japanese strategy depended on their navy and associated bombers attacking the Allied task forces while their merchant fleet brought reinforcements to their island fortresses. The first swing in the balance of the air war came when the Hellcats, Lightnings, and Corsairs took the daytime skies away from the Zeros, allowing the navy and marine dive-bombers to penetrate to the Japanese carrier fleets while protecting their own. The second swing came when the night fighters denied even the night skies to the Japanese, and the final blow came when Allied bombers equipped with air-to-surface radar began to deny the night seas to Japanese shipping, bringing to a climax the nasty and little-known war against Japanese merchant shipping which had been the focus of American strategy in the years following Pearl Harbor.

In those early years of the Pacific war, with our surface navy lying broken on the bottom of the Hawaiian waters and our army and its air corps hopelessly atrophied and out of date, castrated by the isolationism of the 1930s, there was only one way we could strike back at the Japanese—and it was illegal. In the First World War Germany had shown how effective a submarine force can be against merchant shipping, but to be used effectively the force must strike without warning, and in that war and throughout the next two decades the world had agreed that such attacks violate international law. Official American voices loudly agreed with world opinion: it was brutish and barbarian to strike with military arms against unarmed sailors. The surface ships of the Navy could stop such merchant vessels, inspect them for military cargo, take their crews safely aboard, and then sink the ships; but a submarine could do none of these things except the last. As discussed in the next chapter, we cried out with horror when the Germans did this in the Atlantic, but in the Pacific we had no hesitation in carrying this type of warfare to its ultimate limits.

It is a practically unknown phase of our military history, presumably because we have the grace to be ashamed of our cruelty and hypocrisy. But it was our only chance, and we took it and made it work. With England desperately trying to stay alive in Europe and with our own navy destroyed, our air force outclassed, and our armies overwhelmed, the entire Pacific lay open to the Japanese armadas, and they wasted no time in taking it. For the first time since we had entered world affairs we were the underdog, at least until our industrial machine began to roll, and so we turned to the weapon we had wanted to outlaw in the prewar years. The submarine was the guerilla weapon of its day: it was cheap, sneaky, and effective. England and the United States had wanted to ban it for precisely these reasons, but all the underdog nations of the world, who couldn't afford to match our naval expenditures on big surface ships, insisted on its legitimacy and so it remained in the world arsenal.

Now we had nothing else left, and so we used our submarines, turning them loose on the merchant shipping necessary to supply the far-flung islands of the new Pacific empire of the Japanese. The ensuing undersea war was a brutal and deadly one, fought without warning or quarter, the most un-American war we had ever fought up till that time. (Of course, it loses some of its poignancy in this post-Vietnam era.) We lost fifty submarines in the next few years, nearly all of them with their entire crews, but we broke the back of the Japanese Navy, which lay stretched out across those vast Pacific waters. When later in the war our own war machine began to roll up those captured islands one by one, we found an enemy dug in tenaciously and fighting bitterly, but an enemy half-starved, with insufficient reinforcements, low on weapons and ammunition. The newsreels showed us pictures of skinny Japanese soldiers, looking like survivors of Auschwitz; the cameras showed us small tins half full of rice while the voice of the commentators snidely told us how ill-fed the Japanese army was. We were told, and we believed, that all this was proof of how little the Japanese valued human life. These disgusting, decimated little yellow men were contrasted with the beaming, healthy, well-fed GIs—but we were never told and we never thought that the contrast was due not to "the American way of life" but rather to the ferocity and effectiveness of our illegally fought submarine war against the Japanese supply lines.

That submarine war in the Pacific was the converse of the battle of the Atlantic in which the Nazi submarines attempted to strangle the English-American lifeline. It was fought just as bitterly on both oceans, but the Japanese did not have a working radar system, and so in their war the submariners won, while in the Atlantic the decision was barely reversed. When in 1944 the American air forces brought the capabilities of air-to-surface radar to the Pacific, enabling our night bombers to find and destroy the last remnants of the Japanese merchant fleet which were hopping from harbor to harbor at night to escape the eyes of the day bombers and the submarines, the last hope of reinforcing the island strongholds was gone, and with that hope went the war.

This final achievement of radar in the Second World War, air-to-surface capability, in fact, had turned out to be the most important of all when, the year before on the other side of the world, it had finally defeated "the only threat that ever really frightened" Winston Churchill.

Chapter Seventeen
"Das Boot"

> The airplane can no more eliminate the U-boat than the crow can fight the mole....
> *Admiral Karl Doenitz,*
> *Head, German Submarines,*
> *1942*

> The one single weapon which defeated the submarine and the Third Reich was the long-range airplane with radar.
> *Grossadmiral Karl Doenitz,*
> *Chief of State, Germany,*
> *1945*

In 1620 Cornelius van Drebel built for King James I a boat which was sealed and sunk with its crew of twelve in the Thames, whereupon the crew rowed downstream under the water and emerged several hours later safe and sound. But King James had other matters on his mind, and the vessel was not incorporated into the Royal Navy. Not for another century and a half did a "submarine ship" engage in warfare; this occurred in 1776 when the American rebels' *Turtle* slithered across the floor of New York harbor toward the English man-o'-war *Eagle*. The Yankees tried to screw a gunpowder charge, together with a timed fuse, into the *Eagle*'s hull; but it was covered with a copper skin which they couldn't penetrate. Finally they left the charge on the harbor bottom and rowed away. It exploded on time, but without damaging the *Eagle*.

Nearly another century went by before someone tried again. This time it was a Confederate crew, which in 1864 attacked the *Housatonic* in Charleston harbor and sank her with a torpedo lashed to the bow of their submarine. When the charge went off, both the *Housatonic* and the submarine were sunk.

It's easy to see why this exploit did not arouse much naval interest, and although the self-propelled torpedo was invented just a few years later in 1868 (by an Englishman, Robert Whitehead, employed by the Austrian Navy), no one else used the submarine until the Great War of 1914. By this time the advent of the electric motor, gaining its sustenance from storage batteries for operation below the surface, combined with diesel engines for on-surface locomotion (during which time the batteries could be recharged) had turned the submarine into a fully oceangoing vessel. France and England as well as Germany had small submarine fleets with torpedoes that could be sent through the water to explode at a safe distance from the submerged vessel, but it was Germany alone that made of the *Unterseebooten* a dangerous weapon.

At the beginning of the war she had only ten vessels of oceangoing capability, and she didn't know how to use them. Their first mission was to wait submerged outside the entrance to Scapa Flow and to report on the sailing of any British warship from that home base. This never worked out, however, since a submarine could neither send nor receive radio messages while submerged, and when she surfaced to do so she was too vulnerable if close to Scapa Flow, while if she took the time to sneak away under water her messages were delivered too late to be of any value.

The first portent of things to come seemed to be clear enough when, on September 5, 1914, just three weeks after the declaration of war, the U-21 sank the British cruiser *Pathfinder;* this was followed less than three weeks later by the spectacular sinking of the three British cruisers *Aboukir, Cressy,* and *Hogue* off the Dutch coast by the U-9, with the loss of over 1200 British sailors. The U-boat was obviously the equal or master of the surface warship—and yet not necessarily so. She was not fast enough when submerged to catch or even track a modern warship; the sinking of the *Pathfinder* and the *Aboukir* had occurred when they, by luck, crossed the path of the submerged U-boats, and the sinking of the *Cressy* and *Hogue* came when those two cruisers stopped dead in the water and lowered boats to rescue the sailors of the *Aboukir.*

A different sort of portent had appeared on September 15, when a flying boat of the Imperial Austro-Hungarian Naval Air Arm came upon a submarine off the Austrian coast in the Adriatic. The airplane was returning from a bombing mission and had no bombs left on board, but when the crew returned to base and reported what they had seen, two other flying boats took off to look for it. They found it several hours later, submerged but visible in the clear Adriatic waters. They attacked, forced it to the surface, and sank it. Afterwards they landed on the water and rescued the French survivors.

Clearly, then, the future appeared: The submarine was a new force to be reckoned with by the warships of the established navies, and her nemesis would be the airplane. But the future has a way of clouding its path and doubling back unobserved, sneaking off in new directions unforseen by mortal men: The submarines were to prove of little value in armed combat against surface warships, and although several claims were registered and initially accepted, that French submarine *Foucault* was to be not only the first but also the last submarine sunk by an airplane during the Great War.

Merchant shipping. That was the key in both the first and second wars. The only major naval action in the 1914–1918 war was the battle of Jutland, which was more or less of a standoff. The major effort of the Royal Navy was to inderdict war and food supplies to the Central Powers, who replied with a submarine blockade to starve the British in their turn.

The difference between these two efforts was a major distinction of ethical responsibility to the British, but to the Austro-Germans only a technicality. The Royal Navy, with undisputed command of the high seas (since the Kriegsmarine stayed in their home base of Heligoland Bight except for the Jutland episode), ran a traditional naval blockade: Its warships, coming upon a neutral merchantman, would stop it and search it; if its destination was an enemy port and if it carried contraband cargo, its crew would be arrested and the ship would be sunk. (A German or Austrian ship was a priori a legitimate target, of course.) Upon return to Britain the warship would deliver its prisoners up to internment.

The submarines of the Central Powers could not handle things in such a civilized way. If they surfaced and hailed a merchantman instead of sinking it immediately, there was nothing to prevent the target from radioing its position and calling for help. Although it could be sunk before such help arrived, the submarine would then be in danger from enemy destroyers racing to the spot. And, of course, the submarine was not large enough to take on board the merchant crew.

There was only one recourse open to the Austro-Germans, unless they were to renounce submarine blockades entirely. They would have to sink merchantmen on sight, without announcing their presence, and leave the crew to fend for themselves. Since the first announcement of the submarine's presence would be the explosion of torpedoes, the merchant crews could not be expected to have time to radio news of the attack or their position, greatly lessening their chances of survival in small boats or rafts in the open, surging seas.

* * *

On February 2 and 4, 1915, Germany published an announcement that any ship approaching England was liable to be sunk without warning; the effective date of such "unrestricted submarine warfare" was February 18. On February 19 the Norwegian steamer *Belridge* bound for Holland with a cargo of oil was sunk. The submarine war then followed in a series of escalations and drawbacks. The escalations took place as the German high command came to realize how effective the subs were against commercial shipping; the drawbacks came each time they realized how drastically world opinion was horrified.

The first phase lasted from that February through August. In May the *Lusitania* was sunk without warning, and 1198 men, women, and children were killed. The civilized world protested; more importantly, America protested. The event was headline news from coast to coast, and the use of the British term for the *Boche,* the "Hun," spread with all its connotations and reminders of evils once thought to have been buried with the past. Immediately following this, the two American steamers *Gullflight* and *Nebraska* were similarly torpedoed, and America protested so strongly—with much beating of drums and talk of going to war—that Germany retreated and announced that passenger ships would no longer be attacked.

And then on August 19 the *Arabic* was torpedoed, and forty-four passengers died. This time the protest was so strong that the Germans really were afraid of provoking America into war, and they actually did cease submarine attacks. But on land the war was slogging into a muddy stalemate, a war of attrition in which millions would die. As both high commands worried long hours into the nights trying to unravel a plan that would bring them victory, the German Navy kept reminding the Kaiser of how much damage the submarines had done; in February 1916, they were unleashed again.

The slaughter this time lasted barely a month. On March 24 the French ship *Sussex* carrying passengers from England to Dieppe was lost, and President Wilson sent an ultimatum to the German ambassador demanding the cessation of this "present method of submarine warfare against passenger and freight carrying vessels." The German government capitulated, and the second phase was over. The submarines were not idle, however; they traveled to Scandinavia and spent the autumn of 1916 attacking shipping in that area, none of which was American.

This was followed by an error in judgment. With the end of submarine warfare in the Atlantic, the fervor against Germany in the United States began to die down again and was replaced by a resentment against the British, who—although they were not sinking American ships—were blockading Germany and thus interfering with American commerce. As harsh feelings began to shift toward England, Germany made the mistake of overestimating the swing: she began unrestricted submarine warfare again.

This was a calculated gamble. It was a certainty that American resentment would follow, and if the submarine attacks were continued these would be sufficient to bring America into the war on the side of the Allies. (The situation at this time was quite different from that in the initial stages of the Second World War, where America from the first was a natural enemy of Hitler's Third Reich; in 1939 the question was whether or not America would fight, but never on which side might she fight. In 1916–1917, though it seems so difficult to imagine in retrospect, America could have ended up on either side.)

The German naval high command argued that total warfare by the submarine fleet, which had been building up its numbers during the autumn and now had more than a hundred U-boats ready to sail, could be so effective that it would win the war before any effect of American hostility could be felt. This argument won, and on February 1, 1917, the leash was slipped and the subs were sent out to attack anything that moved. Ships of any nationality—warships, merchantmen, ocean liners, and fishing vessels—were to be sunk at sight with no warning given and no attempt made to help the survivors.

The effect was catastrophic. They sank nearly half a million tons of shipping in February, more than a half million tons in March, and over 850,000 tons in April. The climax came on April 19, when nineteen British ships alone were sunk. The loss amounted to more than twenty-five percent of all ships sailing to and from England, a rate that no nation could long endure.

In that same month America—no longer able to accept the attacks on her own merchant navy—came into the war as predicted, but a bit more quickly than the Germans had anticipated. The question was whether England could survive until the American Army and Navy were brought into working order, and whether these overwhelming American forces would be able to travel across the Atlantic in sufficient numbers to influence the outcome of the war. The answer was as simple as it was profound: convoy.

The convoy concept was an old one, dating back to the sixteenth century when Spanish galleons loaded with New World gold sailed in convoy to escape Elizabeth's preying privateers. It was a solution that worked because, for one thing, the ocean was no longer filled with ships waiting to be sunk: Instead of hundreds of ships sailing independently across the waters, so that wherever one looked one was sure to find something sinkable, they were gathered together into a few convoys for which the submarines had to look long and hard. The second reason was equally as obvious, one would have thought: Instead of scattering the too-few antisubmarine destroyers all over the wide Atlantic, the defensive forces could be concentrated around each convoy. They might not find all the submarines hidden under the waters between the Bay of Biscay and the American coastline, but who cared? The only subs that mattered were those that found the convoys—and when they found them, they would also find the destroyers circling round and round, waiting to blast them out of the water with depth charges.

The twofold advantages of the convoy system were obvious, and quite real. The only puzzle is why it took so long to be adopted. But great travails demand great resolutions, and when the American military machine reported that it expected to have its forces in Europe by 1918 at the earliest, and the sea lords of the Royal Navy reported that, at the present rate of merchant loss, England would have to sue for peace by November 1917, it was clear to everyone that something had to be done—even something as innovative as invoking a system that had been around for nearly 400 years.

And so the merchants began to sail in convoy, and the rate of sinking dropped impressively. From the disastrous high in "Black April," the number of sinkings decreased steadily to roughly a third that number by the end of 1917 and barely a tenth that number by October 1918. The number of submarines sunk rose from just two in that dread April to ten in September and seventeen in the following May. The submarine menace was overcome, the American armies sailed "over there," and after one last surge forward on the battlefield, the German armies collapsed under the weight of the combined Allied strength. The war was over.

From the Encyclopaedia Britannica (written in the 1930's):

The exact measures for combating submarines are necessarily confidential, but it may be taken for granted that the lessons of the World War have been taken to heart and that scientific improvements in antisubmarine devices have continued.

Alas. Nothing may be taken for granted. While it is true that scientific improvements in antisubmarine devices did continue, the lessons of the First World War were nowhere near taken to heart; on the contrary, they were discarded along with the khaki uniforms and the transitory romantic entanglements and the selfless thoughts of King and Country: good and bad together, the useful and the useless, the comfortable and the rather embarassingly sordid but sweet to remember, all went out with the next day's trash as the world showered and scrubbed itself clean and went home.

* * *

From a report of the Committee of Imperial Defence, 1937:

We must be prepared for unrestricted attack by submarine against our trade...by Germany, but we do not visualise that the submarine will constitute the menace which it did in 1914–1918.

O sancta simplicitas!

In the years following the First World War, scientific progress in antisubmarine techniques was marked by the operational status of Asdic. Named for the Allied Submarine Detection Investigation Committee, which initiated the idea in 1917, the system was actually a primitive sort of radar: It sent sound waves through the water, and when they encountered a submarine they bounced back and were detected. Radar itself cannot be used for underwater detection because light waves, unlike sound waves, are not really waves at all. A wave is defined as a cyclic disturbance propagating through a medium. If you see a leaf floating on the sea, for example, you note that as waves pass it the leaf rises and falls, but *does not move forward with the wave;* instead the wave moves on and leaves the leaf behind to rise and fall again with succeeding waves. This illustrates that in the wave motion the water itself does not move along with the wave: The wave is *not* the water, but a rising and falling disturbance which moves *through* the water. For such a disturbance, the medium is a necessary constitutent. A wave cannot move through a vacuum, for example, because a disturbance in *nothing* is itself nothing. Maxwell's theory of electromagnetic radiation showed that light propagation follows equations of the same form used to describe wave motion, and so the conclusion followed that such radiation is itself a wave. It was for this reason that the aether was proposed as the (invisible) medium through which the radiation is propagated.

But all attempts to find the aether have failed, and we are forced to recognize that light is not exactly a wave: It travels best through a vacuum rather than through a denser medium, which is behavior exactly opposite to that of true waves. Sound, for example, travels faster and further through a dense medium like metal than it does through air; this is why in Western movies you see the military scout dismounting and bending down and putting his ear against the railroad tracks, then announcing the presence of Indians somewhere down the line. He has heard the hoofbeats of their horses reverberating through the solid iron rails, though they could not be heard through the tenuous air.

Sound, therefore, travels well through water; light, on the other hand, does not. And so a "radar" depending on sound waves is workable for submarine detection, while true radar is not. When the Asdic system was demonstrated to King George in the early 1930s he suggested using radio waves instead, and was brushed off with the reply that it was not technically feasible; no one thought of taking up his suggestion in the matter of airplane detection and thus developing radar.

The system is technically simple, and it works. Indeed, it still works today, under the American name "sonar." (We seem to be better at inventing names of things than the things themselves.) A burst of sound, as from a struck gong, is emitted from a ship cruising somewhere above a submerged submarine. The sound spreads out through the water, bounces off the sub, and an echoing ping is returned and heard through a set of sensitive hydrophones. The hydrophones are directional: As they swivel, the effect is of someone cupping his ear and trying to determine from where in a forest a sound is emanating. Simple, workable, effective.

The main difficulties in using sound detection for airplanes were the surrounding noises, the slow speed of sound waves compared with that of an airplane, and the very limited range of detection. The Asdic system gets rid of the first by supplying its own sound, distinct from and much greater than any ambient sea noises. The second is no problem since the submarine moves so much more slowly than an airplane or sound. The third is a real limitation: The range of detection is so short that the system is useless for simply cruising through the oceans and listening for subs. But it is useful once a submarine is sighted and then submerges. In that case the attacking destroyer can rush to the spot of the sub's disappearance and commence its search there.

The Asdic system proved effective, but in those years after the Great War the Royal Navy forgot everything else about how to fight the submarine menace. In particular, in their eagerness to believe that the horrors of the past world war could never be repeated by civilized, sentient human beings, they joined with the rest of their nation in subscribing to the expressed national policy that there would be no war "within the next ten years"; year by year this policy was extended, and year by year the existing military forces were allowed to dwindle without renewal. Though they remembered how effective the convoy system was, they forgot about the necessity of having warships to shepherd the convoys; when the war came again in 1939 they had only 150 destroyers, half of which were not front-line, to escort a merchant navy which had 2500 ships at sea on any given day.

They forgot that military exercises carried out in peacetime are not likely to be realistic: The Asdic exercises of the 1920s and 1930s, for example, were performed only in conditions of good weather, and were begun with submarine positions known to the destroyers. Even so, the rate of submarine "kills" was less than fifty percent. Somehow the top brass never appreciated this, and continued to believe that the submarine menace had been licked.

There were other problems. The committee members and the *Britannica* author who were so sanguine about submarines didn't seem to realize that while antisub science was progressing, the people building submarines weren't hibernating. The undersea fleet Germany sent out in 1939 was faster and longer ranging than their First World War counterparts, and carried more torpedoes that were more accurate and more effective. They were a formidable menace, made even more deadly by the new science of wireless telegraphy, which enabled *Kriegsmarine Unterseekommand* to keep in touch and direct the movements of each individual sub. This became particularly important when they broke the BAMS (British and Allied merchant shipping) radio code and thus knew precisely when each merchant ship sailed to which destination by what route.

The awful carnage began again.

In August 1939, when it appeared to the British that a war might be in the offing, the Admiralty called for a complete naval exercise, the major point of which was to practice techniques for keeping the submarines holed up in their pens and denying them access to open water. While the English were away exercising, the Germans—who knew damned well that they were about to march into Poland—slipped their submarines unseen out into the open waters of the Atlantic, so that when September 3 dawned there were thirty U-boats sitting astride the commercial lanes waiting for shipping—and nobody in England knew about them. Little wonder that the war opened with a bang at sea. On September 17 a U-29 sank the aircraft carrier *Courageous,* and in the first month of war the U-boats sank forty-one ships totaling more than 150,000 tons, without loss to themselves. As the first group of thirty submarines ran out of fuel and torpedoes and had to return home, the score fell. Thereafter such a large group could not be maintained at sea, being roughly limited to a third that number, as one part was always in transit out and another part on their way home. This took time, but that was all it took. It took no toll of the submarines, as the British had hoped. A large part of their antisubmarine effort was devoted to patrolling the sea routes from the German U-boat pens, hoping to intercept and sink the subs on their way to and fro, before they lost themselves in the great wastes of the Atlantic; but the subs simply stayed submerged by day and cruised unharmed on the surface at night, and both the Royal Navy and the RAF were powerless to find them.

And all over the ocean, ships were going down. Although the convoy system had been immediately introduced, the introduction had been reluctant and inefficient. The merchant captains who had fast ships didn't like the idea of sailing slowly along with a bunch of sitting ducks, and so anyone with a speed of more than fifteen knots was allowed to sail independently. At the other end of the spectrum, any ship that couldn't make at least seven and a half knots was deemed too slow, and was for that reason excluded from convoy and sent off on her own. The navy, for their part, didn't like the idea of confining their attacking forces—the antisub destroyers—to purely defensive duties such as shepherding convoys. Too many of the too few destroyers were sent out into the endless waters on search-and-destroy missions; they searched, but they found few and destroyed less. And meanwhile the convoys were left with insufficient escorts.

In the air the situation was even worse. During the First World War the blimps—named, by one account, according to the plunking sound they made when a finger was snapped against their taut skin, by another named after the legendary British character Colonel Blimp (the most famous portion of whose anatomy they so roundly resembled)—had not sunk a single U-boat but had been remarkably effective in preventing them from reaching their targets. They couldn't sink the subs because they were easier to spot as they floated in the air than the sub was, and even if they did see it before it submerged they moved so slowly that it had plenty of time to disappear below the waves before any bombs were dropped. But the point was that the sub did have to submerge and stay hidden, while the blimp floated around above, and while the sub was hiding it couldn't do any damage and whatever ships the blimp was shepherding could sail safely past and lose themselves over the horizon. In that entire first war, only two ships were sunk while escorted by blimps. But during the 1930s the improvements in regular aircraft were so great that it became clear that the blimp would be a sitting duck for any enemy fighter plane, and so the blimps were phased out in favor of long-range seaplanes assigned to Coastal Command.

Which would have been fine, except that when the war started Coastal Command didn't have any of those wonderful long-range seaplanes. One had been designed and even brought into production—the Short Sunderland—but then Short Brothers designed the Stirling bomber and the RAF decided that they wanted that, so they tore down the manufacturing jigs for the Sunderland and assigned another company—Saro—to build instead its Lerwick flying boat, which then proved impossibly bad. When the order went back to Short to build the Sunderland again, they had to start from scratch, with concomitant delays.

When the war began, Coastal Command had only a few Sunderlands, a few American Lockheed Hudson airliners converted to bombers, the ancient biplane London and Stranraer flying boats, and the Avro Anson, a six-passenger, twin-engine commercial plane acquired by the RAF for reconnaissance duty, which had no business being in the same sky with Messerschmitts, Heinkels, or Dorniers—and which stayed in RAF service until 1968. It was fun to fly (if there were no Messerschmitts around), but it could carry only 100-pound bombs; these were so ineffective that when one Anson attacked by mistake His Majesty's Submarine *Snapper* and scored a direct hit on the conning tower, the total damage done was four broken light bulbs. Little wonder that by the end of 1940 Coastal Command had not destroyed a single sub, while losing over 300 aircraft in the dangerous weather-tossed skies far from land; during this same time the subs sank nearly 600 ships totaling over 2 million tons.

In the beginning, the subs ignored the convoys. There were enough ships sailing independently to keep them busy. When they had sunk so many of these that the pickings became slim, they began to trail the convoys, waiting until the escorts left them (because there weren't enough escorts to take them all the way across the Atlantic); then they would move in and feast on the scattering, undefended beasts. They sank more than a half-million tons of shipping throughout that winter and spring, averaging about 80,000 tons per month: an effective and worrying attack, but one that Britain could bear.

Then in July 1940, with the fall of France, the entire eastern coastline of Europe down to the Spanish border became Nazi property. The U-boats could now be harbored at Brest and other French ports, cutting hundreds of miles off their transit journeys into the Atlantic and dispersing them so that it was even harder for the British to find them. They now began their concerted attacks on the convoys themselves, operating in large Wolf Packs which overwhelmed the strained defenses. This struggle would soon be identified by Churchill as "the battle of the Atlantic." The submariners had a different name for it. They called it simply, "the happy time."

The strategy had been worked out at the end of the last war by Karl Doenitz, then a U-boat captain and now Hitler's *Führer der U-booten*. He had been waiting for the British to be forced into grouping nearly all their sailings into convoys, and now they did so; no longer were the slow merchant ships allowed to go alone, but instead they were formed into special slow convoys. More convoys necessarily meant fewer escorts per convoy. This—together with the shorter transit time allowed by the Bay of Biscay ports, which meant more submarines in action even though they were somewhat fewer in number than at the beginning of the war—gave Doenitz the confidence to challenge the convoy escorts directly.

The essence of his plan was cooperation rather than individual effort on the part of his U-boats. When a convoy was sent out, Doenitz was informed by his *B-Dienst* (Beta Service, the code-breaking establishment), and a patrol line of submarines was strung across its projected path. When one of these sighted the convoy it did not attack; instead the sub let the convoy pass, then surfaced and radioed back to Doenitz the particulars of its course and composition. The sub would then trail the convoy, lounging back over the horizon where it was invisible, keeping in sight the inevitable plumes of smoke which rose above the horizon. Meanwhile, other subs now drew in to surround it. As each reached its assigned position it radioed in, and only when the entire Wolf Pack was assembled did Doenitz give the order to attack.

Sometime that night the first explosion would rip the guts out of an oil tanker. As the two or three escorting destroyers or sloops would turn about and race to the scene, another explosion would light up the sky—from the other end of the miles-long convoy. And then as the harried escorts would dash back and forth, scanning the deep waters with their Asdics for the invisible subs, ship after ship would explode and sink as the subs—*on the surface*—would slip down the long rows of ships and pick out those big enough or important enough or vulnerable enough (oil tankers, for example, rather than those loaded with wood). At night the subs were as invisible on the surface as beneath it, sliding low through the water, and they were much faster and more maneuverable on top; equally important, while surfaced they were invisible to the Asdic!

Their attacks built in intensity as each success built further confidence. British losses tripled, soon averaging more than a quarter of a million tons each month. It was slaughter, bloody slaughter. *The Happy Time....*

But help was on the way.

The problem was severalfold, but linked together through one vital circumstance. On the surface the submarines, with speeds of up to eighteen knots, were faster than the ships they were attacking. The merchantmen were generally capable of not more than ten to twelve knots, and most of the escorts available at that time had a maximum of only sixteen. But below the surface, the subs crept along at only three or four knots. This was important, first, when the convoy was sighted. The spotter sub would slide below the surface as the convoy passed and would then surface and follow it unseen from beyond the horizon several miles away. If it became necessary for the sub to stay submerged, he would not be able to stay in touch with the convoy as it sped away. A later change in direction would enable it to avoid the gathering wolf pack.

The same factor continued to be of importance during the wolf-pack attack, for the Asdic sets were useless in detecting submarines on the surface, and the subs were fast enough and maneuverable enough to avoid the escorts and to slip up and down the convoy lines, shattering one tanker after another. But if they were forced to submerge, they would become "visible" to the searching Asdic beams, and even if they managed to hit one or two ships, the convoy would soon leave them behind. As it was, they were able to continue the attacks throughout the long night; at dawn they would retire submerged beyond the horizon, surface there and trail the convoy along all that day; the following night they would move in once again to attack. Night after night they would attack until the convoy was hopelessly decimated.

What was needed, aside from actually sinking the subs, was the ability to keep them submerged and thus to allow the convoys to escape. What was needed was a means of spotting them on the surface by day and by night. What was needed was radar.

1937. September 4–5.

The Royal Navy had prepared a fleet exercise designed to test the efficiency with which aircraft can detect ships at sea. The newly activated Coastal Command division of the RAF was prepared to demonstrate that it would be able to carry out its assigned role: to scour the North Sea and find any German surface raiders that, in the event of war, might attempt to break out into the Atlantic.

Units of the British fleet were detailed to sail throughout a wide area stretching from the Straits of Dover to Cromarty Firth on the northeast coast of Scotland. Forty-eight aircraft of Coastal Command commenced the search at dawn on September 4. Among them was one Avro Anson, identification number K 6260, carrying the first airborne radar set designed by Taffy Bowen to operate on a wavelength of 1.5 meters. The original objective of the radar program which had brought the operative wavelength down from 25 to 1.5 meters was to produce an airborne set for night fighters, but on August 17, while conducting an unsuccessful test for air-to-air detection, it had been noticed that echoes were obtained from ships several miles away. This had been totally unexpected: At the previously used longer wavelengths, returns from the sea itself had swamped anything that might be seen from ships. This new capability at 1.5 meters overwhelmed the natural disappointment at not yet being able to detect aircraft, and, as Bowen has said, "excitement could not have been higher."

The naval exercise, coming just two weeks later, was too good to miss. Although not officially scheduled to take part, Bowen took off in Avro K 6260 at dawn, with Sergeant Naish as pilot and Keith Wood to help observe.

Within a few hours the weather began to dirty up. It quickly deteriorated below the level at which safe flying could be performed, and the exercise was canceled. But since Avro K 6260 was not an official part of the RAF (it belonged to the Telecommunications Research Establishment at Martlesham, a civilian unit), no one knew about it—and, in any event, it didn't carry a radio! So it flew on, slipping between the clouds and, one by one, the ships of the Royal Navy began to show up on its primitive radar scope. The navy steamed on serenely, undiscovered, as they thought, beneath the gray haze and towering thunderclouds, when without warning the Anson whirled down out of those clouds and sped across them. The aircraft carrier *Courageous* dispatched a flight of Swordfish to practice interception tactics; inside the Anson their joy at finding the fleet was compounded when they got clear blips from the Swordfish—the first time an air-to-air echo had been obtained. More than happy and satisfied with themselves, they turned now and headed for home.

But the weather had become really nasty. They left the navy behind and climbed slowly into a solid wall of gray haze intermittently darkened with black clouds of turbulent water vapor, somewhere behind which was the craggy shoreline of England. Sergeant Naish flew on instruments, while Bowen and Wood crouched over the cathode screen. Ahead of them the radar beam poked through the opaque clouds and returned echoes of nothing but choppy water. Then suddenly they saw something different on the screen. A distinct difference began to show up, and with joy they recognized the coastline clearly enough to give Sergeant Naish directions home to Martlesham.

Meantime the fleet, which had been informed that the exercise had been called off, had radioed that they had been discovered by a lone aircraft, and whose the hell was it? Coastal Command checked with all its squadrons and found that all had returned upon the cancellation order, and that none had come anywhere near the fleet. When finally Avro K 6260 landed safely and TRE notified the RAF that a civilian aircraft equipped with radar had been able to fly out to sea when Coastal Command was grounded by the weather, had found the navy all by itself, and had then returned to base guided by radar navigation through weather that the birds were watching from the ground, they all realized that a new phase in naval warfare had been suddenly introduced.

ASV (Air-to-Surface-Vessel) radar now progressed from a stepchild of Air-to-Air to become a legitimate device in its own right. Within a year a sideways-scanning set had been introduced, so that an aircraft could search simultaneously a wide area on either side as it cruised along, and by the following year's Fleet Exercises, they were able to pick up the *Courageous* at a range of eight miles. Maximum ranges were nearly twenty miles for ships and thirty miles for a coastline.

By September of the next year they were at war and what had been forgotten in the interwar years was suddenly made abundantly obvious: It was not the battleships and cruisers of the German Navy that they had to find, but the submarines. In November 1939, Taffy Bowen was summoned to the telephone; the call was from Admiral Sommerville at the Admiralty offices in London, wanting to know if ASV radar could locate something as small as a submarine. This was "literally the first occasion that submarines had been mentioned as a target for airborne radar," Bowen noted.

The first ASV sets were just then being fitted to the Lockheed Hudsons of Coastal Command squadrons 217 and 220. Bowen borrowed an airplane and flew down to Gosford, where His Majesty's Submarine L-27 had been assigned for trials. Within a few days he had demonstrated that, flying at a height of 5000 feet, he could detect a sub at a distance of five or six miles. The Admiralty—who effectively controlled Coastal Command, though nominally it was an RAF force—approved the test. ASV began to be installed in a wide range of aircraft, and the submarine search began.

And failed. The apparatus simply wasn't good enough for what it was expected to do. Taffy Bowen might be able to detect a sub five miles away, but no one else could. Introduction of the system at this early stage had the advantage of getting the flight crews trained on radar, but the greater disadvantage of souring them on the whole idea. Just as the first night-fighter crews had done with the early Marks of AI, they found ASV useless, and so learned to distrust it. In this case the situation was actually worse, for even when it worked it seemed not to. Picture a Hudson patrolling the North Sea at night, trying to find a submarine sneaking through those endless cold waters into its happy hunting grounds in the Atlantic. Hour after hour the Hudson flies through the murky weather, seeing nothing outside, finding nothing on the cathode screen.

And then suddenly there is a blip. "Target marked," the observer calls out. "Five miles at zero nine zero."

The crew spring to action, gunner to his turret, bombardier to his sight. The pilot pushes the throttles through the wall, drops his starboard wing, and swings her around. The bomb-bay doors open as she swoops on down—and finds nothing. "Where the fuck is she, laddie?" the pilot calls. "I can't see a bloody thing."

The radar operator swallows, curses to himself, and replies: "She's gone. The fucking blip's not there anymore." And he wonders if he ever really saw anything poking up out of the baseline noise or if he imagined it. And the pilot curses softly and says, "Not your fault. That fucking thing's just no fucking use, is it?" And they find their course again and cruise on through the night.

And maybe the operator did imagine the blip, and maybe it was a whale, or maybe it was just a gremlin in the set. But it could have been a submarine. If it was, when the Hudson began its dive into the attack, the conning tower watch would have heard it or seen its distinctive exhaust flames, and within ninety seconds the sub would have been clambering down through the waves to rest unseen a hundred feet below the surface. It would have stayed there an hour or two, until its Kommandant thought it safe to surface again, and then it would have continued its voyage.

And it would have lost that hour or two of safe journey time through the dark North Sea. If enough planes could be equipped with enough radars, the sub might be intercepted several times during the night and thus slowed down enough to double or triple its time of passage. The cumulative effect would be to diminish submarine action in the Atlantic at any given time by a factor of two or three.

But the crew in the Hudson would never know this; all they would know is that the fucking set had given them another fucking false report.

And at any rate there weren't enough airplanes equipped with enough radars to slow the subs down. The very ineffectiveness of these early ASV radars, combined with the other pressures of war, gave them a low priority. In the spring of 1940, just before France fell, opening the Biscay coast to the submarines and introducing "the happy time," the British Admiralty thought they had the submarine menace under control. One of Winston Churchill's last acts before ascending to the prime ministership was to sit at the head of the Night Interception Committee and proclaim that, due to the menace of night bombing, immediate priority should be given to the installation of airborne-interception radar into the rather useless Blenheim fighters, "as it is a more urgent problem," rather than to the fitting of Air-to-Surface-Vessel radar into Coastal Command aircraft. The question of more aircraft for Coastal Command was similarly answered when, upon the urging of Professor Lindemann, the decision was made that priority should be given to Bomber Command—which at the time was dropping nothing more lethal than leaflets on German cities.

We have seen that the German night-bombing menace was exaggerated by British fears—understandably so, for to actually live under a rain of bombs is infinitely more terrible than simply to imagine it. But never were the German night-bombing forces sufficient to decide the outcome of the war. Similarly, not for several more years at least were the RAF bombers strong enough to do much more than provide a fillip to public morale by their own raids on Germany. It was in the Atlantic Ocean that the war could have been won or lost, and yet with no more radar and no more aircraft, the protection of merchant shipping was turned back to the beleaguered, overextended, inadequate convoy system, which itself was under attack by critics in high places. The terrible effectiveness of the wolf packs was overwhelming the defenses, and so it was argued that the system should be dispensed with and the merchant ships dispersed to find their own ways across the Atlantic.

In October 1940, Convoy SC-7 sailed at a maximum speed of seven knots, with thirty-five ships lined up in five columns of four ships on the outside and three columns of five ships in the center, covering an area of several square miles. Its escort consisted of two sloops and a corvette which, after attack by a Wolf Pack of seven submarines, was increased by another sloop and corvette. To no avail: The Wolf Pack cut into them night after night, and by the end of the trip twenty of the thirty-four merchantmen had been sunk. No U-boats were destroyed.

One after another, the convoys were attacked and slaughtered. The Admiralty now lowered the upper speed limit: any ship which could make twelve knots was to proceed on its own. Instead of reinforcing the convoy escorts, more ships and airplanes were diverted to the attempt to search out submarines on their transit through the Bay of Biscay. These attempts were uniformly unsuccessful, and the carnage mounted nearly to the breaking point.

Fortunately, the resources of the U-boats were not infinite; their sailors, under great strain, needed rest as much as did their equipment. By the end of October the number of submarines at sea declined as more and more of them went home for rest and refitting. And in November the terrible Atlantic winter began, making life miserable for the sailors aboard the merchantmen as they pitched and tossed and froze—but allowing them to survive as the U-boats simply couldn't handle such surface conditions. A winter respite set in, but the spring of 1941 approached inexorably, awaited with dread by the Admiralty and with great expectations by the *Führer der U-booten.*

There were two possible ways to defeat the U-boats. One was to increase convoy defenses and beat them in pitched battles as the wolf packs tried to sink their teeth into the merchantmen. The most effective aid would be airpower, but the British were desperately short. An aircraft carrier traveling with each convoy would do the job, but they had too few such carriers to spare. The navy still resented and fought the idea of taking their precious carriers away from their "proper" role—offensive actions against the enemy—and using them for "purely defensive" duties such as convoy protection. They argued instead that long-range shore-based aircraft should be used. Right enough, except that Bomber Command resented and fought the idea of allowing any of those aircraft coming off the production lines to be diverted to such a passive defensive role instead of to their "proper" offensive role, that of bombing Germany. A miniscule part of Short Brothers' efforts went into producing a few Sunderland flying boats, the rest went into the Stirling heavy bomber. A few medium-range Wellingtons were passed to Coastal Command as the larger Lancasters and Stirlings came into service, but these latter long-range bombers were jealously hoarded by Bomber Command.

Fifty overage destroyers were given to England by the United States in return for naval bases overseas, and these would be extremely useful once they were brought up to modern standards. (Aside from being obsolete in respect to modern armaments, they had been mothballed since the first war, and the terrible North Atlantic weather battered them and exposed every rusted fault in their seams.) In the meantime, other escort vessels—sloops and corvettes—were being produced as quickly as they could be, which was nowhere near quickly enough. Primitive radars were being supplied for these ships, but they were nondirectional in character, nonrotatable, and not beamed tightly enough to be of much use in U-boat detection. They provided a useful service in drawing an electronic picture of the convoy and thus allowing ships to keep station with less strain on the officers on the bridge—an important consideration certainly—but what was really needed if the U-boats were to be fought effectively from the escorts was a nimble radar that could sweep around like a lighthouse beam to pick up the subs as they scooted on the surface in the darkness up and down the rows of merchantmen. Such a set was already on the drawing boards and in the lab: it was centimetric radar. But the night fighters had first crack at it, and Bomber Command had second. It looked to be a long time away from the convoys.

So Professor Lindemann leaped into the fray with a new idea for submarine detection, writing to Churchill with the suggestion that

> a large number of small magnets...could be projected from aeroplanes or destroyers to the region in which the submarine was suspected. Any of these striking the submarine would probably stick.... The magnet might carry a suitable device containing...a substance which in contact with the water produces bubbles of gas...thus providing a useful marker. If any magnets hit the submarine they would stick to it, and as it moved away it would betray itself by a track of bubbles.

Churchill for once was disenchanted with his scientific advisor, remarking that the idea was "rather far-fetched," and pointing out the obvious fact that if the airplanes or destroyers were able to hit the sub with the required magnets they would surely do better to hit it with explosives and sink it right off. "I do not think I wish to associate myself with the matter," he wrote in reply.

The other method of attacking the U-boat menace was to keep them from reaching the Atlantic from their French ports. Daytime airplane patrols kept them submerged but were unable to sink them, and they simply surfaced at night to recharge their batteries and continue their voyage. As the radar improved and ASV Mark II went into production, the subs were more often picked up at night, but the sets weren't accurate enough for a bombing run to be carried out. The subs simply thumbed their noses at the blind bombers flying around in circles above them, while the bombers' crews were frustrated at seeing the blip on their scope but not being able to find it visually.

Another crazy idea was proposed for this situation, perhaps even sillier than Lindemann's magnet scheme. Squadron Leader Humphrey de Verde Leigh proposed installing a giant searchlight in the aircraft; when the radar picked up a sub the light would be turned on and, caught in its illuminating beam, the sub could be attacked. The one great difference between Lindemann's idea and this one was that this one would work. But it would take another year to perfect it and install it, and in the meantime there was nothing to be done except to slog it out with the Wolf Packs and try to muddle through.

From January to June 1941, the U-boats sank nearly a million and a half tons of merchant shipping. The first bright light appeared later that summer when the *Audacity* began her first run. She was an experiment in a new kind of aircraft carrier, one designed specifically for convoy escort duties. Converted from a captured German merchant ship by the simple expedient of covering her top deck with a flat metal sheet, she had no hangar space and room for only six fighters which were incapable of carrying bombs. But her fighters shot down or shooed away the German reconnaissance aircraft that tried to find the convoys she sailed with, and when the subs spotted her aircraft in the skies, they dived and stayed submerged and were left behind by the escaping convoys. She pointed the way to the future, and the conversion of other merchantmen into carriers was immediately begun.

But it would be years before enough escort carriers were ready. In the meantime, desperate measures were tried. A catapult was welded onto the decks of a few merchantmen, and fit with a single Hurricane fighter. If a German recco aircraft or submarine sighted the convoy, the Hurri would be shot off to chase away the Focke-Wulf or to frighten the sub into submerging, and the convoy would then race away. The unfortunate Hurricane pilot would then have to bail out and hope one of the escort vessels would be able to fish him out of the water. The idea worked, but obviously only once per convoy.

And the sinkings continued. Over 300,000 tons in May and June, somewhat of a drop in July and August, but up again to nearly a quarter of a million tons in September. Then just when things began to cool down in the winter months, the Japanese attacked Pearl Harbor and opened a brand new door.

The American entrance into the Battle of the Atlantic took Doenitz by surprise. (It was Hitler who declared war on America, not vice versa, and such an action had never been discussed with his commanders in advance.) But the admiral's reaction was quick, and by January 1942 the first U-boats were reaching their new stations off the eastern seaboard of the United States, where they found a Happy Hunting Ground whose luxurious riches they had never dreamed of. Despite watching more than two years of war in Europe and China, America was incredibly unprepared. Not only were the coastal waters filled to bursting with heavily laden steamers traveling alone and unprotected, but nobody even knew whose responsibility it was to protect them! The navy looked to the Pacific and the terrible turmoil of the Pearl Harbor catastrophe, while the Army Air Corps had never even taught its pilots to fly over the sea. The coastal cities of the United States resented and fought against the idea of a blackout, with the result that tankers off the coast of Miami or Virginia Beach or Jacksonville or Boston could be seen easily at night silhouetted against the hazy glare of light in the background. From a bare hundred thousand tons sunk in December 1941, the U-boat score rose dramatically to over 300,000 tons in January, nearly 500,000 tons in February, and over the half million mark in March. In May it topped 600,000 and in June it was over 700,000 tons. When the lights of the cities finally dimmed and the Air Corps finally instituted antisub patrols and the Navy at last organized convoys, the subs simply slunk off to the Caribbean and continued to pick off the unprotected ships there. The entire Allied war effort very nearly collapsed at the outset as too much vital war material and the ships necessary to transport it to the war zones sank to the bottom of the Atlantic shelf.

But the first ray of light also came that summer, quite literally. Before dawn on June 4, 1942, the Italian submarine *Luigi Torelli* was cruising on the surface from her Bay of Biscay port out to her Atlantic patrol position. The officer of the watch was staring out into the darkness when suddenly he was blinded by a fierce white light in the sky. There was no time to dive, no time to do anything but watch helplessly—and then the light went out. There was no attack, no bombs, there was nothing in that dark night but the sound of the *Torelli's* own diesel engines.

Somewhere up above there was an RAF Wellington crew cursing another aborted attack. They had been training for months with their improved radar sets and the Leigh light, and out of long hours of practice had evolved a precision attack routine: After zeroing in on a radar indication, they would come in at a height of 250 feet, and at a precise range indication they would flip on the great light to find the submarine brilliantly bathed in it. The problem was that their aircraft altimeter, like all altimeters, worked by indicating atmospheric pressure which varied from 14.7 pounds per square inch to progressively lower values as altitude increased, but while the principle was valid, the base number of 14.7 was not precisely true; instead it varied according to atmospheric conditions. Before setting off from base, each aircraft set its altimeter *according to atmospheric conditions at that base and at that time.* There was no way of predicting what conditions might be like hundreds of miles away and several hours later, nor was there any way to measure these parameters accurately from the aircraft in flight.

So when Squadron Leader Jeff Greswell of No. 172 Squadron dived his Wellington into the dark night onto the vector given by his radar operator and then leveled out at precisely 250 feet and switched on his Leigh light, he saw nothing but empty waters ahead—because he was not really at 250 feet but actually considerably higher. (He could just as easily have been considerably lower, and could have dived straight into the ocean waters; this was one of the reasons why more than 300 Coastal Command aircraft had been lost by that summer.)

On board the submarine they were not sure what had happened; the only thing they could be sure of was that the light had been from an airplane. But why had there been no bombs, no attack? They had been caught in the fringe of the light, and to their own eyes it seemed they must have been brilliantly lit up; they didn't realize that once the light was turned on, the eyes of the aircraft crew were so blinded that nothing but an object dead center could be seen. Assuming it must have been a friendly aircraft, the *Torelli* dutifully fired off the recognition signals of the day.

In the Wellington circling above there was a burst of amazement when the red and green flares sailed high into the sky, illuminating the submarine. They swung around and came in again, this time catching the sub square in the brilliant white light of their searchlight, and released four depth charges which straddled the *Torelli* and damaged her so badly that she was lucky to be able to limp back to Spain.

The difficulties with the pressure-reading altimeter were soon solved using a radar-altimeter, in which radio waves were directed straight down and bounced off the ground or water, their time of return giving the altitude precisely and without regard to atmospheric conditions. Using this instrument just one month later, an American who had joined the RAF before Pearl Harbor, Pilot Officer W. Howell, made the first confirmed submarine kill at night with radar and the Leigh light. U-502 went down suddenly, with no warning coming out of the black night except that suddenly blinding beam of light, and with no survivors.

Das verdammte Licht, as the Germans soon learned to call it, com-

The Blitz and the "Boot"

bined with radar search and the radar altimeter to have an overwhelming effect. There were so many night attacks that summer that Doenitz ordered a surprising reversal in policy: From now on the submarines were to submerge at night and try to run on the surface in day! This, of course, opened them up to attack in daylight, but at least they'd have a chance of spotting the aircraft themselves rather than having a blinding light suddenly open up out of the darkness. The result was predictable: The U-boats began to be attacked and sunk as their lifeline home across the Bay of Biscay became the most dangerous stretch of their journey. It began to look as if the Battle of the Atlantic was about to be won—

And then suddenly the radar sightings ceased. Or rather the subsequent visual sightings ceased: A radar blip would be seen but would disappear almost immediately, before visual contact could be made. It was almost, the pilots reported, as if the subs *knew* when the radar spotted them, so they could immediately submerge.

Actually, that was precisely the case. In 1941 an ASV-equipped plane had crashed on French soil and the Germans had captured the set intact. They realized now that it was this equipment that was disrupting their submarine service, and their engineers constructed a simple "Fuzz-Buster" set: a radio receiver that picked up the 1.5-meter radar transmissions and warned the submariners that a radar set was tracking them. As Erhard Milch, the German Secretary of State for Air, put it at a meeting of the German High Command: "Our U-boats now know when they are being observed just as surely as a French lady in a Parisian cafe, eyes lowered and demurely sipping her coffee, knows she is being ogled." Since these Metox receivers (named for the French firm that produced them) could pick up the radar beam before the echoes were recognized by the tracking aircraft, the subs disappeared beneath the waves before a proper fix could be obtained.

Once again the tide swept to the German side.

By the end of that summer a new radar device, one capable of sweeping the U-boats from the seas, was in British hands—and those hands were tied once again by Churchill and Professor Lindemann.

In 1937, when Taffy Bowen had first taken a radar set into the air by tying it with rope and wire to the dangling undercarriage of a biplane Heyford bomber, the object of the exercise was to see whether it was capable of detecting other aircraft. It did not succeed, as Bowen later remembered: "The transmitted power was simply not great enough for that. Quite unexpectedly, though, strong echoes were received from objects on the ground, in particular from the wharves and cranes at Harwich which could be distinguished clearly from their surroundings."

When a few months later Bowen flew out to sea to detect the Royal Navy in the autumn exercises, as has been related, he guided the airplane home again through thick weather by identifying the coastline on his radar set.

This feat was remembered in 1941, when Lindemann angered the RAF by identifying a nasty problem. For nearly a year Bomber Command had been hitting back at targets in Germany and the occupied countries, flying and bombing at night. Intelligence concerning the results of their bombing was hard to come by, especially in Germany, but reports from the underground indicated that at least in France they were not hitting their targets as accurately as claimed. Cameras were then installed in some aircraft, synchronized to go off when the bombs hit the ground, and the photographs supported this contention. The chiefs of Bomber Command argued to the Cabinet that the problem lay in the photography rather than in the bombing: Since the aircraft was in motion, the photographs would be necessarily skewed. Lindemann demonstrated conclusively that, because of the nearly infinite difference in the speed of light and the speed of the aircraft, the latter did not affect the situation. He pressed Churchill to assign a member of his staff to make an exhaustive study of the problem, and when this was done the results showed without possibility of argument that the bombing accuracy was roughly a quarter of what Bomber Command had been claiming. Instead of hitting specific factories, they were missing entire cities.

This led to a change in tactics, saturation bombing of urban areas being substituted for pinpoint bombing of individual factories. When it became apparent, in the following months, that even this was not being satisfactorily achieved because of the difficulties of flying across a continent in darkness and finding a particular city, further solutions were searched for. Late in 1941 no solution had yet been found, and Churchill was applying pressure. The headlines in the London papers announcing bomber raids on the Ruhr, on Berlin, on Hamburg and Cologne were the only headlines that talked about hitting back at the Nazi monsters. Bomber Command was the only weapon he had, and he wanted to use it not to bomb empty fields or decoy cities but to blast the Nazis where they lived.

But first the RAF had to be able to find their cities in the dark, and so far they couldn't. "At the time," Bowen wrote later, "people were in a bit of a panic, and somebody in the Air Ministry—I don't know who it was—said that it could not be done by radar. I heard them say it, and I immediately said: 'Yes, it can: I am quite certain that it can be done because I saw it in 1937.'"

Work was immediately started on adapting a radar set to look at the ground as the airplane flew over it, but the first results were not sufficiently precise. The difference between land and sea could be identified, which would be useful as coasts were crossed, but most other landmarks did not show up clearly enough to be discerned by most RAF crews. The next step, obviously, was to try the new cavity magnetron and the centimetric waves it generated. These short wavelengths gave more precise details, as expected, and now forests and hills, towns and rivers, railway lines and major highways could be seen on the radar scope nearly as clearly as on a map.

Bomber Command was thrilled. The new device was named H2S, and to this day nobody is quite sure why, although various stories and personal reminiscences abound. It has been claimed that the name was given by Lindemann, in his sardonic humor the letters standing for the "Home Sweet Homes" that would be destroyed by the bombers, or that he named it for the gas hydrogen sulfide, (chemical formula: H_2S), which provides the stink of rotten eggs and of most bodily gaseous emissions (the idea being that he thought they all stank for not having thought of it sooner); it has also been claimed that the name was given later, again for the noxious gas, when the idea itself began to stink and the RAF bombers fell in flames because of it. (The Luftwaffe night fighters homed in on the radar transmissions and were thus able to track the bombers and strike without warning. But that came much later.)

When the device was first made ready, the only question was who should get it first, Bomber or Coastal Command? Air Marshal Sir Philip Joubert de la Ferte, C-in-C of the latter, claimed it for his own: It could find submarines even when only their periscope was showing, and the Metox radar detectors were set for metric wavelengths rather than centimetric. This meant that the emissions from H2S would be invisible to the subs; with this weapon he could break their backs.

The Admiralty backed him up. Given these sets, or similar ones, aboard their convoy escorts, they'd be able to see the subs as they attacked convoys on the surface at night. Between the two of them— the navy's convoy escorts and the RAF's bombers—they'd lick the submarine menace once and for all.

But, they cautioned, if they were to succeed they must have the sets before Bomber Command. Once the new device was captured by the enemy, they would obviously know what wavelength it worked on; then they could again build detectors to frustrate it. This was the only thing that could wreck the project, for the concept of centimetric wavelengths was so far ahead of anything Germany had that, without such capture, they'd never catch on to what was destroying their subs.

Obviously, the admirals and C-in-C Coastal Command argued, there would continue to be Bomber Command losses; some unfortunate Stirlings and Wellingtons would continue to fall on French and Dutch and German soil. And it had already been proved that the magnetron was so indestructible that even a battered one would reveal its secrets to any competent technician. Therefore they must accept that the centimetric waves would be known shortly after the first bomber raids. (This is, in fact, what happened, and allowed the German night fighters to use the H2S transmissions to find the bombers.) On the other hand, when Coastal Command aircraft went down, they went all the way down—to the bottom of the sea, taking their radar sets with them. In their hands the secret would remain a secret forever.

Clearly then, the obvious order of battle was to give H2S first to Coastal Command and the navy; then once the submarines had been wiped out it could be used by Bomber Command with no loss in effectiveness. If, however, the first sets went to the bombers, by the time they were received for antisub work they would already be compromised.

It was a strong, fair argument. Churchill blistered under its heat, he withered, but he stubbornly shook his head. The submarine war was the only war "that ever really frightened" him, he later wrote—but the bombing war was where his heart lay. Bomb the bastards, destroy their cities, lay all Germany waste!

That was what Air Marshal Sir Arthur "Bomber" Harris was promising him. With a thousand bombers and H2S he could win the war overnight—and at sea surely they could hang on just a little while longer.

Churchill turned to Lindemann, and the Prof told him what he wanted to hear. Air power could win the war, destroying the subs in their factories before they ever reached the water, smashing the oil production and transportation facilities, denying the German armies their weapons of war and rendering an entire nation homeless and sleepless while bringing them to the brink of starvation. The German capability to wage war would be eliminated, and the RAF could do it if only they could find their targets. With H2S they could find anything. Give it to them and turn them loose.

Churchill nodded, but before he could make a decision Tizard and the other scientists who had been fighting with Lindemann since the early days of radar development spoke up. Tizard pointed out that Lindemann's use of statistics was faulty and that the night bombing of Germany was bound to be ineffective until the Americans were able to continue it by day, an effort which was several years away. ("I am afraid...that the way you put the facts is extremely misleading and may lead to entirely wrong decisions being reached, with a consequent disastrous effect on the war. I think, too, that you have got your facts wrong.") P. M. S. Blackett presented his own calculations showing that German deaths by bombing consisted mostly of civilians and were not much higher than British losses of highly trained bomber crews. He thought that Lindemann's numbers were a good "600 percent" too optimistic.

Coastal Command and the Admiralty took this opportunity to strike again. They reminded Churchill that just a few months previously he had written that "the issue of the war depends on whether Hitler's U-boat attack on Allied tonnage, or the increase and application of Allied air power, reach their full fruition first." All would be lost if the U-boats were not stopped. Churchill agreed (as he had to—they were his own words), but he reminded them that in making war "experience shows that forecasts are usually falsified" and he reserved to himself the final decision.

Watson Watt was called in to give his opinion. He presented, in his naturally rather bombastic manner, a paper of exceptional length; stripped of the verbiage, his argument was that the Germans might well already know of centimetric radar since an early set had been lost off Singapore with the battleship *Prince of Wales,* which the Japanese had salvaged. Further, since British Beaufighters had been using the set in their attacks on night bombers over England, the German crews might well have picked up the emissions. Finally, he estimated that once they knew of the existence of microwaves, they would be able to build a suitable detector "within two or three months at the most." They probably had one working already; if they didn't, they would soon, and so all this talk was accomplishing nothing but using up precious time in which the H2S could be working for the bombers.

With Churchill already predisposed to the bombing arguments, this final push carried the day. H2S would go immediately to Bomber Command; when enough sets were available they would be supplied to Coastal Command.

The winter weather at the end of 1942 brought once again a temporary halt to the Atlantic slaughters, but as the spring of 1943 began to calm the waters, the U-boats again stepped up the attack. Doenitz had now more than 200 submarines, and the attack this year was to be overwhelming; the Atlantic lifeline was to be cut, the war was to be won, Britain would sink and America would be isolated.

On the other side, an equal effort was being readied. The Casablanca conference finally, according to Churchill, "proclaimed the defeat of the U-boats as our first objective." A host of weapons was being made ready. The Ultra decoders had long been reading the U-boat wireless traffic, but the decision had always been that Ultra must be defended from discovery: If it provided warning that a wolf pack was lying in wait for a particular convoy, the convoy would not be warned and warships and aircraft would not be sent to the rescue unless there was independent evidence—in order that the Germans should not become aware that their coded messages were being read. Now this restriction was removed; even Ultra would be sacrificed if necessary. High-Frequency Direction Finding (HF/DF, or Huff-Duff) had come into operation as a means of pinpointing the location of wireless emissions. This technique took advantage of Doenitz's wolf pack strategy, which required frequent radio contact between the packs and headquarters. In operation for more than a year from stations along the coast of England, Huff-Duff picked up the wolf pack signals to headquarters and thus was able to warn the convoys that trouble lay along their route. But now the HF/DF sets had been modified to be carried along by the escort vessels, and thus gave correspondingly more precise locations. They could tell the radar operators just where to look. Long-range bombers had been provided to Coastal Command, more and more escort vessels were built for the Admiralty, including small aircraft carriers, and, above all, centimetric radar "had been handed over somewhat reluctantly by our Bomber Command to Coastal Command." Finally, and of utmost importance, Watson Watt was wrong.

On the second of February, 1943, a German Ju 88 night fighter shot down a Short Stirling bomber over Rotterdam. The Stirling had been equipped with H2S, and the significance (although not the precise function) of the copper-valved set was realized by the officer assigned to investigate the wreck; by the end of the week the battered copper device was in Berlin, where General Martini's electronics staff examined it. As predicted, they realized quickly what it must be, and the dimensions of the set itself delineated the wavelengths of its emissions. The secret of centimetric radar was out.

But the Germans were more concerned with the new bombing offensive than with attacks against their U-boats, and most of their effort went into producing a detector and homing set for their night fighters. In this they succeeded, but somehow the British refused to realize it (although they had predicted it): Thousands of RAF lives were lost as the night fighters homed onto the H2S emissions and shot down the bombers.

At sea, however, it was a different story. Incredibly, Doenitz went on believing that Coastal Command would continue to use the 1.5-meter radar whose pulses were detected by his Metox receivers. This belief was inculcated and reinforced by a combination of resources thrown at him by the British, together with a series of circumstances devised by whatever gods might be.

The German electronics countermeasures group quickly put together a Metox-type radar detector for centimeter waves, and this was fitted to several U-boats. It did not, however, work efficiently and the first few subs so equipped were lost by radar-directed night attacks. When one of them escaped and reported that it had been attacked without warning, Doenitz arrived at the conclusion that the British were using something other than centimeter radar; he evidently never considered the possibility that the radar warning equipment was at fault. (Thus do we become prisoners of our national pride: German technical efficiency was an item of faith not to be doubted.)

At the same time, the German engineers made a surprising discovery: The receiving system of the Metox actually radiated an electromagnetic signal. It was a weak emission, but it did exist and if sufficiently sensitive tracking equipment was available, it could be homed in on. Doenitz exploded when he was told. He immediately radioed all his submarines at sea to cease using *all* radar detectors (centimetric as well as metric), and directed his electronics specialists to concentrate on producing an emission-free detector rather than an improved centimetric detector.

Immediately thereafter, this decision was reinforced by an unknown hero. An RAF Coastal Command crew had been shot down and captured by the Germans, and one of them violated standing military orders to give the captors no information other than personal identification by volunteering during interrogation the information that they were finding subs by following their Metox radiations! Coming when it did, this cunning lie was believed without hesitation: Doenitz radioed to his U-boats that it had been confirmed that the English were using their own Metox emissions rather than radar to find them.

We don't know who the unknown hero was. It seems unlikely that he was an intelligence "plant," for no one has owned up to the story since the war; for the same reason, it seems most likely that whoever he was he did not survive the war, never lived to tell his tale. Wherever he is today, God rest him well.

Finally, without knowledge of the events chronicled above, a strong Allied intelligence effort was put into operation to reinforce the notion that centimetric radar was not being used at sea; through infiltrated spy systems the Germans got the message that the RAF was using an American magnetic detector and an infrared scanning system rather than centimetric radar. Together with their own discovery of the Metox emissions and the unknown hero's tale, these served to confuse the German estimation of the situation so that they never put into proper service on their U-boats a reliable centimetric-radar detection system.

The climax of the Battle of the Atlantic approached inauspiciously for the Allies. In March a hundred U-boats were at sea straddling the convoy routes, more subs than had ever been gathered in one ocean at one time, and they sank more than half a million tons of shipping while losing only a dozen of their own number. In April they added another third of a million tons for the loss of seven U-boats. But then in May the Allies unleashed all their forces: Unrestricted use was made of the Ultra intercepts, more escort vessels and aircraft joined the action, and most importantly of all, the ten-centimeter radar sets were fit to an increasing number of aircraft and escort vessels. The combination was devastating.

Ultra told which submarines were being sent where; high-frequency direction-finding wireless equipment picked up the messages each submarine sent back to headquarters and pinpointed its location; and long-range aircraft sent to that location found the U-boats on their radar scopes and dived to the attack, finding them visually by day and with the aid of the Leigh light by night. The Bay of Biscay was effectively closed to U-boat passage by taking advantage of the U-boat's maximum underwater endurance of twenty-four hours, and so scheduling air patrols that every circle of seventy-two-nautical-mile radius was scanned by radar at least once in every twenty-four-hour period. As one of their commanders described it:

> They always found us. We might dive and have enough electric power to proceed under water, but for how long? Taking our cruising speed as 3 knots (under water), that meant we could travel 3 miles an hour for 24 hours at most, enabling us to proceed under water for 72 nautical miles. After that the U-boat would be compelled to surface to recharge its batteries. But a circle of 72 miles radius could be effectively patrolled with the result that the moment a U-boat surfaced, the enemy radar would pick it up even quicker than before.

Indeed it would. Coastal Command's aircraft by this time were equipped with radar that looked sidewards as well as forward. The side-scanning system generally consisted of eight dipoles arranged in four large transmitting antennae on top of the fuselage and duplicated receiving antennae on each side. This was a long-range, poor-resolution system which swept out a large area on each side of the aircraft as it patrolled.

When a blip was sighted on the side scanners, precise location was not possible—but it would be picked up by the receiving antennae on only one side of the aircraft, so direction right or left was immediately and unambiguously clear. The pilot would immediately turn in that direction while the crew was brought to battle stations, and by the time the blip was picked up on the more precise forward-looking radar, every gunner was at his post, the depth charges were ready for dropping, and the plane would be zooming down in a speed-increasing dive. At night the Leigh light would be ready to be switched on when the radar operator marked the correct distance and height, so that—night or day—out of the blackness or out of the clouds or out of the dazzling sun, the Liberator or Sunderland would appear suddenly and right on course, and the thunder of its engines as it zoomed a hundred feet overhead

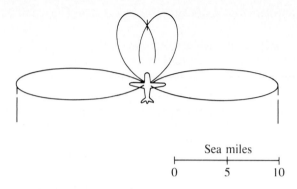

would be followed by the detonation of high explosives—often the last sound the submarine's crew would ever hear.

The side-scanning system increased the effective radius of sight of each patrolling aircraft by nearly ten miles, and made it possible for surveillance to be so intense that every single submarine crossing the Bay of Biscay was sighted and attacked at least once. For many of them, once was more than enough. Forty-one times in that single month of May 1943, air and surface forces found a submarine and sank it, while Allied shipping losses dropped to 200,000 tons. Among the submarines that failed to return home were those that carried Doenitz's two sons, both of whom vanished into the sea without any more warning than a moment's roar of airplane engines.

The wolf packs were pulled back, the Atlantic was abandoned. The U-boats were reassigned to other theaters of operation, but the radar sets found them wherever they were. In June, seventeen more were sunk, while only 22,000 tons of shipping was lost. In that summer of 1943 in which Doenitz had hoped to win the war for Germany, he sank a total of fifty-eight ships—and he lost seventy-two U-boats. For the first time since the *Unterseebooten* took to the seas in the First World War, more U-boats were sunk than merchant ships. When winter finally calmed the seas by roughing the waves, the submarine menace had been irrevocably destroyed.

By the following spring, the Allied buildup for the invasion of Normandy had begun, and by that time the U-boats were so overwhelmingly defeated that not one single American soldier's life was lost due to U-boat action in the North Atlantic, despite the unprecedentedly large numbers of troops transported by ship across that ocean.

When D-Day approached, the remaining U-boats were stationed in a continuous line from Norway to Spain to give advance notice of the invasion flotilla, while every sub in reserve was waiting to sail and destroy it. But on June 6, 1944, the U-boats found they were simply denied the field. Every time they poked a periscope above the water a

Sunderland or Liberator or Catalina or Swordfish came diving down straight at them, and from the prop wash came tumbling a stick of depth charges to send the sub right back down again, often forever.

A new generation of German submarines was being prepared on the drawing boards and in the factories, fitted with *Schnorkel* breathing apparatus and new types of underwater propulsion, but the factories and the drawing boards were being bombed out of existence and these more deadly subs were launched too seldom and too late. Marshal of the Royal Air Force Sir John Slessor, who took over as C-in-C of Coastal Command in 1942, has said that all warfare is a story of muddle and mistakes, with victory going to the luckiest muddler making the fewest mistakes. The mistakes the British made in assigning priority for centimetric radar and long-range bombers to Bomber Command instead of Coastal Command were counterbalanced by Doenitz's failure to recognize how his U-boats were being located so precisely once these factors entered into the war. In a teetering, tottering dance on the very edge of time these two antagonists waltzed to the precipice, and by the grace of God the right one fell over.

Part Five
The Band Plays On

> Casey would waltz with a
>   strawberry blonde,
> And the...
>
> *Old song*

Chapter Eighteen
Armageddon

> Yea, though I walk through the
> valley of the shadow of death,
> I will fear no evil....
>
> *Famous last words
> of the twentieth century*

The dance, unfortunately, is not over; the piper has not yet been paid.

Twenty years ago I got into my Piper Cub to fly from Ithaca to New York. The Cub was a prewar two-seater that had to be started by swinging the prop by hand; the only piece of modern equipment that had been added was a primitive two-way radio. I had just learned to fly, and was taking advantage of this new freedom to fly in to the city for the day.

It was a cold winter morning, with what looked like high overcast. The weather report was for clear skies over New York, but intermittent cloud cover en route. I wasn't rated for instrument flying and should have stayed home, but it looked clear enough here and was certainly clear enough there, and I was young and stupid: I decided to go.

I pulled the prop through a couple of times to clear the lines, then walked around the side of the Cub, ducking under the wing, and reached into the cockpit, flipped on the magneto switches, and advanced the throttle a quarter of an inch. Then I ducked back under the wing and went back to the propeller. Starting the Cub like this was illegal: It was a two-man job, but there wasn't anyone else around at dawn on a freezing Ithaca day. I spun the prop; it caught, the engine coughed, and I scrambled backwards out of the way as the Cub started to move. I skipped sideways to avoid the propeller and ducked under the wing as it passed over me, then ran up to the cockpit and grabbed hold of the door handle. I jumped up and swung myself headfirst into the cockpit as the Cub bounced over the grass. I pulled back on the throttle to slow her down, squeezed into the small seat, got my feet on the rudder pedals, and was back in control.

I didn't call in with a flight plan on the radio because I was only rated for visual flight conditions and was afraid the regional controller would tell me the weather to New York wasn't cleared for VFR pilots. Flight plans weren't mandatory and so I just headed her into the wind, pushed the throttle forward, and took off. She bounced a couple of times and then she was free, clambering up into the air. I held her steady while she climbed a thousand feet, then broke out of the airport pattern and headed southeast.

By the time I reached 3000 feet I was touching the bottom of the overcast: the clouds weren't as high as I had thought. I trimmed her for gradual descent, but had to keep pushing the nose down to stay in clear air. In another ten minutes I was below 2000 feet and the clouds were still pushing me down, and I knew there were hills ahead of me more than 2000 feet high. So I pulled up into the clouds and tried to climb up through.

I wasn't rated for instruments, but I had practiced on them. I climbed at fifty miles an hour and kept the ball and needle centered, and everything went okay for a good fifteen or twenty seconds. Then the ball started slipping. I bent the stick over a bit, but I couldn't keep it from sliding off to the right. I knew I shouldn't try to look out the side window, but finally the temptation was too great and I did. From that moment on I was doomed. I couldn't see anything but white nothing, I couldn't tell up from down from sideways, but somehow as I looked out I was *positive* my left wing was pointing down. The needle-and-ball was telling me I was turning right, but looking out into that white crap I was sure I was turning steeply left, too steeply, about to spin out. I turned right and the needle skidded off the edge of the dial and the ball went bouncing around crazily and I knew I was a dead man.

I'm not sure what happened after that. I swear I was diving but the air speed was dropping off, so I must have been climbing. By all that's holy I should have spun out of that cloud and into the ground but somehow, screaming in terror, I took hold of someone's hand and suddenly popped out on top on the edge of a stall.

As soon as I cleared the cloud, I saw which way was up and which was down, and I settled down into level flight. The speed built up and I was safe again—but everywhere beneath me was nothing but an unending sea of white clouds. I was terrified, because somehow I had to get back down through them.

If I had had a parachute I would have jumped, but I didn't. I didn't know what to do, so I hung around up there for half an hour, flying in circles, looking for a hole below, a break in the clouds.

There wasn't one. As I sat up there looking at my gas gauge run down, I finally remembered my radio. I called Binghamton control.

"Binghamton tower, this is Piper seven six alpha, VFR from Ithaca to New York. I'm at angels five above the clouds, and I think I'm lost."

No one answered, and I began to cry. Then I remembered to turn the radio on. I tried again, and this time a voice broke through the static: "Piper seven six alpha, this is Binghamton tower. Where are you?"

Christ, I didn't *know* where I was. I looked at my watch and saw I had been in the air twenty minutes. I had taken off southeast, but had been flying in circles for a while. I said, "Seven six alpha. I think maybe about twenty-five miles southeast of Ithaca? Over."

"Roger, seven six alpha. Hold on a minute."

About ten hours went by, and then the static crackled again. "Seven six alpha, turn right and then left alternately every ten seconds."

I nodded, then remembered I was alone up here. I pressed the button and said, "Seven six alpha, roger." I bent the stick right as gingerly as I could, afraid by now that any rough movement would send me tumbling back into that white inferno below. I held it for ten seconds, then bent it left, then again right, and again—

"Okay, seven six alpha," the voice burst out suddenly. "We have you on radar. Turn right to one niner five."

I did so, and the radio asked again: "Confirm that you are VFR?"

"Seven six alpha, roger," I answered.

"All right. No need to identify yourself further, we'll be talking only to you from here on in. Now we're going to have to bring you down on through the cloud—"

"No!" I gasped.

"Sorry. There's no break anywhere. But no sweat, just relax and we'll bring you right on in. We've cleared everyone out of your way and you've got nothing to worry about. You're drifting off to the left a bit; come back to one niner five."

I bent the stick a bit more to the right, and watched the needle creep around, then straightened up again.

"Perfect," the radio said. "If you can fly like that you've got no problems. Now start to let down."

I knew he was babying me, but I loved it. He sounded so calm and assured, without a worry in the world. I just wished I could touch him. I reached out and pulled the throttle back a bit, and immediately the nose dropped a couple of inches and the white clouds began reaching up for me—I shoved the throttle forward and bounced up away from them.

"No, no, no. Come on down. You're doing just fine. Slacken off now."

I tried again. This time the throttle stayed back and the nose settled down into the snowy white.

"That's just beautiful. You're drifting left just a bit again. That's it, come around gradually. Lovely. Let's come to two one five now. That's it. You're coming past angels three now, just thirty miles north northeast of the field. Hold that just right and you'll be coming out of the clouds in another few minutes. Let us know when you're VFR again."

I tried hard to believe him. He came on every few seconds, correcting my descent or azimuth, and then suddenly the clouds were gone and I could see the ground. I was a few hundred feet above it and in the clear. I called in, trying to keep the excitement out of my voice. He replied that I was right on course, as if he had never doubted it, and that I should hold altitude and I'd see the field in ten minutes.

He was right. There it was right in front of me. I called in again and he cleared me for a straight-in approach. In another ten minutes I was on the ground.

Radar, in what we with touching hopefulness call the postwar world, has become part of our lives. The ability to guide airplanes through peaceful but still dangerous skies that the Binghamton controller demonstrated to me twenty years ago has been refined to the point where it routinely guides jet airliners into and out of airports that are so busy and crowded that they could not possibly function without it. And it has transformed the black art of warfare.

The transformation can be traced back to the last years of the nineteen-forties, and to two events. The first of these was the decision by Howard Hughes to force his way back into the mainstream of aeronautics. As a young man he had been a champion speed pilot, and in his more mature years he had founded the Hughes Aircraft company, which was an interesting lesson in the futility of dilettantism confronted with professionalism. He spent the war years concentrating the energy of his designers on the *Spruce Goose,* the world's largest airplane. It was an eight-engine, propeller-driven, wooden flying boat, designed and built at just about the time that the rest of the world had abandoned both flying boats and propellers and had discovered jet propulsion. The advantage of the flying boat was that it didn't necessitate the huge expenditure of money to build airfields, which was important in the early days of commercial aviation when the concept had to be proved before any city was willing to build its own airport large enough to take the huge airliners. But by the end of the war there was no longer any doubt about the feasibility of commercial aviation, and it was an accepted truism that a city was not a city without its own gigantic sprawl of runways. (The safety factor that was often advertised, the ability of the flying boat to land anywhere on the water in case of emergency, was little more than an outright lie. It needed the calm, flat waters of a sheltered harbor; the slightest choppiness would give the same effect as if a conventional plane were landing across a plowed and furrowed field.) And the jet engine—which had been brought to the United States by the Tizard mission—changed forever the future of all propeller-driven, long-range commercial aircraft. The *Spruce Goose* flew once—when, with Hughes at the controls for a practice taxiing run, he decided without warning to haul back on the wheel and pull her briefly into the air a few feet above the California waters—and that was the end of her.

Frustrated and annoyed at the condescension with which his plaything was treated, Hughes decided to show them all. The universities were beginning to turn out a large number of high-quality aeronautical and electrical engineers, and this at a time when all the established aviation firms were laying their people off because the war was over. Hughes hired them. He staffed Hughes Aircraft with bright, hard-working, ambitious young men, and he turned them loose—but not on aircraft design. Instead of trying to produce airplanes, he went after components; specifically, he went after radar. And so, when in 1948 the newly organized United States Air Force asked for manufacturing bids on a new level of radar systems for fighter aircraft, Hughes took the contract away from all the established wartime giants. Westinghouse and RCA and General Electric smirked or frowned or laughed outright, but the Hughes idea of a complete weapon-control system convinced the Air Force officers who were skeptical at first.

Just a few months later the Russians sealed off the city of Berlin, and suddenly we weren't sure if we were living in a postwar or prewar world. The Air Force asked Hughes if he would take the chance of investing literally millions of dollars in expanding his facilities before they could promise him any future contracts. He said yes. They asked if he could have his new radar system flying in four months instead of the agreed-upon twelve months, and he said yes.

He and his young group made good on both counts. They beat the deadline by two days, and when the system flew, it found and destroyed a target airplane without the interceptor's crew ever seeing the target with their own eyes. Hughes was overnight the giant of the industry, and the industry was revolutionizing the concept of national security.

In the fighter cockpit the simple radar of the Second World War was transformed into a complete fighter control system, to the extent that in the nineteen-fifties and early sixties the pilot himself was thought to be on the way out. This was because of the revolution in solid-state electronics, coupled with improvements in computer technology—which enabled computers to be built powerful enough to handle incredibly complicated calculations in microseconds and small enough to be installed in a fighter aircraft—and because of the concept of Doppler radar.

The Doppler shift is a well-known wave phenomenon discovered by a Viennese physicist in 1842. It is the reason why an ambulance siren rises in pitch as it approaches and falls in pitch as it recedes. It can be visualized by imagining a machine gun firing bullets at a target at a rate of one bullet per second. If the gun is not stationary but is moving toward the target, each succeeding bullet is fired closer to the target than the preceding one, and so the rate of impact on the target is *greater* than one per second. (The same effect would be seen if the target were moving toward a stationary gun.) If the gun and target were instead moving *away* from each other, the frequency of bullet impacts would be *less* than the true frequency of firing. It also holds true in wave phenomena—for example, in the sound waves of the ambulance. As it approaches, the frequency of the waves seems to increase, and as it recedes the frequency decreases; for sound, an increase in frequency is heard as a rising pitch, a decrease as a descending one.

This principle has been used with electromagnetic phenomena to ascertain that the entire universe is expanding. In the first two decades of the twentieth century the astronomer Vernon Slipher measured the frequencies of light coming to us from stars in distant galaxies and found that they corresponded to the frequencies emitted by hot hydrogen atoms—but that these frequencies were all slightly lowered. He recognized, and it has since been confirmed by Edwin Hubble's extensive work, that this is because the galaxies themselves are all receding from us in a scheme of general universal expansion. This observation was the first indication (though not the last) that the universe originated in a gigantic "Big Bang," creating a blast effect which we are still riding today.

In radar, the Doppler effect means that if the receiving system can not only detect the echo but also measure its frequency (or wavelength, which gives the same information), it can compare this to the frequency of the emitted radiation and see if it has changed, thus determining if the target is stationary or moving toward or away from the system. (We have talked previously as if the radar echo simply bounced off the target; it is actually more precise to say that the target receives the e-m (electromagnetic) pulse, absorbs it, and then reemits it. If the target and radar system are stationary with respect to each other, the reemitted pulse will have the same frequency and wavelength as the original beam, but if they are in relative motion the echo will be Doppler-shifted.) By using a computer to analyze the frequency distribution of consecutive pulses, a Doppler radar system can determine the precise motion of the target relative to the system, and can instantaneously plot its position, its speed, and the direction of its motion—and in that same instant direct the interceptor onto the proper and continuously changing intersecting course. There would be nothing left for the pilot to do but act as a chauffeur: He would drive the bus into the combat area and then turn everything over to the radar-directed control system, which would find the enemy, track him, and fire the guns or rockets at the proper time. It would then ring a bell to wake up the pilot, who would drive the system home.

It didn't work out quite that way. The human brain is a more complex and better-functioning computer than any system we have yet learned how to manufacture, more effective for its weight (even including total body weight, most of which is just a nuisance). So the pilot is still the boss, but the system itself is a marvel. When the radar finds a target, it is displayed at the center of a circle on the scope; the pilot then locks in on it, ordering the radar to follow it automatically. The pilot's job is to fly the plane so that the target remains centered (much like a video arcade game), the computer displaying the centered plot in such a manner as to give him the proper directions no matter what evasive action the target takes. At the same time, the PPI shows the position relative to the fighter of all other aircraft in the vicinity, unidentified, friendly, and enemy, each in a distinctive color for quick understanding and with their directions indicated by a small tail extending from a geometrical symbol. Safe corridors and national boundaries are also shown to help establish the probability of the unknowns being friendly or hostile, and areas of antiaircraft fire are outlined. The radar also indicates which of the bogeys have been assigned for interception to other friendlies, and which radar from other aircraft is tracking which aircraft. His own radar, locked in on the target, gives him a quick readout identifying the enemy by type (Mig 23 or Me 109) and telling him its altitude, direction, and speed. When he draws close enough to fire, he can do so himself or allow the system to fire automatically whatever arsenal he has called for, which may be radar-directed or infrared-controlled rockets, or a cannon, or even a nuclear warhead.

Very sexy. If you ignore the purpose for which the system is built and the pilots are trained—which is to kill people—or if you are sufficiently sanguine (in both senses of the word) to take the longer view in which the purpose is to protect individual liberty and freedoms throughout the world, the image of an F-15 Eagle zooming through the stratosphere as it plucks away the strings of gravity like the coils of a sticky but flimsy cobweb, rolling and soaring like a bird designed by a god who knew what he was doing instead of relying on the random processes of evolution—that image is perhaps the most beautiful one the technological mind can conceive. But when you think about it, there is nothing there that was not predestined; the Eagle is simply the natural result of steady improvements in the basic AI systems and jet propulsion concepts which began in the Second World War. Totally different and much more important, if a lot less sexy, is the AWACS aircraft.

In the 1970s the Air Force asked Congress for funds to build two and a half dozen Airborne-Warning-and-Control-Systems aircraft, at a cost of more than 100 million dollars each. The Spitfires that saved England had cost a few tens of thousands of dollars apiece, and Congress and the General Accounting Office and everybody trying to scrape up a few dollars for Medicare or slum renovation or cancer research were understandably suspicious of the price of modern jet fighters, which was up over a few million dollars each. So this request sparked an acrimonious debate.

We already had in place or in construction the Ballistic Missile Early Warning System (BMEWS): radar stations in Thule, Greenland; Clear, Alaska; and Fylingdales Moor, England, which together keep an overlapping vigilance spread out across the North Pole and the Great Circle routes from Russia, from which a missile attack might come. We were safe enough, many claimed. But the Air Force responded that new missile-launching submarines meant that an attack might come from anywhere, and that guerilla-type limited wars erupting in remote areas of the globe seemed to be more likely than an all-out assault. Their concept was to take a Boeing 707 airliner and turn it into a mobile but complete radar-tracking control station. From the outside it looks like a standard airliner except for the absence of windows and the addition of a thirty-foot pancake on stilts sitting astride the fuselage. The pancake is a rotating radome housing a series of antennae, and the fuselage is crowded with thirteen radar and computer specialists, nine radar consoles, and one dedicated data processor. The system is basically a Doppler radar with built-in IFF (Identification-Friend-or-Foe), with linking antennae that can send the signals directly to other stations around the world. It's the most expensive—and the most useful—airplane ever built.

It is, in effect, a complete defensive and offensive command post, able to fly at more than 500 miles an hour to wherever in the world it's needed. From its position five or six miles high it can look out with radar to a distance of more than 200 miles, seeing everything that happens from the ground on up, within a circle of nearly 150,000 square miles. Its data is processed by the on-board computer and displayed on the nine consoles, while being simultaneously transmitted in any desired form via computer-to-computer links to other command posts anywhere in the world.

In the early airborne radar systems the chief obstacle was ground clutter, the return of echoes from the ground which—the ground being infinitely bigger than any airplane—swamped the target echo and rendered the target invisible. In the early days, any airborne system attempting to look down on an airplane beneath it would have no chance of finding it in the massive ground clutter. But if the pulses could be made sufficiently short and sharp, the ground returns could be differentiated since the target is necessarily closer to the spotter aircraft than is the ground, and its return echo would come at a different time.

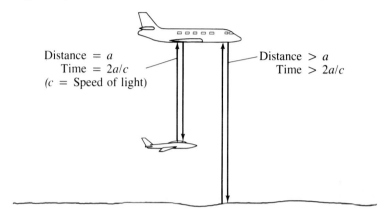

Stimulated by Hughes's electronic successes, other major corporations were by now rapidly increasing their research efforts, and Westinghouse beat out Hughes for the AWACS radar contract. Their system has pulses shorter by many orders of magnitude than those used in the first AI sets; augmented by use in the Doppler mode to differentiate moving from stationary target echoes, it has a clear and sharp "look-down" capability. In a 1975 exercise an Air Force fighter, trying to avoid detection, flew hugging the ground, using ground-avoidance radar which enabled it to slide up hills and down valleys at full speed. The AWACS plane picked it up as a supersonic automobile, and, of course, the operators realized immediately what it was.

That exercise was designed to test the AWACS control capabilities and its vulnerability to enemy fighters. The former turned out to be immense, the latter to be virtually nil. In that test the AWACS aircraft had two F-15 Eagles on its side, against nine attackers. All nine of the attackers were picked up as soon as they entered AWACS air space—the 150,000-square-mile circle extending from the ground straight up to space—and the two defenders were directed so perfectly that all nine were shot down before they got within a hundred miles of the AWACS.

It can, in fact, keep track of nearly 1000 different aircraft, missiles, ships, and ground targets at the same time, plotting not only their positions but their speeds and directions. Every system is redundant in case of failure, and uses a variety of techniques to counter enemy jamming. The system has proved so useful that it has spawned the world's first multinational air force: the NATO Airborne Early-Warning Force (NAEWF), created in 1980 and currently based in Geilenkirchen, West

Germany, with forward operating bases in Greece, Turkey, Italy, and Norway, commanded by Major General Klaus Rimmek of the Luftwaffe, flying eighteen Boeing NE-3A Sentries (the modified 707) and eleven British Nimrods, staffed by flying and ground personnel from the armed forces of twelve NATO countries (the opters-out being Iceland and France).

NAEWF's mission is aerial surveillance in peacetime, fighter control if war should break out. Three aircraft flying coordinated patterns delineate moving ground circles that effectively blanket all of central Europe from the ground up, so that no low-flying Cruise missiles or aircraft can sneak through; at higher altitudes they can see as far as the Moscow air space. If an attack should come, the group would coordinate and direct aerial responses, giving the superiority that was shown in the 1975 air exercise when the two AWACS-directed fighters outfought nine attackers. In the 1982 Falklands war, the British forces went into action without AWACS support, and the HMS *Sheffield* was sunk by a low-flying Exocet missile. Had an AWACS been available, that missile would have been spotted, plotted, and destroyed before it got anywhere near the ship.

These awkward-looking, pancake-bearing airplanes have become symbolic of our first line of defense and of our presence everywhere in the world. The Air Force's 552 AWACS Wing has picked up aircraft smuggling marijuana into Florida, and when two Iranian Air Force F-4s attacked an oil tanker in Saudi Arabian waters, they were shot down by AWACS-directed F-15s. When President Park of South Korea was assassinated in 1980 and tensions built up there, when conflict seemed about to erupt between North and South Yemen the year before and later the same year in eastern Europe, when Sadat was assassinated in 1981 and when rebel forces rose up in the Sudan in 1983, when the Libyans bombed Khartoum in 1984 and when the Korean Airlines 747 was shot down by Russia in the Sea of Japan in 1983, our first response was the AWACS. It is the single most important airplane we have.

The look-down capability of short-pulsed radar is also incorporated on surveillance satellite systems. The photo-reconnaissance capabilities of satellites are based on television, which of course uses visible light, and on infrared and ultraviolet electromagnetic detectors, as well as radar. But visible light is not available in the night hours, and of all these systems only the comparatively long wavelength radar radiation can penetrate through clouds. So while the radar images do not have the resolution of visible television pictures, they fill in the otherwise inescapable void of long polar nights and northern European winters: Neither snow nor sleet nor dead of night, as they used to say, can keep them from their appointed task.

The resolution of the satellite radar is the limiting factor in its usefulness, but once again the technological tricks of the trade provide a drastic improvement over what had been thought possible a few years ago. The resolution is defined by the wavelength, as noted before in the attempts to provide a workable airborne World War II radar, and this has been shortened as far as possible. But the resolution is also inversely proportional to the size of the antenna used; and this can be made to assume a magnitude beyond its natural proportion.

The technique, known as synthetic-aperture radar (SAR), processes the reflected waves electronically in such a manner that, combined with the motion of the satellite relative to the target, the waves are "fooled" into thinking that the receiving antenna is larger than it actually is. In a large-antenna system, echoes from many points on the target are simultaneously received; their interference patterns contain information about the size and/or structure of the target. In SAR the smaller antenna is in motion, and so it can pick up echoes from different points at successive intervals as it moves through space; these are recorded and subsequently analyzed to give the same sort of information. It's a complex system, but it works. The resolution is improved drastically; it is not sufficient to pick up a loose basketball bouncing around the suburbs of Moscow (as the television cameras can do), but quite good enough to spot a missile being launched or a single submarine at sea.

A satellite surveillance system has two related problems. First, it can't be steered or flown like an airplane to look at whatever is of interest at the moment. It cruises on its orbit over the surface of the earth, looking down as it passes over a predetermined swath of the earth's surface, or it is put into a geosynchronous orbit in which it stays stationary over one particular spot. (In the latter case it is revolving around the earth, of course, but at the same speed as the earth revolves.) A geosynchronous satellite can be put up over any spot of interest to us, such as the Plesetsk or Tyuratam missile-firing stations in Russia, but Newton's laws of motion and gravitation insist that such a satellite remain roughly 35,000 kilometers above the surface of the earth, and that leads to the second problem: At that great distance the sensitivity and resolution aren't as good as we would like. To spot that loose Moscow basketball at night or in a snowstorm, one further type of radar is used: the phased-array system. This forms the basis of extraordinarily powerful ground-based stations with fail-soft and over-the-horizon capabilities that keep a never-blinking eye on everything that moves from South America across the Atlantic to Europe and up through the Arctic into the depths of Russia and down again on the other side across the Pacific. Three pyramids, one isolated in the ice of Alaska, another on the Atlantic coast at Cape Cod, and the last in California, direct an unbroken electronic curtain which hangs forever across those vast reaches.

Watson Watt's original concept was expressed in terms of just such a curtain through which any German bomber would have to penetrate in order to reach the English shores, but the ubiquity of that system was lost with the introduction of rotating "searchlight" beams of short-wave radar in which the curtain consisted of interlocking beams generated by a revolving antenna which continually swept over its assigned region of space. Between rotations, of course, the curtain vanishes: It would be theoretically possible for a very fast bomber to slip through such a sporadic curtain undetected, in the manner of a small child jumping past a circularly spinning water sprinkler without getting wet.

In practice, in Watson Watt's war, this radar deficiency was not important. The beams rotated a few times per minute, and with a range of several miles they would come back to sweep over the area before any bomber flying at speeds of only a few hundred miles per hour could get through. But today, with Mach-3 fighters and ICBM missiles, such gaps are as permeable as a revolving door at Macy's the day before Christmas. A "real" curtain is needed, one that is never drawn aside for an instant.

Phased-array radar. Instead of sweeping the beam across an area by mechanically steering the antenna, it relies on electronic sweeps, which in turn rely on computer control and which travel back and forth virtually instantaneously.

The concept is based on the phenomenon of constructive/destructive self-interference of waves. If two machine guns situated side by side and pointed in the same direction fire separate parallel streams of bullets, there is no interaction between the two fusillades: Each bullet travels its own path unaffected by the others. But if two electromagnetic waves are generated by adjacent sources, they *will* interact: constructively if they are "in phase"—with each wavelength crest and trough matching—or destructively if they are "out of phase." If the waves are perfectly in phase, the result will be as if a single beam of twice the amplitude (intensity) has been generated; if they are out of phase, they will self-destruct.

(In the diagram two different forms of electromagnetic wave representation are used. There is no single way of drawing an electromagnetic wave in reality, for its essence does not correspond to any reality that we can understand. The best we can do is to draw various aspects, to emphasize the behavior to which we want to draw attention.) Now imagine a vertical array of pulse-generating antennae. A pulsed beam is

Armageddon

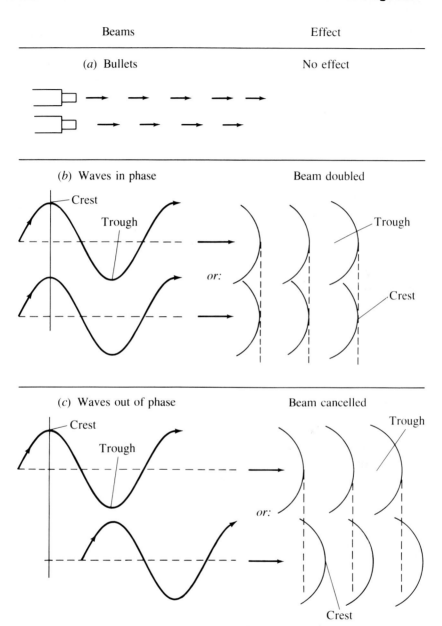

sent out from the uppermost antenna. One-millionth of a second after generation of the first pulse, a beam is sent out from the next antenna. This second beam is slightly out of phase with the first because of the time lag. And so on, down the array.

The beams interfere with each other and die out in the horizontal direction. But they do not, in fact, totally disappear, for although the

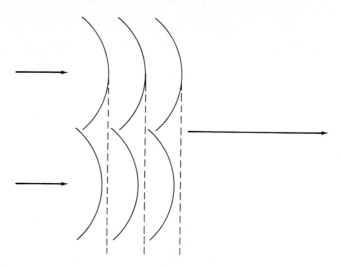

waves are out of phase in the horizontal direction, they are *in* phase at a slightly skew angle. The beam is therefore reinforced in that direction while diminished in the horizontal direction; the effect is exactly as if the beam had been steered slightly away from the horizontal.

In a phased-array radar system there is a huge bank of such parallel antennae, pulsing millions of times every second, with every succeeding pulse slightly out of phase with the preceding one by a continually changing infinitesimal amount. The effect is that though the antenna array never moves, the beam sweeps out an area just as if it were directed by a revolving antenna. The difference between this system and

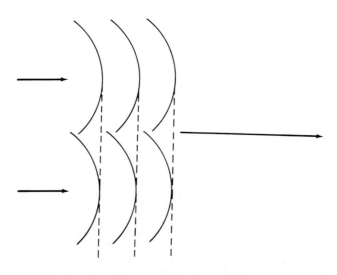

the early radars is that the sweep speed is limited by electronic circuitry and the speed of light rather than by resistance of a mechanically moved large antenna, and so the sweep travels so fast as to be virtually instantaneous (it is measured in microseconds). This means that the system is not blinded in one direction while its antenna is pointing elsewhere; like a many-headed Hydra it can look everywhere at once and keep track of everything that is happening within its total operational area.

The two main systems being used today are Cobra Dane on the Aleutian island of Shemya, and Pave Paws on Cape Cod and California. Cobra Dane is a single-sided slab which looks out over a 2000-mile corridor across the Bering Sea and the Kamchatka Peninsula deep into Siberia. Without moving, Cobra Dane sweeps out an areal coverage of 120°, watching every Soviet missile test launch in that area, giving precise measurements of new Soviet satellites, and providing early warning of any surprise ICBM attack. Pave Paws, looking like a hundred-foot-high truncated pyramid, consists of two slanted, tilted walls lined with thousands of eight-inch antenna elements, staring out from the Cape across the Atlantic and from California across the Pacific. Each pyramid covers 240° and can pick up a ten-square-meter target at a range of 3000 miles. Two more stations currently under construction will provide surveillance southeasterly from Georgia and to the southwest from Texas. The Pave Paws system will be extended in a complete line across the top of the world by new updates to the original BMEWS line. Phased-array systems are replacing the old antennae in the BMEW system station in Thule, Greenland; in Clear, Alaska; and in Fylingdales Moor in England. Taken together, this system will provide unblinking coverage of ballistic missiles launched from sea or land from any conceivable direction of attack.

Cobra Judy is a similar system operated on board a ship which can take it to the limits of international waters off the Russian coasts. From these locations it can observe missile tests, satellite launches, and aircraft testing deep within the Soviet Union. The Soviets, for their part, are now constructing a large and sophisticated phased-array radar at Krasnoyarsk in central Siberia, some 4000 kilometers inside the Soviet border. The United States has pointed out that this is in violation of the 1972 antiballistic missile treaty, which permits phased-array radars only on the edge of each country's territory. The Soviets claim that its purpose is merely peaceful satellite-tracking, and that in any event our own Thule and Fylingdales installations are equally in violation of that treaty, by the same reasoning. One can't logically quarrel with that line of argument. The Russians have offered to decommission Krasnoyarsk if we do the same with the Thule system, which is nearly complete, and agree not to go ahead with Fylingdales. The Reagan administration has thus far rejected the idea, maintaining that somehow our far-flung radars are perfectly legal, and that Krasnoyarsk should be unilaterally dismantled. In any event, our claims are rather more moot than mute, considering the Strategic Defense Initiative (discussed in the following chapter).

The phased-array system is "fail-soft" in the sense that it is powered by many individual tubes or transistors; if any one of them should fail, the total efficiency is only infinitesimally diminished, in contrast to a normal radar which, if it fails, fails completely. The range of the system is also in tremendous contrast to that of traditional radar: When Cobra Dane looks outwards for 2000 miles and Pave Paws stretches its electronic fingers out for 3000 miles, they are obviously reaching over the horizon—by methods which are still classified. Two possibilities which have been discussed are Skywave and Seawave.

Skywave bounces the signals off the ionosphere, in the same manner that Marconi's original transatlantic radio broadcasts worked. But anyone who has listened to such radio broadcasts is familiar with the frequent static and the continually rising and fading signal. The ionosphere is an active region, providing in effect a moving target; the details of the computer-controlled system which achieves continuously uninterrupted surveillance through this medium are currently classified.

The other system, Seawave, depends on a phenomenon sometimes encountered in junior-high-school science classes. If a straight stick is inserted at an oblique angle into a pool of water, it appears to "break" at the water surface and go off at a sharp angle. This is because water has an index of refraction different from that of air. The stick has not changed shape, of course, but light reflected from it changes direction when passing from one medium (water) to another (air) of different refractive index. The air itself is not a homogeneous medium; over any one spot on the earth's surface, the atmosphere varies with height, and the index of refraction varies with it.

If the stick should be inserted not from air into water but from air of one refractive index into air with an index only infinitesimally different, it would appear not "broken" at the interface but rather "bent" ever so slightly. So it is with radar waves passing through the atmosphere. The effect of the continuously-upward-varying index of refraction is to bend the waves concavely downward: If the effect were strong enough, the radar waves would follow the curve of the earth so that, in effect, they would never see a limiting horizon.

The effect is *not* strong enough, however; under normal circumstances it is a very small effect indeed, and does not increase the range significantly. But at the very surface of the oceans the effect increases dramatically because of the saturation of water vapor in the surface layer of the air. This water-vapor saturation obviously must decrease rapidly with height as it diffuses upward, and since water has such a different index of refraction (than air), this leads to a large and continuously varying change of index with height for the first few tens of feet above the sea surface.

It then turns out that if the wavelength of a beam of radiation is carefully matched to the change in index, it can be made to reflect at an angle equal to that of its incidence. In this way the highly water-saturated layer of air lying over the ocean forms a "vent" through which the beam passes and is confined, bending with it around the curve of the earth and thus passing without interruption over the horizon. The precise details of the system, in which not only the transmitted but also the reflected signal must be kept within the vents, are secret and difficult to visualize. It is said, however, that Cobra Dane can tell whether the security guards at the Soviet missile launching stations are tossing around a basketball or a soccer ball.

Even allowing for a slight exaggeration there, the system is certainly capable of noticing anything as untoward as a sudden launch of missiles. Together with BMEWS, AWACS, and satellite survey systems, it provides a secure electromagnetic curtain more impermeable than that made of iron, behind which we can find the time, it is hoped, to design a lasting peace.

Unfortunately, one further development has now risen phoenixlike from the ashes of scientific disdain. Like the Ghost of Christmas Past, it is now beginning to hover over us, haunting us, driving us closer to the edge on which we find ourselves still precariously dancing.

Chapter Nineteen
The Death Ray: Finale

> The most incomprehensible fact about the universe is that it is comprehensible.
> —*Albert Einstein*

> The most comprehensible fact about human behavior is that it is incomprehensible.
> —*David Fisher*

When Watson Watt asked Wilkins to calculate whether it might be possible to construct a death ray in 1935, the answer was a clear and resounding *no*. During the next few years the concept finally died out in the minds of the military and was sloughed off back to the pages of cheap science fiction stories. Today, fifty years later, the government of the United States is spending many billions of dollars to perfect the death ray and put it up into space. In a development typical of the muddled and muddied path along which this entire scheme has proceeded, the idea which turned the impossible nonsense of 1935 into the military-technology conversation piece of 1987 was published way back in 1916.

In the most productive dozen years any scientist has ever had, Albert Einstein between 1905 and 1917 worked out the explanation of the photoelectric effect, which was the first concrete demonstration of the reality of the new quantum mechanics and for which he won the Nobel Prize; explained the heat capacities of solids at low temperatures by this same quantum mechanics; provided the first proof of the existence of atoms by explaining the effect known as Brownian motion; and bracketed all this with his special theory of relativity in 1905 and his general theory in 1917 (for which he did *not* win a Nobel Prize because most people couldn't quite understand or believe it). As a rather minor contribution during those revolutionary years, he pointed out in 1916 the concept of the *stimulated emission of radiation.*

Atoms absorb energy by jumping to higher energy states, and can then reemit the energy in the form of electromagnetic radiation by dropping down again to their former level. This is how a normal light bulb works: An electric current passing through the wire heats it, bouncing the atoms into higher energy states from which they spontaneously emit radiation in the form of visible light as they drop down again. This was explained in 1913 by Niels Bohr, who visualized the atom as being composed of a central nucleus surrounded by electrons in circular orbits. In the "ground" state of a normal atom, all electrons are in orbits of the lowest possible energy, circling as close to the nucleus as they can get. When the atom absorbs energy, one or more of the electrons jump ("quantum-jump") to the next higher level, further away from the nucleus. Since they are unstable there, they drop back down again, emitting as they do so electromagnetic energy equal to the difference in energy levels of the two orbits. Using an equation derived by Max Planck, which relates the energy of radiation to its frequency or wavelength, Bohr was able to explain the previously puzzling unique distribution of wavelengths observed from such excited atoms.

In 1916 Einstein pointed out that one photon could stimulate emission of an identical photon from an atom in an excited state if the stimulating photon's energy matched exactly the difference in energy levels of the atom. Ordinarily, atoms are in their lowest possible energy states. This does not mean, however, that *all* atoms are. At any temperature above absolute zero some (small) proportion of atoms or molecules will have absorbed thermal energy and quantum-jumped to a higher state, and some such states are "metastable," so that the atoms stay in them for a finite time, which may range from nearly infinitesimal fractions of a second to geological eras. Luminescence, for example, is a phenomenon in which atoms or molecules absorb energy from external sources and remain in a metastable state for some time, emitting light when they drop back down to their true "ground" state. Thermoluminescent effects lasting for thousands of years have been measured in meteorites.

Einstein's prediction was that if a photon strikes an atom in such an excited state, and if its energy exactly matches the energy of the radiation that the atom might eventually emit when dropping down to a lower energy state, then that incoming photon will in effect tickle the excited atom and cause it to emit its energy immediately. The end result will be that one photon comes in and two come out, exactly matched in energy and therefore also (according to the laws of quantum mechanics) matched in frequency and wavelength. This is not, as at first glance it might appear to be, a violation of the concept of energy conservation: Although more light energy comes out than goes in, that extra energy first had to be pumped into the atom by an unrelated mechanism such as heat or light or electricity or chemical reaction, and then stored in the metastable excited state until the appearance of the stimulating photon.

The idea was not thought to be important. Theoretically it was an interesting concept, but not in the world-shaking class of Einstein's other discoveries of those years; as far as practical matters went, no one could foresee any use for it since most atoms in any given situation are not in excited states and therefore any incoming photon would be much more likely to strike and be absorbed by ground-state atoms with no subsequent stimulated emission of matching photons. The effect would necessarily die out; it seemed to be as useless in any practical sense as nuclear energy, another discovery of those early years of this century, of which Lord Rutherford—one of the greatest nuclear physicists of them all—said at the time: "The idea of harnessing nuclear energy is just so much moonshine!"

So much for the Nostradamus effect.

The reason Wilkins found the death ray to be impossible was the inverse square law, which says that radiation intensity diminishes with the square of the distance between illuminating and illuminated object so that, for example, the planet Mercury, which is less than half the earth's distance to the sun, is hot enough for metals such as lead and zinc to melt on its surface while Pluto, a hundred times further away, is so cold that gases such as methane and ammonia are frozen solid there.

This is because of a simple and unalterable property of the geometry of space. Since energy is conserved and since there is little interplanetary material to absorb the sun's radiated energy, whatever heat and light leave the solar surface must continue outward to reach each of the planets without being significantly diminished. If we could surround the sun with a sphere lying just above its surface, we could collect and measure all the energy it emits. The result would be a large but finite and definite number (it is actually 3.9×10^{23} ergs per second). Since the area of a sphere is given by $4\pi r^2$ and the radius would be just about the radius of the sun (7×10^5 kilometers), its area would be 6×10^{12} square kilometers, and one square centimeter laid out on such a collecting sphere would receive just $1/(6 \times 10^{22})$ of the total energy, or 6.5×10^{10} ergs per second. If we now expand that sphere so that it includes the earth and its orbit around the sun, its radius must increase to the sun-earth distance of just about 300 million kilometers, and its area is now

1.1×10^{18} square kilometers, or 1.1×10^{28} square centimeters; the same one square centimeter on its surface would now receive only $3.9 \times 10^{23}/1.1 \times 10^{28}$, or 0.000035 ergs per second of solar energy. A further expansion of the sphere to the distance of Pluto would give it a radius of about 10 billion kilometers and therefore an area of about 1.25×10^{31} square centimeters; the undiminished solar energy at that point is spread out so that one square centimeter receives only 3×10^{-8} (0.00000003) ergs per second.

So although the sun's radiated energy travels through the solar system without appreciable diminution, it spreads out geometrically so that the same object held at the orbit of Mercury would receive six times as much heat and light as it would on earth, while on Pluto it would receive only one-thousandth as much. This dispersion of electromagnetic radiation, which means that its local intensity must diminish with the square of the distance traveled, is a characteristic of all natural emissions. But what if one tries to artificially direct a beam of radiation, as in a flashlight? The technique involves placing a convex mirror behind an ordinary bulb. An electric current passing through the wire in the bulb heats it, the atoms in the wire are excited to higher energy levels, and as they drop back down they emit radiation which comes out spherically symmetrical, emitted in all directions. The radiation which is directed backwards hits the curved mirror and is reflected and focused dead ahead, so that all—or at least most—of the light is directed forward in a narrow beam.

But it doesn't stay in a narrow beam for long. The atoms in the wire can be excited to a multitude of energy levels, each with its own characteristic energy, and so when they drop back to the ground state they emit photons with a variety of energies and so the radiation necessarily is composed of a variety of wavelengths. These propagate outwards and interfere with each other, as mentioned in the last chapter's discussion of phased-array radar systems. The result is that some of the energy directed straight ahead dies out and is redirected at various angles from the initial direction; the difference between this and a phased-array system is that the latter is controlled, while this is a random process. The final effect is that the beam, no matter how well collimated at the outset, necessarily diffuses radially outwards as it travels, spreading its energy over increasingly wider boundaries and thus diminishing its concentration at any one point. (It is generally thought that this spreading effect is due to scattering by atoms in the air through which it passes, but while this does, of course, take place, it is a minor reason for the beam diffusion: Even in empty space, a beam of ordinary light would spread out and be lost. Thus no ordinary searchlight stationed on earth or even placed on a satellite in orbit above the atmosphere could illuminate for us a spot on the dark surface of the moon.)

This is the vital consideration that made the death ray an impossibility. It was certainly possible in 1935 to generate enough radiant energy to hurt or kill; something as innocuous and simple as an ordinary light bulb radiates enough energy from its central wire to burn your hand if you hold it for more than a few seconds, but just a few inches away the energy intensity dies off to nothing more than a gentle warmth. A much greater source of energy could radiate enough to kill, but the intensity would die off so rapidly with distance that it was quite impossible to use any such weapon to damage aircraft or crew at any reasonable range.

Neither Wilkins nor Watson Watt nor anyone else at that time thought to use Einstein's effect in such a weapon, but it could have two major results. First, the incoming photon generates another photon, so that where there was one there subsequently are two: the radiation is amplified rather than diminished by being absorbed. Second, the emitted photon has precisely the same direction and wavelength as the incident one, so that the two are necessarily in phase; i.e., they interfere constructively with each other in the direction of their motion and destructively in all other directions, the result being that they do not disperse radially.

We have here the beginnings of a method of simultaneously increasing radiation intensity and confining it to a narrow beam. What we have here, in effect, is the beginnings of a workable death ray.

What we have is a laser. The acronym stands for Light Amplification by Stimulated Emission of Radiation, and actually we don't have one yet. There are two problems that have been glossed over. First, the atomic population of the lasing material must be energy-inverted. In any given situation most atoms are in their lowest energy state, with successively fewer atoms inhabiting states of higher and higher energy. A photon entering such an atomic region is more likely to be absorbed by one of the silent majority of ground-state atoms than to encounter one in the precise energy state it has the capability of stimulating into emission. Another way of looking at this is to realize that energy must always be conserved: A photon cannot enter a laser and provoke emission of light with the subsequent generation of greater energy, *unless that energy has already been entered into the system*. The laser is not a free-energy machine; it is simply a means of converting other forms of energy into radiation (and collimating it).

So first a quantity of energy must be pumped into the system sufficient to raise most of the atoms to a particular energy state, and not all energy states are available for the process: Only metastable states, in which the atom can retain its excitation for a finite time, can be used. If energy is pumped into an atomic (or molecular) system in which such a metastable state exists, random excitation processes will raise the atoms to a variety of energy levels from which they will drop down into the metastable state—where they are momentarily held. Then when a photon of the proper energy is sent in, it will have a good probability of encountering one of these excited atoms and tickling it into releasing a duplicate photon. Then the two of them will cause the further release of four photons, which will in turn release eight, and very quickly a cascade of photons all of the same wavelength can be generated.

The second problem is that though the photons are of the same wavelength they are not emitted at precisely the same instant, and so they will be not quite in phase. If they are not put into precise phase they will, of course, interfere with each other and the beam will disperse.

The energy can be pumped into the laser by any one of a variety of techniques, including chemical, electrical, or optical means. The phase forcing is accomplished in a manner analogous to the way the cavity magnetron generates microwave radar beams. In that case, which is itself analogous to a penny-whistle, the electrons reverberate within what is called a resonant cavity; this is a box of dimensions equal to the wavelength desired. When wavelengths equal to the length of the box are produced, they are said to resonate within it: they bounce back and forth continually. Radiations of different-sized wavelengths hit the walls of the cavity unequally and set up self-interferences which reverberate out of phase and thus self-destruct. For lasers, the wavelengths involved are on the order of 10^{-8} centimeters, much too small for any box to be manufactured; but mathematical relationships allow larger dimensions of particular relationship to the wavelength desired.

A simple laser can be constructed of a rod made of artificial ruby, composed of aluminum oxide doped with atoms of chromium, with one wall mirrored and the opposite wall half-mirrored. A fluorescent light in contact with the rod is flashed, and its light energy serves to pump up the ruby laser optically so that the chromium atoms are excited and within immeasurably short times fall into a metastable state. Some of them immediately begin to emit radiation of the frequency corresponding to the ground-state transition, while others remain in the metastable state, so that the emitted photons then tickle other metastable atoms to emit more such photons. Those emitted lengthwise to the rod encounter the mirrored ends and bounce back and forth, with each passage releasing more and more photons in the same direction; those emitted in different directions pass out of the rod without further interactions and thus without building up to high intensity. Since the lengthwise photons are bouncing back and forth with the speed of light, and since the rod has dimensions on the order of inches, all this is virtually instantaneous. The length of the rod is prescribed by the wavelength of the light generated in order to produce a resonant cavity to put all the photons in phase. While they are reflected by the full-mirrored end, some fraction of them will pass through the half-mirrored end to produce the laser beam—a unidirectional beam of photons of identical wavelength, perfectly in phase.

With the laser the death ray becomes possible, but not yet practical. Since the beam is in phase, it does not spread out with the inverse square of the distance, and therefore practically all the energy generated at the source can be delivered to the target. But two problems remain, and they exacerbate each other. First there is the problem of the atmosphere: the air we breathe is not perfectly permeable to visible radiation. We see this in the scattering of sunlight which causes the sky to be blue and the setting sun to be red, and also in the fact that we can see a flashlight beam in a dark room. If the beam penetrated the dark air perfectly, we would see only the glow of the bulb and the reflected glow where the light hits the wall. In fact, we can see the beam itself as it crosses the room, and this is because a part of it is being scattered out of its path by dust and other particles in the air.

It is necessarily the same with a laser, and so the beam is scattered and absorbed as it penetrates the air and thus loses energy with distance. This loss is much smaller than the loss of normal light, and so the laser can bore holes in a razor or cut through human tissues in a hospital, but in these cases we are talking about travel distances on the order of inches rather than the miles which would be necessary for a useful weapon. The second problem is again related to the question of generating sufficient energy to kill: When a bullet hits, it delivers all its destructive power at the moment of impact, but a laser beam delivers its energy relatively slowly. If the beam is directed against an airplane, for example, at the very instant it hits there is little effect; then as the beam is held on the target, the energy transfer builds up and the airplane is heated until its gas tanks are punctured and the fuel explodes. But meanwhile, of course, the plane is moving, and so the trick is to keep the beam fixed on the moving target while it takes evasive action.

Furthermore, the laser energy is directed strictly along the line of sight and therefore is of no use if it misses its swerving target. A cannon shell, on the other hand, can be set to explode and destroy the target if it gets anywhere near its vicinity, even for a moment. Clearly, for most purposes the death ray is inferior to more conventional weapons. A greater number of enemy troops can be killed by napalm or gas or antipersonnel bombs, all of which spread out from the point of impact to cover a large area, than by spritzing them with a laser beam which would have to be held at least momentarily on each soldier before moving on to the next. Airplanes can be brought down more easily with guided exploding missiles which only have to come close—and if the warhead is nuclear they don't have to come very close at all. Tanks were thought at first to be a legitimate laser target, but even if the problem of atmospheric attenuation were solved, there is still no intrinsic advantage of lasers over missiles.

The armed forces were frustrated: They had here a handy-dandy, high-tech gadgetry weapon from the very pages of science fiction, and no one could figure out a use for it.

And then on March 3, 1983, President Reagan took warfare out into space—where there is no absorbing, scattering atmosphere.

When the intercontinental ballistic missile replaced the long-range bomber as the principal delivery system for the hydrogen bombs of the world's two superpowers, international politics became a whole new ball game. It became clear that such a weapon could not be stopped, and the policy of mutually assured destruction became, in the absence of any viable alternative, the somewhat less-than-secure system of security under which the world has since attempted to survive. But some scientists, engineers, and soldiers familiar with the ordeals of the 1930s never forgot Lindemann's lesson that no weapon had ever failed to produce its own counterweapon; he had said then that he failed to see why the bomber should be supposed to be any different, and this group of disciples throughout the 1960s and 1970s failed to accept that the intercontinental missile should be any different.

But it is. Armed with pods of multiple nuclear warheads which can scatter over large areas and destroy a city apiece, the ICBM has proved indestructible. One site of an antiballistic missile (ABM) defense system was approved by Congress in 1969 and built 100 miles northwest of Grand Forks, North Dakota, at enormous expense and over the protestations of most scientists who pronounced it unworkable. It was installed in 1974, and dismantled in 1975 after 7 billion dollars had been spent on it. It simply couldn't do the job, and wasn't worth the money to keep it operating.

The basic problem is that in order to shoot a person or animal or mechanical contrivance you have to have a bullet that goes faster than the person or thing can run. This is not altogether true—for example, torpedoes in World War II were only marginally faster than some warships—but by and large it is. The smaller the difference in speeds between the target and projectile, the less efficient will be the projectile (for example, the faster warships could often dodge the torpedoes). And so the great speeds of the ICBMs, coupled with their great number, with the possibility of producing vast crowds of decoy missiles, and with the terrible destructive power of each of their warheads, made the ABM an anachronism at the moment of its conception.

But the laser beam strikes at the speed of light. Virtually instantaneously with the electronic pull of the trigger, its power can flash across the world if it is unhindered by the atmosphere. And the ICBM on its terrible journey must of necessity rise above the atmosphere before plummeting down again on its target. The implication is clear.

It was very clear to Edward Teller, widely known as the "father of the hydrogen bomb." Although Teller has been in official semiretirement from the Lawrence Livermore Radiation Lab (LLRL), which was founded expressly for him (when he disagreed with the then-director of Los Alamos, J. Robert Oppenheimer, on the feasibility of building a hydrogen bomb), his is the kind of mind that can never retire, neither from scientific problems nor from the concomitant political struggles to turn scientific feasibilities into political or military realities. In the winter of 1982–1983 I heard him tell an informal audience that, though his tongue was tied because of official secrecy, he could assure us that it was possible and practical to build an umbrella over the United States through which no nuclear weapon could penetrate. He evidently was able to convince the President, for just a few months later the Strategic Defense Initiative was announced.

The nuclear-pumped x-ray laser is Teller's pet project, and the heart of the SDI program. It is not the only potential weapon in the "Star Wars" arsenal but it is, according to Teller, "the most novel and potentially the most fruitful," while other experts in the field describe it as the central item in the system, the one most likely to be most effective at the earliest time. The expectations are that it will be able to destroy tens or perhaps even hundreds of missiles at one shot, within a few minutes of their launch. It gains its tremendous energy from the detonation of a nuclear bomb: The x-rays from the explosion strike the laser rods and excite the atoms to the desired metastable state, creating the laser action with a reasonable fraction of the original nuclear energy and generating an intense beam of coherent high-energy x-rays. A tiny fraction of a second later, of course, the entire weapon is destroyed by the shock waves of the explosion, but by that time the x-ray beam is on its way.

There are other possible weapons in the Star Wars arsenal: beams similar to or variants on the x-ray laser, such as the electron laser or charged particle or neutral rays; missile projectiles which would fire homing or directed rockets at 20,000 miles an hour; and giant electromagnetic "rail guns" which would unleash what are basically cannon shells at incredibly high velocities of more than 50,000 miles an hour. But these latter weapons have problems that seem to be insuperable, at least on the time scales we are interested in, since a missile defense system that comes into operation a hundred years from now is likely to be too late to save us. The rocket-fired homing projectiles currently weigh in at nearly 10,000 pounds and the rail guns at over 100 tons;

their weights would have to be brought down to a few pounds for enough of them to be put into orbit to destroy the thousands of Soviet missiles that are the targets.

The laser-beam weapons are our hottest choice. In the autumn of 1985, the Defense Department released video pictures of a test of the weapon: A Titan intercontinental missile more than 100 feet long and weighing nearly 200 tons was caught in midflight by a laser beam. Its skin was heated momentarily to thousands of degrees, as hot as the surface of the sun; the metal evaporated, setting up a shock wave that tilted the missile out of its flight path, and a moment later it burst into flames and fell back to earth.

It was an awesome demonstration. General James Abramson, Director of the Strategic Defense Initiative, described the test as a "world-class breakthrough," one of many recent "exciting" and "incredible" progressions in our capabilities.

Unfortunately, the test was not really incredible, nor even exciting. It was not a breakthrough of *any* proportions, really; it wasn't even a test. It was, in the best of all possible worlds, merely a demonstration; in the worst of all possible worlds, it was a fraud.

It wasn't a test since it involved neither anything new and previously untested nor anything with the remote capabilities of doing what it purported to do. The laser used was ten years old; it represented nothing new. Its capabilities were well known, as were its incapabilities: It is useless against modern ICBMs under combat conditions, having a wavelength too long to be useful at the long ranges which would be necessary in reality. The video "test," according to Dr. Kosta Tsipis, a leading laser expert and professor at MIT, was typical of "the kind of bamboozling of the public the SDI program is undertaking."

In fact, the x-ray laser in particular and the entire Star Wars program in general are thought of by most scientists as, at best, unworkable, and, at worst, a fraud and a deadly danger to all of us.

While the most optimistic view is that the laser could destroy up to hundreds of missiles, this claim is seen by most workers as highly extravagant. If it could destroy tens of missiles, that would be a magnificent achievement—but an attack would involve thousands of missiles, which means that hundreds of such lasers would have to be used. And all of them would have to work perfectly, or at least tens of missiles would get through, representing a catastrophe beyond imagination.

The energy of each laser comes from a nuclear explosion, which has all the radiation and blast effects of such detonations. In particular, the explosions would almost certainly destroy our own low-flying surveillance and fire-control satellites, leaving us helpless against further attacks.

The entire complex would have to be controlled by computer, and according to most experts the computer software is simply impossible to achieve. Not only impossible under today's technology, but impossible with any currently foreseeable improvements. Of course, the unforeseeable is always possible, but to set the world's security upon a foundation as flimsy as the unforeseeable is unforgivable.

The system would have to be triggered within seconds or, at most, minutes of an enemy attack, which reduces the time for rational consideration of alternatives to below the human limit. We would have to place our faith in a computerized system which would spy the missiles taking off, evaluate the situation, and set off our response—all without any human being taking part in the process.

The nuclear weapons needed to trigger the x-ray lasers would either have to be launched at the moment of enemy attack, reducing the time interval between initial radar spotting of the attack and our reaction time to virtually nothing, or would have to be placed in orbit ahead of time. Placing nuclear weapons in orbit cannot possibly be viewed by any other nation as a purely defensive move, for they could just as easily be dropped on enemy heads with no warning as be used to trigger the defensive laser. Even with the best will in the world, it is remarkably difficult to identify this system with the "nonnuclear defensive system that would threaten only offensive missiles, not people" described by President Reagan.

An alternative to the space-based laser is a mountain-top laser that would fire its beam through the atmosphere to an orbiting mirror which would redirect it to another mirror orbiting over the enemy target. But aside from the problem of generating enough energy through the atmosphere to the mirror above, we have no mirror capable of reflecting the beam without melting. And even if we should attain that capability, the placing of a mirror satellite in geosynchronous orbit directly over Moscow or any other target would be obviously just as deadly as placing a hydrogen bomb on a thread of Damocles over the heads of those people below. Whatever our views of the good intentions of the people involved, we cannot honestly describe such a system as defensive or stabilizing, or as a step towards removing the world from the threat of nuclear destruction.

On the contrary, all such Star Wars weapons deployment would only trigger Soviet countermeasures and thus escalate the current tensions and weapons expenditures. We see this clearly by our own actions: In the 1960s Russia initiated an ABM defense system surrounding Moscow, to which we responded by simply putting in place more Moscow-targeted ICBMs in order to swamp their defenses.

Lindemann's notion that every weapon has its counterweapon is in a sense true, but that sense includes counterweapons to the counterweapons ad infinitum so that never can there be a weapon which would be the ultimate, the Holy Grail that would render us invulnerable. The more effective any nuclear umbrella might be, the more nuclear missiles any nation that sees itself threatened by us or inimical to us would certainly build.

And we must not forget that ICBMs are by no means the only weapon in the modern arsenal. The Cruise missile is to the ICBM as the German V-1 was to the V-2, and is in reality no less deadly. Armed with a nuclear warhead and launched by submarine, it would rise up out of the waters and skim low over the ground to find our cities and centers of war, never once rising above the atmosphere, *and no space-based defensive system could touch it*. So if we were to put up our multiple-billion dollar Star Wars system and the Soviets were to switch the burden of their weapons to Cruise missiles, what would we have gained?

Add to this the difficulty of properly testing the Star Wars components; which is just about impossible because of the nature of the threat the system is set up to safeguard against, a massive nuclear strike launched without warning. The testing of weapons *under realistic battle conditions* is a most necessary stage of their development—necessary before even the simplest weapon is ever introduced into use, so much more necessary for the unbelievably complex computerized system under discussion. Without such testing we have a system founded on hyperbole and hope rather than on a concrete foundation of workable and reliable components. But how can one test—not merely once or twice as an initial guide to suitability, but continually as insurance against any conceivable breakdown—a system whose operational imperative it is to work perfectly on a few seconds notice after years of inactivity, and whose extent encompasses not only the radar warning system but also the launch of hundreds of nuclear lasers and secondary backup weapons? The economics alone are mind-boggling: The last underground test of just one weapons laser cost 30 million dollars (and was a failure because some of the measuring equipment had been improperly calibrated beforehand). The politics of testing are stultifying: Since the laser system is nowhere near a usable weapon, it will necessarily involve much testing over the next decade, and to test the nuclear component obviously involves nuclear weapons testing, both underground and eventually in space. With this in mind, our negotiations toward reaching a comprehensive test ban with the Soviets are obviously less than sincere. If we truly want to outlaw nuclear weapons—and particularly nuclear weapons in space—we can't possibly keep the antimissile laser; and without it there seems practically no likelihood at all of ever establishing even a marginally effective umbrella.

Some of our experts see the Star Wars program as a way to force the Soviets into trying to catch up with our high technology and thereby ruining themselves economically. But at an estimated cost to us of up to 500 *billion* dollars, who is ruining whom?

And worst of all, our military leaders are neither stupid nor ignorant—it is only we who are. They must understand clearly such implications, which leads to the conclusion that the Star Wars program is an offensive, first-strike weapons program, and nothing else.

Consider. It is extremely unlikely that within any foreseeable future the system will be capable of providing the puncture-proof umbrella President Reagan is selling us. Comments range from those of the head of the program at the Livermore National Laboratory, Dr. George Miller, who says flatly: "We can't do what the President asked for," to those of George Rathjens of MIT who says: "The President is ill-informed on military matters, perhaps out of touch with the scientific community," and those of Noel Gayler, former Director of the National Security Agency and former Deputy Chief of Naval Operations, who says: "What we are observing is the will to believe, and it is irreducible. People will believe in hopeful things. Cancer sufferers still go to Mexico and get a shot full of laetrile, and this is that kind of operation."

Any deployment of the system, so that it would be ready to go into operation at the first sign of an attack, would itself be a provocative move, and might trigger the attack it is designed to prevent. But on the other hand, the system could be built and held on earth, and while not 100 percent effective, it would certainly be powerful. We could then launch the laser detonators at the same time we launched a first strike against the enemy, and they would be in position to destroy a large number of the enemy's retaliatory missiles. Even more effectively, the lasers could be used to wipe out the enemy radar and infrared sensor satellites moments before our missiles were launched. The nation being attacked would know something was happening as they went blind, but they wouldn't know what; they wouldn't see our missiles coming until it was too late. If we were to send off an initial attack without warning, this would have to be a most useful adjunct to our offensive arsenal.

Whether, in fact, we would ever do such a horrendous thing or not, we can hardly expect the rest of the world to depend on our innate goodness. If we develop this capability, the world must expect that some day we might be tempted to use it. It must be a source of further instability in world tensions, it must provide a temptation to other countries to strike first when some inevitable confrontation arises, it must be and it is a deadly threat to us all.

The lessons of history are difficult to learn. It is not enough to know that after the First Great War Germany was reduced to the status of a feudal village and held in fiefdom by the Treaty of Versailles and that the civilized nations of the world set up an international body of law administered by a League of Nations, and that all this body of law failed and we were nearly destroyed by another Great War more terrible than the First. The lesson one might learn from this is that international law is not to be trusted, we must rely not on paper guarantees to secure our freedom, but must arm ourselves and resist aggression with force.

But *intra*national law was also ineffective at first. When the first justices of the King were traveling sporadically throughout the original counties of England, administering the King's Justice, foolish was the man who relied on law rather than on his own strength to secure justice and freedom. Yet who would now want to live in a nation without laws, without police, without courts?

Clearly international law and cooperation must be our only ultimate salvation. But clearly also, we do not yet live in a world with such an effective system. For the immediate future we must rely reluctantly but firmly on military strength for our protection. In deciding which courses of military preparedness should be followed, the lessons of history must be learnt. But they are distinctly difficult.

"The bomber will always get through," Baldwin said in 1932. He wanted to base Britain's national defense thinking on that concept, he wanted "the man in the street" to realize that in any future war those streets would be blasted into rubble and each man and woman and child would be bombed and their homes set afire.

But Lindemann said no. In his letter to the *Times* of August 8, 1934, he said: "That there is at present no means of preventing hostile bombers from depositing their loads of explosives, incendiary materials, gases, or bacteria upon their objectives I believe to be true; that no method can be devised to safeguard great centres of population from such a fate appears to be profoundly improbable."

And Wilkins and Watson Watt said no. Starting with the initial funding of 10,000 pounds provided by Dowding, they and a very few fellow scientists began the work that led to radar and the defeat of Nazi Germany.

But the lesson is not as clear as it seems to be, for Baldwin was right and Lindemann was wrong. The bomber *did* always get through; there was throughout that war never any method devised to safeguard great centers of population. London suffered a conflagration worse than the Great Fire of 1666, Coventry and Berlin and Cologne and Warsaw were destroyed, and Hiroshima was wiped from the face of a weary earth.

The war was won, but "the man in the street" suffered as badly as Baldwin had warned. Radar provided defenses strong enough to take a toll of the attacking bombers sufficient finally to turn them away, but not before they had set their loads of fire and destruction on the heads of the huddled populace.

This, ultimately, is the lesson. Yes, every weapon has its counterweapon. Yes, we can build at a cost of hundreds of billions of dollars a system that will destroy a large fraction of enemy missiles. But counterweapons also have their own counterweapons, and no defensive system will be even close to 100 percent efficient. We cannot build a system that will destroy *every* enemy missile.

If we seriously want to destroy any other country in the world, we can do so. And we can provide ourselves with a sort of protection, so that any damage to ourselves will not be as great as that we inflict on others. But is that what we want? For the overwhelming majority it is not, and yet quite obviously for some few it is. There is no other rationale for spending well over 100 million dollars next year on nuclear-pumped x-ray laser research—on a weapon that is tremendously useful for first-strike capabilities but has not the potential to provide the defensive umbrella claimed as its raison d'être.

Sixty-five million years ago the dinosaurs became extinct, and so did half the genera of all marine invertebrates. This is the ultimate lesson of history: It is not inconceivable for an entire form of life to be wiped out, to vanish forever, leaving behind nothing but a few bones trapped in silicified ash layers or in beads of amber. On the contrary, this is the normal and usual fate of living creatures on this earth. We mammals are not indestructible; *Homo sapiens* is not necessarily predestined for glory everlasting.

We are still dancing gingerly, nervously, dangerously out of balance, on the slippery, slithery edge of unfriendly time.

Then said I, Lord, how long? And He answered,
Until the cities be wasted without inhabitant,
And the houses without man,
And the land be utterly desolate...

Isaiah 6.11

Appendix

The development of German and Japanese radar has been virtually ignored in this book, because it was not important. That may seem a dreadful sentence to those who remember fathers and brothers who died aflame in their falling Lancasters and Liberators in the skies over Berlin or the Pacific islands; nevertheless, the Axis radar was *not* important in the sense of being a decisive weapon.

Certainly the Germans in particular developed radar into a formidable weapon—but too late to decide the fate of the war. In 1940 they had their own radar sets, but scorned to use them as the British did. The British system was a *defensive* one, and the Germans in general and Goering in particular sneered at it. Their motto was that of the old French (when they had been, in German eyes, an admirable, warlike people): *Élan, élan, toujours élan!* The Nazis saw how the French had retreated from this offensive stature in the years between the wars and had built and relied upon the defensive Maginot Line, and they saw how poorly this had served them as the Luftwaffe and the panzers rolled around and through it in the spring of 1940.

The First World War had been a defensive war, dominated by extensive lines of trenches defended by artillery, gas, and machine guns, in which any attempt to mount an offensive resulted in fields littered with gasping, bleeding, shattered bodies of men dying in the pockmarked mud. But the Second World War exploded over the map of Europe with the concept of blitzkrieg, war like the flash of a lightning bolt, with Stukas destroying defensive positions over areas of hundreds of square miles and panzers rolling over machine-gun nests and never stopping till the battle was won.

And so the Germans developed radar, but solely as part of their offensive spirit. They had radar equipment set up on the coast of France, looking out over the Channel to search for convoys, helping them to dispatch their deadly bombers. But radar as a defense against those bombers was regarded as worse than a useless waste of technical man-

Appendix

power: It was a sign of decadence, of a return to those outmoded years of 1914–1918. And so Goering did not worry too much about those high, flimsy towers lining the coast of England; to him they were a symbol of weakness rather than of strength and determination.

Later in the war, when the Allied air fleets were in their turn ravaging the German cities, the Luftwaffe was turned of necessity into a defensive weapon. And it was a deadly one, and the Allied bombers fell out of the skies nearly to the point where the bomber offensive would have had to be called off.

Nearly.

If the Luftwaffe had developed their defensive radar sooner, if they had had it ready to deal with the first few, halting, blundering waves of Wellingtons and Flying Fortresses who were feeling their way uncomfortably across the skies of France and Germany, they could have devastated them and forced the British and American air forces to call off their attacks on the Fatherland. They could have, but they didn't. By the time they went over to the defensive it was too late; by the time the German radar was an effective force picking out bombers in the night and above the clouds, there were hundreds upon hundreds of those bombers, escorted by Mosquitos at night and by Mustangs in the day, and Germany died shattered and broken amid the rubble of factories that could have turned out the jet fighters and bombers, the V-1 and V-2 rockets, and the radically new submarines that could have won the war for her.

In this final and ultimate sense, German radar was not important. It could have been, it nearly was—but this book is about what *was,* not what *nearly* was.

The story of Japanese radar is roughly similar. We were all terribly wrong in our estimates of the Japanese at that time, regarding them as a technically inferior people. It's incredible, knowing what we do of them today, to think that we could have been so naive, so ignorant.

But again, although the Japanese scientists and engineers independently developed successful radar systems, the military tended to ignore them at the beginning as their armies flooded over the Pacific and through Asia, and by the time they needed them and turned to them, it was too late. When their radar was operative and integrated into a unified system of defense, when it picked up the blips of PBYs and B-29s and Mitchells and Marauders and Hellcats and Helldivers, there were no longer enough Zeros with sufficiently trained pilots to shoot them down. When the *Enola Gay* cruised high over the Japanese coastline on its leisurely way to Hiroshima, it was surely seen on Japanese radar—but they did not have forces to waste chasing one or two solitary bombers on what must have appeared to be a reconnaissance or sight-seeing mission.

The focus of forces and events came together in the tight glare of history on one spot and at one time: on the island of England in the last years of the fourth decade of this century, when radar was conceived and the world we live in was born.

Notes

References are to books listed in the Bibliography.

Introduction

The story of the Mongol campaigns is based largely on the account in the *Encyclopaedia Britannica,* in which the phrase "indignities worse than death" was used in all seriousness.

Chapter 1: In God We Trust

Details of the zeppelins are from the *Britannica;* of the Martini mission and early bombardments from Wood and Dempster, Clark, Mason, Watson-Watt (*Radar*), M. Smith, Kinsey (*Bawdsey, Orfordness*), and Everett; and the Geneva conference from M. Smith.

Chapter 2: Death Ray

The discussion of the death ray is based on Jones and on Clark, and that of the atomic bomb is based on Hyde. The Churchill quotes are from Manchester and R. W. Thompson; "Makin' mock o' uniforms..." is from "Tommy" by Rudyard Kipling; and the discussion of Londonderry is largely from Cross. Details of the sound detectors are given by Rowe and by Clark, and the story of the army's and navy's rejection of aircraft is from Slessor. Wilkins's involvement is based on his own unpublished account, on private correspondence with his wife, on informal talks with several people involved in radar development, and on Kinsey (*Orfordness*), Clark, Swords, and Terraine; Wimperis's consultation with Watson-Watt is from Terraine, Clark, and Rowe.

Chapter 3: Radio

The development of the telegraph is taken largely from the *Britannica*. The account of Farraday and Maxwell is taken from several physics texts, most notably Richtmyer. The history of the early pioneers (Marconi, Tesla, et al.) is from Swords and assorted physics and engineering texts. The story of Dr. Crippen is from several contemporary records and from the marvelous account given by Priestley. The King George anecdote is presented by Watson-Watt

(*Radar*); the Milch story is in Kinsey (*Bawdsey*). The Wilkins/Watson-Watt controversy is discussed further in Kinsey (*Orfordness*) and Terraine.

Chapter 4: Beginnings

A detailed history of American radar development can be found in Page and that of German development in Swords. The relationships between Dowding, the Tizard Committee, Wimperis, and Watson-Watt are discussed by Clark and by Gunston; the test at Daventry is described by Clark, Rowe and Kinsey (*Orfordness*). The Blucke discussion is from Clark.

Chapter 5: Opposition

Churchill's involvement is from the *Britannica*, Manchester, Tuchman, Gilbert, Snow, and Soames; Lindemann's from Snow, Birkenhead, and Colville. The halibut anecdote was given by Priestley, the Fisher by Bacon, quoted in Tuchman.

Chapter 6: Barking Creek

This account was pieced together from stories in contemporary newspapers and the various bits and pieces in Slessor, Wallace, Revie, Townsend, Shaw, Deere, and Kinsey (*Bawdsey*).

Chapter 7: Mulberry Bush

The Lindemann references are from Birkenhead and Snow; the Tizard are from Clark and Snow. Unemployment statistics in Britain are given by Mowat. The story of Z is in Read and Fisher. Lindemann's involvement with the Tizard Committee is described by Churchill (*Gathering Storm*), Cross, Jones, Birkenhead, Clark, and R. W. Thompson. The development of infrared detection is from the account by Jones.

Chapter 8: 1935–1940

The first radar experiments are described by Swords; the difficulties in direction and height finding by Swords and in Taylor and Westcott; the air exercises by Rowe, Watson-Watt (*Radar*), and Clark.

Chapter 9: Spectre and Terror

The Dowding material is from Wright and Deighton. The first months of war are described by Nicolson, the May 7 debate by Mowat; the Maastricht debacle by Shaw, the May 16 debate and letter by Churchill (*Finest Hour*) and by Wright. Ian Jacob is quoted by Wright. No. 1 Squadron's trip to France is from Richey.

Chapter 10: Night

43 Squadron's victory is described by Townsend. Baldwin's vision is described by Thompson. The Dowding quotes are from Dowding (*Lychgate, Many Mansions, God's Magic, Twelve Legions*).

Notes

Chapter 11: The Human Element

The Bader story is recounted by Lucas and by Brickhill; the organization of Fighter Command and the Luftwaffe is from Wood and Dempster, Deighton, Townsend, and Mason. The Battle of Britain descriptions are from Mason, Terraine, Townsend, Wood and Dempster, and personal accounts. The 609 Squadron anecdotes are from Ziegler; 601 Squadron from Moulson, quoted by Lyall.

Chapter 12: *Adlerangriff!*

Details of the Battle are from Mason, Mosley, Terraine, Townsend, and Wood and Dempster. The story of Bader and the "big wing" controversy has been pieced together from personal accounts and from Lucas, Brickhill, Terraine, Mason, Townsend, Wood and Dempster, and Johnson. Day-by-day episodes are from Mason; Biggin Hill's involvement is from Wallace.

Chapter 13: London

Further details can be found in Mason, Terraine, Townsend, and Wood and Dempster.

Chapter 14: Pride and Petulance

The story of Dowding's dismissal is subject to varying interpretations. The account given here is largely from Wright; but see also Terraine, Deighton, Townsend, and Allen. The Churchill episode is from Churchill (*Finest Hour*).

Chapter 15: Blitz

Technical aspects are discussed by Page, Ridenour, Swords, Bowen, and Taylor and Westcott. Details of the first airborne work are given by Bowen; of the magnetron by Swords, Clark, and Brittain. The first AI kill has been described by Mason and by Gunston; the description of "the thing" by Rawnsley and Wright; early missions in the blitz are based on accounts by Brandon, Braham, and by Rawnsley and Wright. ("The magician was still kneeling...." is from Rawnsley and Wright.)

Chapter 16: Microwaves

Microwaves and the Tizard mission are from Clark. The Hellcat and the Butch O'Hare incident comes from Anderton.

Chapter 17: *Das Boot*

Activities of submarines and antisub warfare in World War I are from the *Britannica* and Price; World War II accounts are from Macintyre, Price, Churchill (*Closing the Ring*), and private correspondence with Bowen. The *Luigi Terelli* story is told by Price; the story of the U-boat commander was told by Schaeffer, quoted by Slessor. For more details of the 1943 activities, see Brown; for the German reaction to centimetric radar, see Price.

Chapter 18: Armageddon

Information on the AWACS aircraft came from air force sources and from Dickson. The satellite and phased-array radars are described by Hafemeister et al., and by Brookner, synthetic aperture radar by Elachi; and some details of the Seawave theory can be dug out of Ridenour.

Chapter 19: Finale

The Star Wars scenario has already generated a literature of its own. I have relied largely on R. Smith (1983), Hecht, Yonas, and Panofsky. Reagan was quoted in R. Smith (1985).

Bibliography

Allen, H. R.: *Who Won the Battle of Britain?*, Barker, London, 1974.

Anderton, David A.: *Hellcat,* Crown, New York, 1981.

Bacon, Sir Reginald Hugh: *The Life of Lord Fisher of Kilverstone,* Hodder and Stoughton, London, 1929.

Birkenhead, Earl of: *The Prof in Two Worlds,* Collins, New York, 1961.

Bowen, E. G.: *Development of Airborne Radar in Great Britain—1935–1942,* Institute of Electrical Engineers, London, 1985.

Braham, J. R. D.: *Night Fighter,* Norton, New York, 1962.

Brandon, Lewis: *Night Flyer,* Kimber, London, 1961.

Brickhill, Paul: *Reach for the Sky,* Collins, New York, 1954.

Brittain, James E.: "The Magnetron and the Beginnings of the Microwave Age," *Physics Today,* July 1985.

Brookner, Eli: "Phased-Array Radars," *Scientific American,* February 1985.

Brown, Anthony Cave: *Bodyguard of Lies,* Harper & Row, New York, 1975.

Clark, Ronald W.: *The Rise of the Boffins,* Phoenix House, Los Angeles, 1962.

Churchill, W. S.: *The Gathering Storm,* Houghton Mifflin, Boston, 1949.

———*Their Finest Hour,* Houghton Mifflin, Boston, 1949.

———*Closing the Ring,* Houghton Mifflin, Boston, 1949.

Colville, John: *Winston Churchill and His Inner Circle,* Wyndham, London, 1981.

Cross, J. A.: *Lord Swinton,* Clarendon Press, Oxford, New York, 1982.

Deere, Alan C.: *Nine Lives,* Hodder and Stoughton, London, 1959.

Deighton, Len: *Fighter,* Jonathan Cape, London, 1977.

Dickson, Paul: *The Electronic Battlefield,* Indiana University Press, Bloomington, 1976.

Dowding, Lord: *God's Magic,* Spiritualist Association, London, 1962 (Reprint).

———*Lychgate,* Rider, New York, 1945.

———*Many Mansions,* Rider, New York, 1943. (Reprint, Cedric Chivers, Portway, Bath, 1976.)

———*Twelve Legions of Angels,* Jarrolds, London, 1946.

Elachi, Charles: "Radar Images of the Earth from Space," *Scientific American,* New York.

Encyclopaedia Britannica. University of Chicago Press, Chicago, 1947.

Everett, Susanne: *World War I,* Hamlyn, London, 1980.

Gilbert, Martin: *Winston S. Churchill,* Houghton Mifflin, Boston, 1975.

Gunston, Bill: *Night Fighters,* Scribner's, New York, 1976.

Hafemeister, David, Joseph J. Romm, and Kosta Tsipis: "The Verification of Compliance with Arms-Control Agreements," *Scientific American,* March 1985.

Hecht, Jeff: *Beam Weapons,* Plenum Press, New York, 1984.

Hyde, H. Montgomery: *British Air Policy Between the Wars,* Heinemann, London, 1976.

Johnson, J. E.: *Full Circle,* Ballantine, New York, 1964.

Jones, R. V.: *The Wizard War,* Coward, McCann & Geoghegan, New York, 1978.

Kinsey, Gordon: *Bawdsey,* Terence Dalton, Suffolk, 1983.

———*Orfordness,* Terence Dalton, Suffolk, 1981.

Londonderry, Marquess of: *Wings of Destiny,* Macmillan, New York, 1943.

Lucas, Laddie: *Flying Colours,* Hutchinson, London, 1981.

Lyall, Gavin (ed.): *The War in the Air,* Morrow, New York, 1968.

Macintyre, Donald: *The Battle of the Atlantic,* Batsford, London, 1961.

Manchester, William: *The Last Lion,* Little, Brown, Boston, 1983.

Mason, F. K.: *Battle Over Britain,* Alban, London, 1969.

Mosley, L.: *The Battle of Britain,* Time-Life Books, New York, 1977.

Moulson, Tom: *The Flying Sword,* MacDonald, London, 1964.

Mowat, Charles Loch: *Britain Between the Wars, 1918–1940,* Beacon, Boston, 1955.

Nicolson, Harold: *Diaries and Letters, 1939–1945,* Collins, New York, 1967.

Page, Robert Morris: *The Origin of Radar,* Doubleday, New York, 1962.

Panofsky, Wolfgang K. H.: "Strategic Defense Initiative: Perception vs. Reality," *Physics Today,* June 1985.

Petersen, Stefan: "Multinational AWACS in Europe," *Defence Update International,* No. 64, Eshel Dramit, Cologne, W. Germany, 1985.

Price, Alfred: *Aircraft versus Submarine,* Jane's, London, 1973.

Priestley, J. B.: *The Edwardians,* Harper & Row, New York, 1970.

Rawnsley, C. F., and Robert Wright: *Night Fighter,* Collins, New York, 1957.

Read, Anthony, and David Fisher: *Colonel Z,* Viking, New York, 1985.

Revie, Alastair: *The Bomber Command,* Ballantine, New York, 1974.

Richey, Paul: *Fighter Pilot,* Batsford, London, 1941.

Richtmeyer, F. K., E. H. Kennard, and T. Lauritsen: *Introduction to Modern Physics,* McGraw-Hill, New York, 1955.

Ridenour, Louis N.: *Radar System Engineering,* McGraw-Hill, New York, 1947.

Rowe, E. P.: *One Story of Radar,* Cambridge, London, 1948.

Schaeffer, Heinz: *U-Boat 977,* Kimber, London, 1952.

Shaw, Michael: *Twice Vertical,* MacDonald, London, 1971.

Slessor, Sir John: *The Central Blue,* Cassell, London, 1956.

Smith, Malcolm: *British Air Strategy Between the Wars,* Clarendon Press, Oxford, New York, 1984.

Smith, R. Jeffrey: "The Search for a Nuclear Sanctuary," *Science,* July 1 and 8, 1983.

Smith, R. Jeffrey: "Summit Ends with Exchange Agreements," *Science,* December 6, 1985.

Snow, C. P.: *Science and Government,* Harvard University Press, Cambridge, Mass., 1961.

Soames, Mary: *Family Album,* Houghton Mifflin, Boston, 1982.

Swords, Sean S.: *A Technical History of the Beginnings of Radar,* Trinity College, Dublin, 1983.

Taylor, A. J. P.: *English History, 1914–45,* Oxford, New York, 1965.

Taylor, D., and C. H. Westcott: *Principles of Radar,* Macmillan and Cambridge Press, New York, 1948.

Templewood, Viscount: *Nine Troubled Years,* Collins, New York, 1954.

Terraine, John: *The Right of the Line,* Hodder and Stoughton, London, 1985.

Thompson, Laurence: *1940,* Morrow, New York, 1966.

Townsend, Peter: *Duel of Eagles,* Editions Robert Laffont, 1969.

Tuchman, Barbara W.: *The Proud Tower,* Macmillan, New York, 1966.

Wallace, Graham: *RAF Biggin Hill,* Putnam, New York, 1957.

Watson-Watt, Sir Robert: *Man's Means to His End,* Potter, New York, 1961.

———*Three Steps to Radar,* Dial, New York, 1959.

Winterbotham, F. W.: *The Ultra Secret,* Harper & Row, New York, 1974.

Wood, D., and D. Dempster: *The Narrow Margin,* Paperback Library, New York, 1969.

Wright, Robert: *The Man Who Won the Battle of Britain,* Scribner's, New York, 1969.

Yonas, Gerold: "Strategic Defense Initiative: The Politics and Science of Weapons in Space," *Physics Today,* June 1985.

Ziegler, Frank: *The Story of 609 Squadron,* MacDonald, London, 1971.

Index

Aboukir (ship), 282
Abramson, General James, 341
Acrobatics, air, 172–175
Adler Tag, 198, 199, 200, 202, 225
Adlerangriff, 198–217
  most massive attack, 205–210
Aerial mines, 102–103, 107, 257, 259
Aeronautical Research Committee, 29, 98,
Aether, the, 243, 287
AI (Airborne Interception), 251–254, 257, 261, 263, 321, 323
  (*See also* Blenheims, radar-equipped; Bristol Beaufighters)
Air Defense Exercises, final, 126
Air Defense Experimental Establishment, 23–24; 46–47, 54
Air Defense Forces, Soviet, 52
Air Defense Research Committee (*see* Committee on Air Defense Research)
Air Group 6 (U.S.), 277
Air Staff for Research and Development, 120, 131, 134
Air-to-surface radar, 280, 293–297, 303–306
  interception of, 302
  (*See also* Submarines)
Airacobras, 271
Aircraft:
  Airacobra, Bell P-39, 271
  Avenger, Grumman TBF, 276–278
  Anson, Avro, 291, 293–294
  AWACS, 322–324, 331
  Battle, Fairey, 145–146
  Black Widow, Northrop P-61, 275, 278
  Blenheim, Bristol, 175, 179, 186, 187–188, 199, 251–254, 255–256, 258, 261, 262, 264, 296

Aircraft (*cont.*):
  Beaufighter, Bristol, 257–261, 262–265, 273, 274, 275, 306
  Bulldog, Bristol, 12, 172–174
  Catalina, Consolidated, 311
  Corsair, Chance-Vought, 275
  Defiant, Boulton-Paul, 175–176, 179, 186, 257, 262
  Eagle, F-15, 321, 323
  Flying Fortress, Boeing B-17, 268, 271, 274, 350
  Gladiator, Gloster, 144, 175
  Gamecock, Gloster, 172
  Gotha, bomber, 10, 97
  Halifax, Handley-Page, 271
  Heinkel He-111, 53, 126, 144, 147, 157, 161, 179, 185, 205, 206, 216, 222, 223, 230, 252, 264, 291
  Hellcat, Grumman F6F, 275–278
  Hudson, Lockheed, 179, 291, 295, 296
  Hurricane, Hawker, 14, 85–86, 87, 88, 89–90, 126, 134, 135, 144, 146–152, 155–157, 159, 161, 175, 176, 179, 185, 186, 192, 198, 199, 200, 202, 203, 206–207, 209, 210, 213, 216, 224, 227, 228, 229, 235–236, 257, 259–260, 262, 299
  Junkers Ju-87 Stukas, 53, 145, 147, 179, 181, 185, 198, 200, 202, 205, 207, 209, 236, 252, 307, 349
  Kondor, Focke-Wulf, 13, 299
  Lancaster, Avro, 271, 298, 349
  Lerwick, Saro, 290
  Liberator, Consolidated B-24, 310, 311, 349
  Lightning, Lockheed P-38, 275, 278

INDEX

Aircraft (*cont.*):
  Messerschmitt Me-109 and Me-110, 89, 90, 126, 147, 161, 175–176, 181, 186, 189, 190, 198, 199, 200, 201, 203, 205, 206–207, 216, 218, 222, 223, 224, 227, 228, 229, 230, 237, 268, 271, 291
  Mitsubishi bombers, 277–278
  Mosquito, de Havilland, 273–274
  Mustang, North American P-51, 268, 271
  Spitfire, Supermarine, 14, 32, 45, 84, 88–89, 90, 119, 126, 127, 132–133, 134, 135, 137, 146, 151, 157, 161–162, 163, 172, 175, 176, 177–178, 179, 185, 186, 187, 197, 198, 199, 200, 201, 202, 203, 204, 205, 206–207, 208–209, 210, 214, 218, 220, 223–224, 226, 227, 228, 229–231, 237, 259, 268, 271, 322
  Stirling, Short, 271, 290, 298, 304, 307
  Sunderland, Short, 290–291, 298, 310, 311
  Swordfish, Fairey, 294, 311
  Tigercat, Grumman F7F, 275
  Tomahawk, Curtiss-Wright P-40, 53, 271
  Wildcat, Grumman F4F, 53, 275, 276, 277
  Zero, Mitsubishi, 53, 275, 276, 277, 278, 350
Aircraft Radio Laboratory (U.S.), 50
Air power, early advances in, 26–27
Allied Expeditionary Air Force, 236
Allied Submarine Detection Investigation Committee, 287
  (*See also* Asdic)
Allison engine, 271
Altimeters, pressure-reading, 301–302
Altitude (*see* Height)
Amery, Leopold, 144
Ampère, André Marie, 34
Anacostia Naval Air Station, 50
Anderson, Maxwell, 218
Andover, 203, 207
Andrade (*see* Costa)
Antiballistic missile (ABM) defense system 339, 342

Arabic (ship), 284
Archimedes, 29
Armament Experimental Flight, 116
Army Cooperation Command, 145
Asdic, 43–44, 267, 287–289, 292–293
Ashfield, Flying Officer G. "Jumbo," 252, 254, 255, 257, 262
ASV (*see* Air-to-surface)
Atmospheric permeability, 338
Atmospherics (static), 112–116, 124, 330
Atomic bombs, x–xi, 8, 15–16, 20, 139, 151, 268, 339, 340, 350–351
Atoms, atomic energy, 332–339
Audacity (carrier), 299
Avengers, 276–278
Avro Anson planes, 291
  K 6260, 293–294
AWACS aircraft (Airborne-Warning-and-Control-Systems), 322–324, 331

B-Dienst, 292
Bacon, Roger, 33
Bader, Flying Officer Douglas, 171–175, 178, 190, 191–192, 193, 198–199, 206, 211, 212, 215–217, 224, 228, 229, 234–235, 236, 237
Baldwin, Stanley, 3, 8, 11, 25, 28, 72–73, 81, 82, 98, 100, 101, 102, 133, 159, 224, 345
Balfour, Arthur James, Earl of, 66
Balfour, Harold, 235
Ballistic Missile Early Warning System (BMEWS), 322, 329, 331
BAMS radio code, 289
"Bat" teams, 276–278
Battle of the Atlantic, 280, 291–293, 296–297, 299–302, 306–311
  and America, 300
  and night attacks, 302
Battle of Barking Creek, 53, 84–91, 109, 120, 126, 138, 175
Battle of Britain, 45, 53, 54, 60, 84, 88, 108, 109, 119, 132, 133, 149, 150, 152, 155–238, 239, 250, 255, 267
  Adlerangriff, 198–217
  aftermath, 233–238
  and big-wing action, 216–217, 224, 228–231, 234–235
  and bombing of Berlin, 212–213, 218–219, 233

INDEX

Battle of Britain (*cont.*):
  bombing of London, 212, 219–233
  Ch stations, 182–184
  culmination, 203–208
  Dowding's report on, 236
  Fighter Command pivotal formation, 176–177
  first stages, 155–170, 175–176
  night kill, 253–254
  "penny packets," 163–164, 211, 214, 235
  plotting problems, 184
  radar and the human element, 171–195
  settling in, 182–195
Battle of France, 141–142, 146–152, 156, 157, 176, 180, 190, 191, 192, 203, 235–236, 291, 296
Bawdsey Manor, 4–5, 123–125, 136–137, 182–183, 246
  and airship reconnaisance, 5–8, 125
Beaverbrook, Lord, 149
Becquerel, Antoine Henri, 20
Bell, L. H. Bainbridge, 112
Belridge (ship), 284
Bentivegni, Hauptmann, 10
Bentley Priory, 7, 85, 88, 122, 132, 135–136, 138, 148, 149, 151, 156, 161, 176, 183, 195, 201, 203, 206, 208, 219, 220
Berlin, bombing of, 212–213, 218–219, 233
Big Bang theory, 320
Biggin Hill aerodrome, 23–25, 46, 54, 85–86, 167, 188, 200, 201, 207, 215, 217, 219–220, 225
Bilney, Air Vice Marshal C.H.N., 89–90
Birkenhead, Frederick E.S., Earl of, 73, 75, 79–80
Black Widow (P-61), 275, 278
Blackett, P. M. S., 96, 102, 107, 108, 305–306
Blake, William: "The Tyger," 241
Bleinheims, 175, 179, 186, 187–188, 199, 251–254, 258, 262
  radar-equipped, 251–254, 255–256, 259, 264, 296
Bletchley Park, 181, 196, 197, 229, 231
Blimp, Colonel, 290

Blimps, antisubmarine, 290
Blitz, the (*see* London, bombing of)
Blucke, Air Vice Marshal R. S., 62–63
BMEWS (*see* Ballistic Missile Early Warning System)
Boadicea, 85
Bofors repeating cannon, 269
Bohm, Dr., 51, 52
Bohr, Niels, 32, 333
Bomber Command (*see* RAF)
Bombers, bombs and targets, 8–14, 22, 23–26, 73, 81–83, 96, 99, 101, 120–121, 133–135, 170, 212, 224, 303, 305, 345
  Luftwaffe, 241–242
  saturation bombing, Allied, 241, 303, 345, 350
  (*See also* Battle of Britain; Blitz)
Bonar Law, Arthur, 71–72
Boot, Henry, 249–250, 256, 268, 272
Bore War, 126
Born, Max, 75
Bowen, E.G. "Taffy," 1, 46, 117–118, 245–246, 251, 252, 261, 272, 293–294, 295, 302–303
Bracken, Brendan, 100
Brand, Air Vice Marshal Sir Christopher Q., 176
Brantes, Captain, 86
Braun, Wernher von, x
Breit, Gregory, 48, 112
Brinkley, Charlie, 4
Bristol Beaufighter, 257–261, 262–265, 273, 274, 275, 306
Bristol Bulldog, 12, 172–174
Broglie, Louis Duc de, 55, 73
Brooke-Popham, Sir Robert, 82
Brownian motion, 332
Brundrett, Frederick, 248–249
Bruning, Heinrich, 143
Bryan, William Jennings, 69
Bulldog fighters, 12, 172–174
Bullock, Sir Christopher, 13
Butement, W. A. S., 54

Campbell-Bannerman, Sir Henry, 67
Canaris, Admiral Wilhelm, 3
Canewdon, 124–125
Carlow, Lord, 251
Casablanca conference, 306
Cascariolo, Vincenzo, 17
Catalinas, 311

INDEX

Cathode tube, rays, screens, 17–18, 52, 59–63, 113–116, 134–135, 137, 156, 255, 256, 261–262, 264, 274–275, 294
Cavendish Laboratory, 36, 79
Cavity magnetrons, 250, 256–257, 261, 269, 271–273, 304, 337
Centimetric radar (*see* Radar: centimetric)
Chain Home (CH) stations, 118, 124–126, 142, 160–161, 182–184, 201, 204, 225, 226, 228, 229, 242, 243, 245, 246–247, 252
Chain Home Low (CHL), 142, 161
Chamberlain, Neville, 16, 73, 84–85, 140, 141, 143, 144–145
Chamberlain, Wing Commander, 252–254
Chatfield, Lord, 85
Cherwell, Lord (*see* Lindemann)
Churchill, Clementine, 74, 75, 81
Churchill, Lord Randolph, 64, 65
Churchill, Winston S., 15–16, 23, 25, 26, 28, 29, 60, 99–100, 127, 140–141, 142, 145, 146, 148–152, 160, 168, 179, 180–181, 182, 187, 194–195, 212–213, 231, 251, 270, 280, 291, 298–299, 302–303
  and bombing of Berlin, 212–213, 219, 233, 305–306
  books, 65, 151, 233
  and Dowding, 127, 133, 148–152, 213, 235–236
  earlier career, 64–67, 68–73, 145
  The Gathering Storm, 30
  insight into Hitler, 141
  and Lindemann, 73, 83–84, 91, 96, 100–102, 106–109, 298–299, 302–303, 305–306
  and Norway, 143–144
  opposes radar, 73, 83–84, 91, 99–102, 106–109
  and Roosevelt, 269, 270, 271
  and scientific mission to America, 267–268
  Their Finest Hour, 147–148
Clarendon Laboratory, 75, 79, 248
Clark, Ronald, 46
Coastal Command (*see* RAF)
Coastal defense (CD), 266
Cobra systems, early warning, 329, 330, 331

Code breaking, 181, 185, 190, 196, 197, 220, 229, 231–232, 234, 289, 292, 306–307, 309
Colney Heath, 44, 53
Coltishall, 191–192, 206, 224
Colville, Sir John, 74, 75, 168
Commercial aviation, 318–319
Committee on Air Defense Research (Swinton Committee), 101–102, 106, 107–108, 118
Committee of Imperial Defense, 101, 287
Committee for the Scientific Survey of Air Defense (Tizard Committee), 28–29, 59, 83, 96, 98–99, 100–108, 117
Computer technology and weapons systems, 319–321, 322, 326, 341–342, 343
Constantine of Greece, 72
Convoy system, convoys and escorts, 285–286, 288, 290, 291–292, 296–297, 298, 299, 304, 306–307, 308–309, 349
  SC-7, 297
Corsairs, 275
Costa Andrade, Edward Neville da, 21
Courageous (carrier), 289, 294
Courtney, Air Vice Marshal Christopher, 131
Coward, Noel, 95, 108
Cressy (ship), 282
Crippen, Hawley Harvey, 39–41, 68
Cromwell, Oliver, 144
Cross, F. A., 108
Cruise missiles, 343
Cunliffe-Lister, Philip, 101–102 (*See also* Swinton, Lord)
Cunningham, John "Cat's Eye," 45, 262
Curie, Marie, 20, 120
Curie, Pierre, 20

Danby Beacon, 156
Dark Star (Dowding), 166
Darley, Squadron Leader, 189
DAT (*Défence Aérienne du Territoire*), 53
Daventry, 61–63, 99, 102, 112, 134–135
David, Pierre, 53
Davies, W. T., 62

Davy, Sir Humphrey, 34
Dawn Patrol (film), 130
"Death rays," 4–5, 96, 169, 257, 267, 332–344
 and inverse square law, 334–336, 338
 lasers, 336–344
 and origins of radar, 15–31, 44–45, 59, 91
Deere, Alan, 88–89
Defensive spirit, 349–350
Defiants, 175–176, 179, 186, 257, 262
De Havilland Mosquitos (*see* Mosquitos)
Department of Scientific and Industrial Research, 29, 98, 107
Department of Terrestrial Magnetism (Washington), 272
Descartes, René, 247
"Detection and Location of Aircraft by Radio Methods" (Watson Watt), 59
Detling, 203
Dew, Inspector, 40–41
Diechmann, *Oberst* Paul, 204–205
Direction finding, 119, 121–122, 125–126, 183, 198–199
 high frequency, 307
Directive No. 16 (German), 186
Directive No. 17 (German), 196–197
Directorate of Scientific Research, Air Ministry, 23–26
Dirigibles:
 antisubmarine, 290
 reconnaisance, 5–8, 125, 200
Distance, measuring, 119, 125–126
Doenitz, Admiral Karl, 281, 291–292, 297, 300, 306, 307–308, 310, 311
Dominik, Hans, 43, 50
Doppler shift, effect, 319–321, 322, 323
Dornier bombers, 126, 181, 185, 193, 202–203, 205, 208, 220, 222, 223, 230, 252, 254, 291
Douglas, Air Vice Marshal Sholto, 85, 131, 236
Dover, 124–125
Dowding, Clarice, 122, 130, 158, 159–160, 165–168
Dowding, Derek, 130, 158
Dowding, Hilda, 122, 163, 164

Dowding, Air Chief Marshal Sir Hugh Caswall Tremenheere ("Stuffy"), 14, 24, 27, 60–61, 63, 84, 88, 89, 112, 122, 124, 125, 126, 133–135, 136, 138, 139, 146–147, 148–151, 161–170, 171, 175, 179–180, 184, 185, 189–190, 196, 198, 200, 203, 204, 208, 211, 213, 215, 216, 218, 220–221, 226–227, 229, 233, 251, 276, 345
 books, 166–168, 234, 236
 character and career, 28, 60, 127–133
 and death of wife, 122, 130, 158, 159–160, 162–166, 234, 236
 insanity, 60, 133, 168–170, 215
 relieved of command, 234–237
 report on Battle of Britain, 235
Drebel, Cornelius van, 281
Duel of Eagles (Townsend), 157
Dunkirk, 108, 132, 151–152, 178–179, 181–182
Duxford, 177–178

Eagle (ship), 281
Early warning and security, xi, 322, 323–324, 329
 (*See also* CH stations)
Eastchurch, 202–203
Echoes, echolocation, 110–120, 124, 126, 136–137, 161, 182, 320
Ede, Chuter, 11
Eden, Anthony, 181
Edward VII, 67, 69
Eighth Air Force (U.S.), 268
Einstein, Albert, 36, 54, 55, 56, 151, 247, 332–334
 quoted, 332
Eisenhower, General Dwight D., 133
Electromagnetism, 33–37, 51–52, 243, 287–288, 333, 335
 and Doppler shift, 320
 interaction of waves, 326–329
 (*See also* Wavelengths)
Elektrische Luistertoestel, 52
Elizabeth I, 285
Ellington, Air Chief of Staff Edward, 131
Elliott, Maxine, 67
Elmore, Belle, 40
Encyclopedia Britannica, 286, 288
English History, 1914–1945 (Taylor), 149
Enola Gay, 350–351
Enterprise (carrier), 277

INDEX 364

Expanding universe, 320
Extinction of species, 346

F-15 Eagles, 321, 323
Falklands war, 324
Faraday, Michael, 33–35
Farren, Warren, 76–77
Fermat, Pierre de, 21
Fighter Command (*see* RAF)
Fighter Interception Unit (FIU), 251, 252, 258–259
Fighting Services Economy Committee, 100
Filter Room, 135–136, 138, 156, 183
Fink, Kommodore Johannes, 181, 185, 202–203
Fisher, David:
  airborne adventure, 315–318
  quoted, 332
Fisher, Sir John, 69, 70–71
Fiske, Billy, 209
Fliegerkorps 2, 205
Fluorescence, 17–20
Flying and air pressure, 173
Flying boats, 318–319
Flying Fortresses, 268, 271, 274, 350
Focke-Wulf bombers, 13, 299
Folkes, "Tiger," 157
Foucault (submarine), 283
Freya radar system, 3
Frisch, Otto, 268
"Fruit machine," 183
Fylingdales early warning system, 329

Gallipoli, 70–71, 145, 176
Gamelin, General M.G., 147
Garland, Flight Lieutenant "Judy," 145–146
Gayler, Noel, 344
GCI (ground-controlled interception), 262, 263, 265
GEMA (*Gesellschaft für Elektroakustische und Mechanische Apparate*), 51–52
Geneva Conference (world disarmament), 10–14, 24
George V, 43–44, 51, 288
George VI, 88, 89, 157, 220
Gilbert, William, 33
Gilbert, Sir William Schwenck, 28
Gladiators, 144, 175
Glorious (carrier), 144
Gloster Gamecock, 172

God's Magic (Dowding), 166
Goebbels, Joseph, 213
Goering, Field Marshal Hermann Wilhelm, 5, 14, 32, 54, 178, 179–180, 181, 182, 185, 186–187, 190, 197, 199, 200, 202, 203–204, 205, 207, 208, 209, 211, 213, 214, 218, 219, 220, 221, 225, 226, 227, 229, 231, 349, 350
Gort, Field Marshal Lord, 147
Gosewisch, *Oberstleutnant*, 7–8
Gossage, Air Vice Marshal Ernest Leslie, 177
Gotha bombers, 10, 97
GPO (General Post Office) Report No. 233, 44–45, 47–49, 50, 52, 53
Graf Zeppelin, 5–8, 85, 125
Graham, Flight Lieutenant, 185
Great Bromley, 124–125
Greswell, Squadron Leader Jeff, 301
Grey, Sergeant, 146
Ground clutter, 322–323
Guinness, Loel, 188
Gullflight (ship), 284

H2S, 304–306, 307
Haig, Field Marshal Lord Douglas, 20
Halifax bomber, 271
Halifax, Lord, 143, 159, 194, 196
Hallowes, Sergeant H.J.L., 157
Hankey, Sir Maurice, 98, 101
Harris, Air Marshal Sir Arthur "Bomber," 305
Hawkers, 134
Heaviside, Oliver; Heaviside layer, 39, 47–48
Height, determination of, 119–120, 121, 125–126, 142, 161, 182, 183, 216, 227
Heine, Heinrich: "Die Lorelei," 50
Heinkel bombers, 53, 126, 144, 147, 157, 161, 179, 185, 205, 206, 216, 222, 223, 230, 252, 264, 291
Heinkel, Ernst: Heinkel 119, 274
Heisenberg, Werner, uncertainty principle, 55
Hellcat (F6F), 275–278
Hendon air shows, 172, 192
Henry II, 133
Henry V, 14, 28, 110
Hertz, Heinrich, 36, 42

INDEX

HF/DF (High-Frequency Direction Finding—Huff-Duff), 307, 309
Hill, A. V., 27, 29, 96, 102, 106, 107, 108, 267
Hindenburg (airship), 6
History, learning lessons of, 344–346
Hitler, Adolf, 3, 14, 15, 16–17, 22, 54, 75, 91, 100, 139, 140, 141–142, 143, 160, 178, 179, 180, 181, 182, 186–187, 194, 195–196, 197, 211, 212, 213, 214, 218–219, 220, 225, 226, 227, 231, 269, 272, 275, 285, 291, 300, 306
HMS L-27 (submarine), 295
HMS *Sheffield*, 324
HMS *Snapper*, 291
Hogue (ship), 282
Hook, Robert, 247
Horace, 33
Home, Sir Robert, 72
Housatonic (ship), 281–282
Howell, Pilot Officer W., 301–302
Hubble, Edwin, 320
Huff-Duff, 307, 309
Hughes, Howard; Hughes Aircraft, 318–319, 323
Hull, Albert W., 247–248
Hulsmeyer, Christian, 42–43, 50
Humbert, King, 37
Hurricanes, 14, 85–86, 87, 88, 89–90, 126, 134, 135, 144, 175, 259–260, 262, 299
  in Battle of Britain, 155–157, 159, 161, 176, 179, 185, 186, 192, 198, 199, 200, 202, 206–207, 209, 210, 216, 224, 227, 228, 229, 257
  in Battle of France, 146–152, 156, 157, 176, 203, 213, 235–236
  engine, 268
  pilots, 127, 161–162, 163, 164–165
Hydrophones, 288

ICBM missiles, 326, 329, 339
  and SDI, 340–346
Identification of blips, 126, 198–199
IFF (Identification Friend or Foe), 46, 126, 161, 185, 262, 322
Infrared detectors, 103–105
Intercontinental ballistic missile (*see* ICBM)

International law, 345
Inter-Services Committee for the Co-ordination of Valve Development, 248–249
Invasion Alert No. 1, 220, 225
Ionospheric layer, 39, 47–48, 112, 330
Isaiah 6:11, 347

Jacob, General Sir Ian, 149
James I, 281
Jet propulsion, 268, 271, 318, 319, 321–322
Joad, Cyril, 142–143
Joan of Arc, 14
Jodl, General Alfred, 196
Jones, Reginald V., 16–17, 103–105
Joubert de la Ferte, Air Marshal Sir Philip, 304
Junkers, 53, 202, 205, 207, 236, 252, 307
Jutland, 283

KAL 747, 324
Karinhall, 204
Keitel, Field Marshal Wilhelm, 196, 227
Kendall, Captain, 40–41
Kenley, 207
Kennedy, Joseph, 143
Kennelly, A. E., 39
Kesselring, Field Marshal Albert, 181, 201, 220
Khartoum bombing, 324
Kipling, Rudyard, 66
  quoted on Tommy Atkins, 22
  "Recessional," 233
Krasnoyarsk, 329
Kuhnhold, Rudolph, 51–52

Lancaster bombers, 271, 298, 349
Lasers, 336–339
  and "Star Wars," 340–344, 346
Law, Andrew Bonar, 71–72
Law, international and intranational, 345
Lawrence Livermore Radiation Laboratory, 340, 344
League of Nations, 10–11, 12, 13, 270, 344
Leigh, Humphrey de Verde, 299
Leigh lights, 300–302, 309–310

INDEX

Leigh-Mallory, Air Chief Marshal Sir Trafford, 133, 176, 177, 190, 199, 206, 208, 211, 212, 215, 216, 217, 224, 234, 236, 237
Le Neve, Ethel, 40–41
Leopold of Belgium, 141
Lerwick flying boat, 290
Le Sage, M., 33
Lester, "Pingo," 178
Leyden jars, 33
Leyland, Sergeant R. H., 252, 254
Liberators, 310, 311, 349
Life after death, 158–160, 163–168, 208, 215, 236
Light, composition and nature of, 55–57, 287–288
Lightnings (P-38), 275, 278
Lincoln, Earl of, 188
Lindemann, Frederick Alexander (Lord Cherwell), 73–83, 91, 95–96, 97, 98, 99, 100–101, 102–109, 248, 257, 259, 298, 302–303, 304, 305–306, 339, 342, 345
   character and behavior, 74–79, 95–96, 98, 105–106, 107
Livermore National Laboratory, 340, 344
Lloyd George, David, 66, 68, 71–72, 73, 100, 140, 144
Location, measurements, 118–126
Lockheed Hudsons, 179, 291, 295, 296
London, bombing of:
   Battle of Britain, 212, 219–233, 296
   blitz, 241–242, 255, 259–261, 262–265, 296, 345
London flying boat, 291
Londonderry, Lord, 27, 28–29, 82, 96, 101, 236
Loomis, Alfred, 272
Lorentz, Hendrik, 247
Lorzer, General Bruno, 204–205
Lothian, Philip, 143
Lucas, Laddie, 216
Luftflotte 1, 181
Luftflotte 2, 181, 201
Luftflotte 3, 181
Luftflotte 5, 181, 190, 199, 203–204, 205–207, 208
Luigi Torelli (submarine), 300–301
Luminescence, 333

Lusitania sinking, 284
Lychgate (Dowding), 166, 167

MacArthur, General Douglas, 127, 133
McComb, Group Captain James, 188
MacDonald, James Ramsay, 28, 73, 82, 100–101, 102
Macdonald, Peter, 235, 236
Mach-3 fighters, 326
Maginot Line, 139, 141–142, 349
Magnetrons, 247–250, 256–257, 261, 268, 269, 271–273, 304, 305, 337
Magnets, magnetism, 33–37
   anti-submarine, 298–299
Malan, "Sailor," 89
Manchester, William, 65, 67
Mannerman, Sir Henry, 66
Marconi, Guglielmo, 21, 36–39, 50, 52, 330
Margaret Rose, Princess, 157
Mark IV radar, 257, 261, 262–265
Marlborough, John Churchill, Duke of, 64
Martini, General Wolfgang, 3, 5, 6–7, 199, 200, 307
Massey, Raymond, 139
Matthews, Grindell, 20
Maxwell, James Clerk, 32, 36, 42, 48, 55–56, 243–245, 287
Meitner, Lise, 15, 268
Merlin engines, 268, 271, 273
Messerschmitts, 89, 90, 126, 147, 161, 175–176, 181, 186, 189, 190, 198, 199, 200, 201, 203, 205, 206–207, 216, 218, 222, 223, 224, 227, 228, 229, 230, 237, 268, 271, 291
Metox receivers, 302, 304, 307–308
Micropup transmitter, 261
Microwave Committee (U.S.A.), 272
Microwaves, 266–280, 295, 299, 306, 337
   American cooperation and aid, 267–273, 275
   American developments, 274–280
   and first airborne radar set, 273, 274
   (*See also* Radar: centimetric)
Middle Wallop, 203, 207
Milch, Erhard, 5, 45, 302
Military preparedness, problematics of, 345–346
Miller, George, 344
Mines, aerial, 102–103, 107, 257, 259

INDEX

MIT Radiation Laboratory, 272
Mitchell, Reginald, 134
Mitsubishi bombers, 277–278
Montgomery, Field Marshal, 133
Morale, peculiarities of, 259
Morris, Pilot Officer G. E., 252, 254
Morse, Samuel F. B., 33, 35–36
Mosquitos, 273–274
Musschenbroeck, 33
Mustang fighters, 268, 271

Naish, Sergeant, 293–294
Napoleon, 187
National Physical Laboratory (NPL), 104–105, 123
NATO Airborne Early-Warning Force (NAEWF), 323–324
Nebraska (ship), 284
Nelson, Admiral Lord Horatio, 161
New York Times, The, 86
Newall, Air Chief Marshal Cyril, 85, 131, 148, 151
Newcastle, Duke of, 188
Newton, Sir Isaac, 34, 36, 55, 247, 325
Nicholas II, Czar, 67
Nicolson, Harold, 139, 140–141
Night bombers, night fighting, 245–246, 248, 251–265, 274, 275, 293, 295, 296, 298, 304, 305–306, 307, 308
(*See also* Microwaves)
Night Interception Committee, 296
Nimrods, 324
Nobel-Baker, Mr., 11
Norman, Frederick, 16–17
North Weald, 215–216
NPL (*see* National Physical Laboratory)
Nuclear energy, usable, 334
Nuclear weapons tests, 343

Observer Corps, 135, 184, 193, 202–203
Oersted, Hans Christian, 34
Offensive spirit, 349–350
Office of Scientific Research and Development (U.S.A.), 269, 272
O'Gorman, Mervyn, 77–78
O'Hare, Lieutenant Commander Edward H. "Butch," 277–278
Okabe, professor, 52
Oliphant, Mark, 248–249

One Story of Radar (Rowe), 30
Operation Sea Lion, 197, 214, 225–226, 227, 228, 231–232, 237
Oppenheimer, J. Robert, 340
Orfordness, 116–118, 123
Oshchepkov, Pavel, 52

P-38s, 271, 275–278
P-39s, 271
P-40s, 53, 271, 275
P-51s, 268, 271
P-61s, 275, 278
Pacific war, 276–280
   submarines, 279–280
Park Chung Hee, 324
Park, Keith, 176–177, 185, 201, 206, 210–211, 215, 223, 228, 229, 230, 234, 236–237
Pathfinder (ship), 282
Patton, General George, 127, 133
Pave Paws system, early warning, 329, 330
Pearl Harbor, 32, 53, 268, 278, 300, 301
Peel, Robert, 35
Pennywhistle, concept of, 249–250, 337
Pile, General Sir Frederick, 133
Planck, Max, 55, 333
Poincaré, Jules Henri, 38
Poincaré, Raymond, 72
Pollard, P. E., 54
PPI (plan position indicator), 257, 261–262, 274, 321
Predictor gun sight, 269
Preparedness, military, difficulties of, 345–346
Pretty, Air Marshal Sir Walter, 7, 85
Prince of Wales, 306
Proximity fuse, 269
PVO (*see* Air Defense Forces, Soviet)

Quantum theory, mechanics, 55–57, 332, 333–334
Quilter, Sir Cuthbert, 4, 123, 124

Radar:
   acceptance by services, 138
   air-to-surface (*see* Air-to-surface radar)
   altimeters, 301–302
   American, 32, 268, 269, 271–272, 274–278, 319

INDEX 368

Radar (*cont.*):
  and attacks on installations,
    201–202, 209–210, 210–211,
    213–214, 217, 227
  at beginning of war, 138, 142–143
  beginnings, early stages, 50–63,
    131, 182–184, 243, 247
  centimetric (small sets, fine tuning)
    243–265, 266, 269, 293–294, 296,
    298, 304–306, 307, 309, 311
    German countermeasures, 307–309
    (*See also* Microwaves)
  concept and technical considerations, 110–126, 243–250,
    272–273, 287–289, 318–331
  and "death rays," 4–5, 15–31, 59,
    91, 169, 267
  early malfunctions, 84–91
  final commitment to, 125
  German, 3, 199–200, 268, 349–350
  and guiding of airplanes, 317–318
  invention and importance of, ix–xi
  Japanese, 280, 349, 350–351
  mobile stations, 210
  and night bombing, fighting, 163,
    243–265
  opposition to development of, 73,
    83–84, 91, 95–109
  phased-array, 326–331, 335
  and radio-wave development,
    32–49, 100–109, 111–116
  SAR, 325
  short-pulsed, 322–325
  submarine detection (*see* Air-to-surface; Asdic)
  surveillance satellites, 324–326, 341
  and transformation of warfare,
    318, 319
  weapons-control system, 319–331
  (*See also* Battle of Britain; RDF;
    Submarines)

Radiation stimulated emission of,
    333–336
  and lasers, 336–339
Radio; radio waves, 32–49, 111–116,
    330
  direction finding (*see* RDF)
  pulses, 112–116
Radio Direction Finding Unit, 116–117
Radio Research Board, 29
Radio Research Laboratory
    (Slough), 29, 30, 46, 47,
    112–113, 115–116, 117, 123

Radioactivity, radium, 20, 120
Raeder, Admiral Erich, 196
RAF Bomber Command, 197,
    212–213, 218–219, 296, 297–298,
    303, 304–306, 311
RAF Coastal Command, 290–291,
    293, 294, 296, 298, 301, 304–306,
    307, 308, 309–310, 311
  Squadron 217, 295
  Squadron 220, 295
RAF Fighter Command, 14, 84, 85,
    88, 90, 118, 122, 126, 127, 131,
    138, 146, 148, 150, 157–158, 161,
    175, 176–177, 182, 183, 186, 190,
    196, 199, 204, 205, 207–208, 210,
    213, 216, 220, 225, 226, 228,
    231, 236, 260, 265
  Groups:
    10, 176, 230
    11, 45, 176–177, 187, 188–190, 191–
      192, 201, 206, 208, 211, 212,
      215, 216, 217, 219, 222, 223,
      224–225, 228, 230, 234, 236
    12, 133, 145, 176, 177–178, 185,
      189–195, 207, 208, 211, 215,
      217, 223, 228, 234
    13, 176, 206, 207, 208
  Squadrons:
    1, 146, 152
    17, 207
    19, 177, 216
    23, 171–174
    25, 251, 259
    29, 259
    41, 206
    43, 155–157
    49, 206
    66, 216
    72, 185, 206
    79, 206
    172, 301
    219, 259
    242, 190–194, 198–199, 206, 212,
      215–216, 224
    266, 208–209
    303, 224
    600, 187, 251, 275
    604, 259
    605, 206
    609, 189
    610, 200–201
  (*See also* Battle of Barking Creek,
    Battle of Britain; London,
    bombing of)

RAF Volunteer Reserve, 187
Randall, John, 249–250, 256, 268, 272
Rathjens, George, 344
Rawnsley, Sergeant C. F., 252, 262
Rayleigh, Lord John, 38
Rays, mysterious, 17–20
  (*See also* "Death rays")
RDF (radio direction-finding), 45, 46–49, 116, 121, 125, 271
  RDF 1.5, 246
  RDF 2, 246–247, 251
Reagan, Ronald; Reagan administration, 329, 339
  and SDI, 340–344, 346
Red Baron, the, 180
Refraction indexes, 330–331
Reynolds, Leading Aircraftsman, 146
Rhodes-Moorhouse, Willie, 188
Ribbentrop, Joachim von, 15–16
Richey, Pilot Officer, 87
Richey, Flying Officer Paul, 146
Richthofen Squadron, 180
Right of the Line, The, (Terraine), 44–45
Rimmek, General Klaus, 324
Roentgen, Konrad Wilhelm; Roentgen rays, 17–20, 36, 55, 113, 340–346
Rogers, Buck, 16
Rommel, Field Marshal Erwin, x, 237, 275
Roosevelt, Franklin D., 140, 143, 151, 267, 269, 270, 271
Roosevelt, Theodore, 67
Rowe, Albert Percival, 23–24, 25–26, 27, 29, 30, 59, 62–63, 96, 118, 125, 137
Royal Auxiliary Air Force, 187–189, 251, 275
Royal Institution, 34
Royal Society, 73, 267
Runge, Dr., 51, 52
Rutherford, Ernest, 20, 75, 79, 96, 334

Sadat, Anwar, 324
SAR (*see* Synthetic-aperture radar)
Satellites, surveillance, 324–326, 341
Saul, Air Vice Marshal Richard, 176, 206
Scanners, airborne, 273, 294
  side scanners, 309–310

Scharnhorst, 144
Scherl, Richard, 43
Schneider Trophy, 134
Science and technology, ix–xi, 17, 23, 29
Scientific cooperation, British and American, 267–269, 270–273, 275
Scott, Robert F., 84
SDI (*see* Strategic Defense Initiative)
Searchlights, anti-submarine, 299, 300–302, 309–310
Seawave, 330
Secret Intelligence Service (SIS), 16
Sentries (Boeing), 324
Shakespeare, William, 168
  Henry IV, 155
  Henry V, 110
  Macbeth, 93
  Sonnet 147, 171
Shaw, Napier, 29
Short Brothers, 290, 298
Signals Experimental Establishment (SEE), 53–54
Sinclair, secretary, 148
Skywave, 330
Slessor, Air Marshal Sir John, 26, 27, 85, 149, 311
Slipher, Vernon, 320
Slonim, Captain Gilven M., 159
Snow, C. P., 73, 74–75, 97–98, 98–99
Solid-state electronics, 319
Sommerville, Admiral, 295
Sonar, 288
  (*See also* Asdic)
Soper, Sergeant, 87
Sopwith planes, 97, 129
Sound location, 23–24, 43–44, 50–51, 52, 110–111, 287
  (*See also* Asdic)
Soviet Air Defense Forces, 52
Space, geometry of, 334–336
Sperrle, Hugo, 181
Spitfires, 14, 32, 45, 84, 88–89, 90, 119, 126, 134, 135, 137, 146, 151, 172, 175, 237, 259, 271, 322
  Battle of Britain, 157, 161, 163, 176, 178, 179, 185, 186, 187, 197, 198, 199, 200, 201, 202, 203, 204, 205, 206–207, 208–209, 210, 214, 218, 220, 223–224, 226, 227, 228, 229–231
  engine, 268

Spitfires (*cont.*):
  pilots, 127, 132–133, 161–162, 163, 177–178, 210
Spruance, Admiral Raymond, 159
Spruce Goose, 318–319
Stalin, Josef, 52, 140
Stalingrad, battle of, x, 237
"Star Wars" (*see* Strategic Defense Initiative)
Static (*see* Atmospherics)
Stirling bombers, 271, 290, 298, 304, 307
Strahlenzieler, 43
Stranraer flying boats, 291
Strategic Defense Initiative (SDI— "Star Wars"), 329, 332, 340–344, 346
Stuka dive bombers, 145, 147, 179, 181, 185, 198, 200, 205, 209, 349
Stumpff, Hans-Jürgen, 181
Submarines:
  and air defense, 282–283, 290–291, 292–299, 300–302, 304–311
  between the wars, 286–289, 293–294
  and convoys (*see* Convoy system)
  detection; anti-submarine devices, 43–44, 51, 245, 248, 267, 286–290, 292–297, 304–311, 325
  early, 281–283
  missile-launching, 322
  snorkels, 311
  torpedoes, 282, 289, 339
  World War I, 282–286, 290, 291, 310
  World War II, 145, 238, 269, 289–311, 350
    American against Japanese shipping, 279–280
    Battle of the Atlantic, 280, 291–293, 296–297, 299–302, 306–311
    German wolf packs, 242, 291–293, 297, 299, 306–307, 310
  (*See also* U-boats)
Sunderland flying boats, 290–291, 298, 310, 311
Supermarine, 134
Surveillance satellite systems, 324–326, 341
  two problems of, 325–326
Sussex (ship), 284

Swinton, Lord, 98, 102, 107–108
Swinton Committee (*see* Committee on Air Defense Research)
Swordfish, 294, 311
Synthetic-aperture radar (SAR), 325

Tangmere, 146, 199, 209, 225, 251, 252
Task Force 50, 2, 277
Taylor, A. H., 50–51
Taylor, A. J. P., 149
TBF Avengers, 276–278
Technology, influence of, ix–xi, 9
Telecommunications Research Establishment (Air Ministry), 4, 294
Telefunken, 51, 52
Telegraph, electric, 33, 35–36
Telegraphy, wireless, 36–37, 289
  (*See also* Radio)
Telemobiloscop, 42–43
Teller, Edward, 340
Tennyson, Alfred: "The Charge of the Light Brigade," 1, 146
Terraine, John, 44–45
Tesla, Nikola, 42, 43
Thermal noise, 115
Thermoluminescence, 333
Thomas Dylan, quoted, 266
Thompson, J. J., 20
Thule early-warning system, 329
Tigercats, 275
Times (London), 86–87, 87–88, 345
Titanic, 38
Tizard, H. T., 29, 59, 60, 79, 83, 96–99, 102, 107, 108, 138, 245, 249, 251, 305
Tizard Committee (*see* Committee for the Scientific Survey of Air Defense)
Tomahawks (P-40), 53, 271
Torpedoes, self-propelled, 282, 289, 339
Townsend, Peter, 157
Trenchard, Lord Hugh ("Boom"), 25–26, 28, 60, 128, 129, 130–131, 132, 163, 187
Trunk telephone lines, 135–136
Tsipis, Kosta, 341
Tucker, W. S., 46
Tukachevsky, Mikhail, 52
Turtle (submarine), 281
Tuve, Michael, 48, 112

Tweedsmuir, John Buchan, Lord, 26
Twelve Legions of Angels (Dowding), 236

U-boats:
   U-9, 282
   U-21, 282
   U-29, 289
   U-502, 301–302
Ultra interception and decoding, 181, 185, 190, 196, 197, 220, 229, 231–232, 234, 306–307, 309
United States National Defense Research Committee, 272
Usworth, 156
Uxbridge, 219

V-1 and V-2 rockets, 343, 350
Vacuum tubes, 247–248
Vansittart, Sir Robert, 140
Ventnor, 202, 209–210, 225
Versailles Treaty, 11, 22, 100, 270, 344
Victoria Regina, 8, 65
Von Weiler, J. L. W. C., 52

Walker, Flight Lieutenant Johnny, 87
Watson, 33
Watson, Dr., 35
Watson Watt, Robert Alexander, 29–31, 44–49, 50, 53, 54, 57–60, 61–63, 83, 91, 99, 106, 112, 113, 114–116, 117–118, 119, 120, 121, 123–126, 136–137, 182–183, 245, 246, 247, 248, 261, 272, 306, 307, 326, 332, 336, 345
   most important contribution, 183
Wavelengths, wavelength choice, 55–59, 112–120, 243–247, 261–263, 266, 271–272, 304, 325, 330–331, 337–338
Wavell, Field Marshal Archibald, 149
Weapons and counterweapons, 339–346
Wellington bombers, 298, 300–301, 304, 350
Wells, H. G., 20, 95
   Things to Come, 139
Westminster, Duke of, 73, 74
Weygand, General Maxime, 160

Whitby, 157
Whitehead, Robert, 282
Wildcats, 53, 275, 276, 277
Wilhelm II, Kaiser, 8, 9–10, 67, 68, 284
Wilkins, Arnold F. ("Skip"), 30–31, 32, 44–49, 50, 53, 54, 57–59, 61–63, 70, 91, 99, 112, 117–118, 119, 120, 123–124, 126, 137, 245, 332, 334, 336, 345
Wilson, Woodrow, 270, 284
Wimperis, H. E., 15, 20, 23, 26, 27–29, 30, 58–59, 60, 61, 63, 83, 96, 105, 116, 267
Wings of Destiny (Londonderry), 236
Wireless telegraphy, 36–37, 289
   (*See also* Radio)
Wizard War, The (Jones), 103
Wolfe, General James, 23
Women as plotters, 137, 183, 201, 204, 221
Women's Air Transport Auxiliary, 260
Wood, Keith, 293–294
World Disarmament Conference (Geneva), 10–14, 24
World War I:
   air power, 6, 8–10, 172, 176–177, 180
   defensive war, 349
   submarines, 282–286, 290, 291, 310
World War II:
   beginning, 138–152, 285
   D-Day, 188
   (*See also* Battle of Britain; Bombers; Submarines)
Wright, Orville and Wilbur, 8, 25, 26
Wright, Robert, 149–150

X-rays, 17–20, 36, 55, 113
   and lasers, 340–346

Yagi, professor, 52
Yeats, William Butler: "The Second Coming," 153, 196
Young, Leo C., 50–51

Zeppelin, Count Ferdinand von, 6
Zeppelins, 5–8, 85, 125, 200, 290
Zeros, 53, 275, 276, 277, 278, 350